DNA Repair Disorders

Chikako Nishigori • Kaoru Sugasawa
Editors

DNA Repair Disorders

Springer

Editors
Chikako Nishigori
Department of Dermatology
Graduate School of Medicine, Kobe University
Kobe
Japan

Kaoru Sugasawa
Biosignal Research Center
Kobe University
Kobe
Japan

ISBN 978-981-10-6721-1 ISBN 978-981-10-6722-8 (eBook)
https://doi.org/10.1007/978-981-10-6722-8

Library of Congress Control Number: 2018959112

This Springer imprint is published by the registered company Springer Nature Singapore Pte Ltd.
The registered company address is: 152 Beach Road, #21-01/04 Gateway East, Singapore 189721, Singapore

Preface

Xeroderma pigmentosum (XP) is an autosomal recessive hereditary photosensitive disease, in which patients display extreme hypersensitivity to ultraviolet radiation (UVR) because of the deficiency in the ability to repair the UVR-induced DNA lesions. Although the existence of the disease had been known since the first case report on XP by a dermatologist, Kaposi, in 1883, the cause of XP was at length discovered in 1968, 85 years after the first case report. This year marks the 50th anniversary of the discovery of the cause of XP, a deficiency in nucleotide excision repair (NER), by James E. Cleaver. NER is an indispensable DNA repair mechanism for all living things on earth to remove various forms of DNA lesions from their genomic DNA, including UVR-induced DNA lesions, such as cyclobutane pyrimidine dimers and (6-4)photoproducts. In this sense, NER involves in an essential mechanism for living things and recently it has been shown that NER is closely involved in the biologically fundamental role such as transcription and replication. Therefore the deficiency in NER results in a disastrous condition. In this book we focused on the clinical aspects of DNA repair disorders. We would like to delineate the outcome of the deficiency of DNA repair so that we will come to know the essence of the DNA repair mechanisms. The authors are experts in this subject, and the publication of this book is timely because a Nobel Prize was given to the scientists who discovered the mechanisms of the NER, and the readers may be interested in what will become of individuals who are deficient in DNA repair.

Kobe, Japan — Chikako Nishigori
Kobe, Japan — Kaoru Sugasawa

Contents

Chapter 1
Molecular Mechanism of DNA Damage Recognition for Global Genomic Nucleotide Excision Repair: A Defense System Against UV-Induced Skin Cancer

Kaoru Sugasawa

Abstract Nucleotide excision repair (NER) is a versatile DNA repair pathway responsible for removal of ultraviolet light (UV)-induced DNA photolesions from the genome. In mammals, NER operating throughout the genome decreases the risk of UV-induced mutagenesis arising due to DNA translesion synthesis across photolesions on template DNA strands and thereby contributes to suppression of skin cancer. Lesion recognition for global genomic NER relies on multiple xeroderma pigmentosum (XP)-related protein factors, XPC, UV-DDB, TFIIH, and XPA, each of which probes for a different aspect of abnormal DNA structure. A combination of diverse strategies is likely required to achieve the broad substrate specificity, efficiency, and accuracy of this DNA repair system. To regulate this elaborate system *in vivo*, post-translational protein modifications, such as ubiquitination, and higher-order chromatin structures also play important roles.

Keywords Nucleotide excision repair · Xeroderma pigmentosum · DNA damage recognition · XPC · UV-DDB · Transcription factor IIH (TFIIH) · XPA · Ubiquitination Chromatin · Histone

1.1 Introduction

Among the complex clinical symptoms associated with xeroderma pigmentosum (XP), the predisposition to skin cancer is an important diagnostic hallmark [1]. Mutagenesis following ultraviolet light (UV)-induced DNA damage is a fundamental cause of skin cancer. In patients with XP, the risk of mutagenesis is elevated tremendously by a hereditary defect in one of two biological processes, nucleotide excision repair (NER) or DNA translesion synthesis (TLS) [2].

K. Sugasawa
Biosignal Research Center, Kobe University, Kobe, Hyogo, Japan
e-mail: ksugasawa@garnet.kobe-u.ac.jp

C. Nishigori, K. Sugasawa (eds.), *DNA Repair Disorders*,
https://doi.org/10.1007/978-981-10-6722-8_1

NER is an important DNA repair pathway that can remove various types of structurally unrelated DNA lesions from genomic DNA [3]. In many species, including humans, NER is exclusively responsible for repair of major UV-induced DNA lesions (*i.e.*, dipyrimidinic photolesions). Consequently, a defect in NER can result in accumulation of unrepaired photolesions in genomic DNA of skin cells. Such lesions interfere with normal processes of DNA replication, transcription, and other aspects of DNA metabolism. Mutations can then arise when TLS incorporates incorrect nucleotides opposite photolesions on the template DNA strand [4, 5]. According to this widely accepted model for skin carcinogenesis, global coverage of the genome by NER functions as a defense system against UV-induced skin cancer by decreasing the frequency of collisions between DNA replication forks and photolesions. Sustained DNA lesions block transcriptional elongation but can be removed from transcribed DNA strands by a specialized NER sub-pathway called transcription-coupled NER, which is reviewed in another chapter of this book.

By the end of the twentieth century, the causative genes for all known genetic complementation groups of XP had been identified; among them, seven groups (XP-A through XP-G) are associated with defective NER. Extensive studies of these gene products have contributed to our understanding of the basic molecular mechanisms of NER [6]. It has been also revealed that NER is subject to elaborate regulation, which involves post-translational protein modifications and alteration of chromatin structures. This chapter reviews our current knowledge on the mechanism and regulation of mammalian global genomic NER (GG-NER), especially the DNA damage recognition step.

1.2 Mammalian NER Pathways

Among the various DNA repair pathways, a remarkable characteristic of NER is its extremely broad substrate specificity. Typical examples include the following: (1) UV-induced photolesions, such as cyclobutane pyrimidine dimers (CPDs) and pyrimidine (6-4) pyrimidone photoproducts (6-4PPs); (2) intrastrand crosslinks caused by bifunctional chemical compounds (*e.g.*, cisplatin); and (3) bulky base adducts induced by various chemical carcinogens, including benzo[*a*]pyrene and acetylaminofluorene (AAF) [3]. Although most of these lesions are induced by environmental factors, NER can also handle certain types of endogenous oxidative DNA damage, such as 5′,8-cyclo-2′-deoxypurine lesions [7, 8], albeit with relatively low efficiency. In addition, NER is thought to participate in repair of highly detrimental interstrand crosslink lesions [9] that can be induced not only by cancer chemotherapeutic agents such as cisplatin and mitomycin C but also by aldehydes from endogenous and exogenous sources [10, 11].

Mammalian NER is an elaborate system involving more than 30 polypeptides, which can be dissected into several reaction steps (Fig. 1.1).

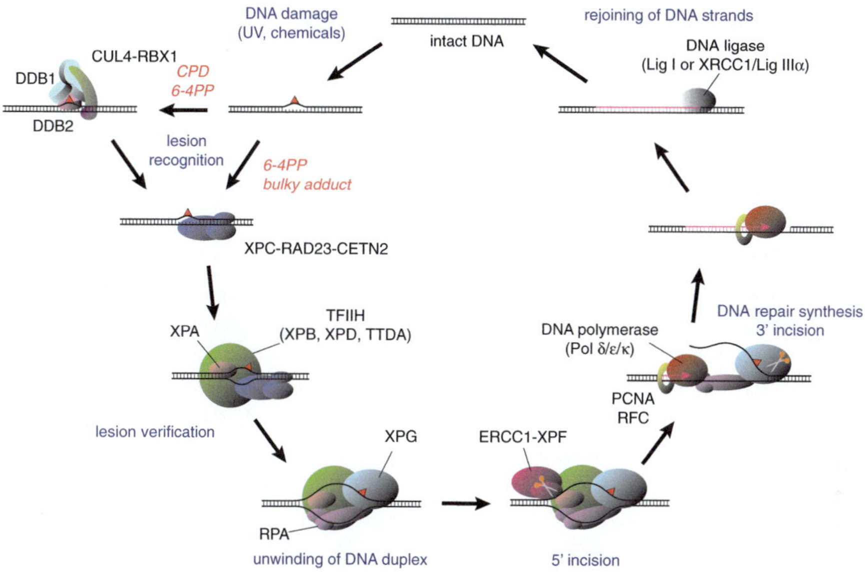

Fig. 1.1 Model of the molecular mechanism of mammalian global genomic NER. See text for details

1.2.1 Lesion Recognition

To initiate a repair reaction, it is crucial to sense the existence of a relevant lesion and determine its precise location within the DNA. In mammalian GG-NER, the protein complex containing the *XPC* gene product plays a central role in this key step [12–14]. The XPC protein complex has the potential to interact with various types of abnormal DNA (see below), thus serving as a highly versatile lesion recognition factor. On the other hand, a more specialized damage recognition pathway has evolved to ensure efficient repair of UV-induced photolesions. The UV-damaged DNA-binding protein complex (UV-DDB) exhibits exceptionally high binding affinity and specificity for photolesions (both CPDs and 6-4PPs) and promotes recruitment of XPC to the damaged DNA sites [15–17].

At transcriptionally active gene loci, lesions generated on transcribed DNA strands can be sensed by elongating RNA polymerases as a result of blockage of their translocation, efficiently triggering a specific sub-pathway of NER called transcription-coupled NER (TC-NER) [18]. Because RNA polymerase functions as the primary lesion sensor, both XPC and UV-DDB are dispensable for TC-NER. By contrast, some gene products implicated in Cockayne syndrome and UV-sensitive syndrome (CSA, CSB, and UVSSA) are required specifically for TC-NER, but not for GG-NER.

1.2.2 Lesion Verification and Demarcation

Despite the diversity of lesion recognition modes, the subsequent repair reaction is processed by a common set of protein factors. One of the key players in this process is the transcription factor IIH (TFIIH) complex, which is also essential for initiation of basal transcription [19]. In GG-NER, TFIIH is most likely recruited through a direct interaction with XPC [20–22]. Among ten subunits of TFIIH, the XPB and XPD proteins possess DNA-dependent ATPase/helicase activities that unwind the DNA duplex into a single-stranded state [23–25]. During unwinding, the presence of a relevant lesion is verified to decide whether the repair reaction should proceed toward dual incisions (see below). The XPA protein, recruited by TFIIH, stimulates the ATPase/helicase activities [26] and is thus vital to lesion verification.

Following enlargement of the unwound segment of DNA, a heterotrimeric complex of the single-stranded DNA-binding protein replication protein A (RPA) binds to the undamaged DNA strand [27], guided by interactions with XPA [28–31]. The configuration of this complex containing the partially unwound DNA demarcates the lesion for the subsequent dual incisions.

1.2.3 Dual Incisions

Two structure-specific endonucleases, the ERCC1-XPF complex and the XPG protein, are responsible for the dual incisions, which excise an oligonucleotide containing the lesion and surrounding (intact) nucleotides [32, 33]. Both enzymes introduce a single-strand cleavage near a junction between the double- and single-stranded DNA segments but with opposite polarities: ERCC1-XPF and XPG make incisions at the 5′ and 3′ ends of the unwound DNA region, respectively [32, 34, 35].

XPG is incorporated into the pre-incision NER complex, presumably through a strong interaction with TFIIH [20, 36, 37]. ERCC1-XPF is subsequently recruited mainly through an interaction with XPA [38–40]. Although biochemical studies revealed that the 5′ incision by ERCC1-XPF precedes the 3′ incision [25, 41], the presence of XPG in the pre-incision complex is required for the 5′ incision, independent of its catalytic function [42, 43].

1.2.4 DNA Repair Synthesis and Ligation

The excised DNA strands need to be resynthesized by DNA polymerases, a process involving the DNA polymerase clamp, proliferating cell nuclear antigen (PCNA), and the clamp loader ATPase complex replication factor C (RFC) [44–46]. PCNA interacts with XPG [47], and DNA repair synthesis may start following the 5′ incision but prior to the 3′ incision [41]. As a result of the strand displacement synthesis, 5′-flap structures are formed and subsequently removed by XPG.

Various DNA polymerases could be redundantly involved in the DNA repair synthesis of NER. Both PCNA and RFC were initially identified as accessory factors of replicative DNA polymerases δ and ε, suggesting that these enzymes are involved in NER, particularly in proliferating cells [44–46]. Following the DNA repair synthesis and 3′ incision, the DNA strands can be rejoined by DNA ligase I, which is also involved in Okazaki fragment maturation. However, other enzymes, such as DNA polymerase κ [48, 49] and DNA ligase IIIα/XRCC1 [50], are also implicated in the late steps of NER, presumably depending on cell type and other conditions.

1.3 XPC Recognizes a Broad Range of Substrates for NER

As mentioned above, GG-NER can handle an extraordinarily large repertoire of DNA lesions. These substrates cannot be stereotyped by any common chemical structure but have in common a relatively large distortion induced in the DNA helices in comparison with substrates for another major DNA excision repair pathway, base excision repair. This versatility of GG-NER is attributed to the DNA-binding specificity of the heterotrimeric XPC protein complex.

1.3.1 Architecture of the XPC Complex

The human *XPC* gene encodes a basic protein containing 940 amino acids (calculated molecular mass: 106 kDa), the largest subunit of this complex [51–53]. Evolutionarily conserved functional domains have been identified mostly within the C-terminal half of the protein [54]. The transglutaminase-homology domain (TGD) and three β-hairpin domains (BHD1, BHD2, and BHD3) search for a lesion on DNA and form a stable repair initiation complex at the affected site (see below). Because of an amino acid change at the predicted active site, the XPC TGD does not have transglutaminase enzymatic activity [55] but instead provides interfaces for interactions, not only with DNA but also with another subunit of the complex, RAD23 [54, 56].

RAD23 is a NER protein originally identified from the budding yeast *Saccharomyces cerevisiae* [57]. Mammalian cells express two RAD23 orthologues, RAD23A and RAD23B, which have redundant functions in the XPC complex [52, 58]. In the absence of RAD23, the XPC protein is destabilized and degraded by the ubiquitin-proteasome system (UPS) [59, 60]. RAD23 possesses a ubiquitin-like (UBL) domain at its N-terminus and two copies of ubiquitin-associating (UBA) domains and is therefore implicated in regulation of UPS-mediated protein degradation, independent of its functions in NER [61, 62]. Consistent with this dual role, RAD23 proteins are expressed at much higher levels than XPC [63]. The XPC-interacting site has been mapped between the two UBA domains [64].

The third component of the XPC complex is CETN2 (also called centrin-2 or caltractin-1) [65]. CETN2 is a calmodulin-like calcium-binding protein that stably binds to an α-helix near the C-terminus of XPC [66]. Although the XPC-RAD23 heterodimer is functional in a cell-free NER reaction [12, 45, 67], the interaction with CETN2 substantially increases its binding affinities for both damaged and undamaged DNA, thereby stimulating GG-NER [66]. CETN2 was originally identified as a component of the centrosome and is thought to play a pivotal role in the microtubule organization [68, 69]. Whether such a "non-NER" function is related to GG-NER has yet to be clarified.

1.3.2 XPC Indirectly Senses DNA Lesions

In a cell-free system recapitulating the mammalian GG-NER reaction, the XPC protein complex is the first factor that interacts with DNA lesions, thereby serving to initiate NER [12, 13]. Biochemical and physicochemical analyses revealed that XPC exhibits specific binding affinities, not only for DNA with relevant NER substrates, such as UV-induced 6-4PPs and AAF-dG adducts [70–73], but also for DNA containing only mismatched bases [74]. These findings suggest that XPC does not recognize any specific chemical feature of DNA lesions but rather more general aspects of DNA with structural abnormalities.

This notion was further corroborated by structural studies of the *S. cerevisiae* XPC homologue RAD4 bound to damaged DNA [54]. In this structure, TGD and BHD1 of XPC/RAD4 interact with the intact double-stranded DNA segment 3′ of the lesion, whereas the β-hairpin of BHD3 is inserted into the major groove at the lesion site, so that BHD2 and BHD3 together interact with the undamaged DNA strand. As a result, the lesion *per se* is flipped out of the DNA duplex and does not contact the protein. Biochemical studies with cell-free NER systems also support the notion that XPC must interact with the undamaged DNA strand to induce productive dual incisions [75, 76]. Moreover, recent single-molecule imaging analyses suggest that XPC/RAD4 may engage in both three-dimensional and one-dimensional diffusion on the DNA duplex to search for its target sites [77]. It has been proposed that the BHD1–BHD2 region may serve as the dynamic scanning module, enabling the protein to rapidly interrogate the intactness of the DNA duplex [78]. As a prerequisite for formation of the stable DNA–protein complex, the DNA duplex needs to be thermodynamically destabilized by the presence of lesions and/or base mismatches, lowering the free energy barrier sufficiently to allow BHD3 insertion during the rapid scanning of the DNA duplex [79]. Thus, XPC/RAD4 adopts an indirect mode of damage sensing, providing the molecular basis for the unprecedentedly broad substrate specificity of the GG-NER system.

1.4 TFIIH and XPA Ensure Fidelity of GG-NER

Although the XPC protein complex can interact with mismatched DNA devoid of any lesion, biochemical studies revealed that such damage-free DNA substrates do not give rise to productive dual incisions by the NER system [73, 80]. This observation indicates that the presence of a relevant lesion must be somehow verified before the NER protein complex makes the decision to proceed with the dual incisions. TFIIH and XPA play crucial roles in this damage verification process.

1.4.1 Two-Step Damage Recognition and TFIIH ATPases/Helicases

Among the known substrates of GG-NER, UV-induced CPDs are associated with only a modest DNA helix distortion [81]. Because of their relatively weak destabilization of the DNA duplex, these lesions tend to escape direct recognition by XPC, and another specialized damage recognition factor is involved in efficient repair of CPDs *in vivo* (see below). Although the repair efficiency of CPDs in the cell-free NER system is consequently much lower than that of 6-4PPs, introduction of mismatched bases opposite a CPD substantially enhances recruitment of XPC and susceptibility of the lesion to dual incisions [73]. This stimulatory effect is also observed when the mismatched bases are 60–150 bases apart from the lesion [76], providing evidence for spatial separation of the initial DNA binding by XPC and the subsequent dual incisions at the lesion site. Importantly, to observe such stimulation of CPD repair, the mismatched bases serving as a XPC-binding site must be 5′ of the lesion. Based on these findings, we propose that certain NER factors may search for a lesion by scanning a DNA strand in the 5′–3′ direction (Fig. 1.2).

Among known NER factors, the XPD ATPase/helicase subunit of TFIIH is the only one known to be able to move on a DNA strand with such a specific polarity [82, 83]. Biochemical and structural studies with yeast and archaeal XPD homologues suggest that when XPD translocates along a DNA strand containing a bulky lesion, it is likely to be detained at the damaged site [84, 85]. More recently, this notion has been corroborated for the recombinant whole TFIIH complex: its helicase activity is blocked by a bulky cisplatin adduct, but not by a non-bulky abasic lesion, on the XPD translocating DNA strand [26]. Given that XPC interacts with the DNA strand opposite a lesion, recruited TFIIH may be loaded such that XPD binds to the damaged strand immediately 5′ of the lesion. Although this model could explain how XPD detects and locates the lesion as soon as it starts translocation [86], further studies are required to understand how the DNA duplex is unwound and lesion verification is eventually accomplished. For instance, XPB, another

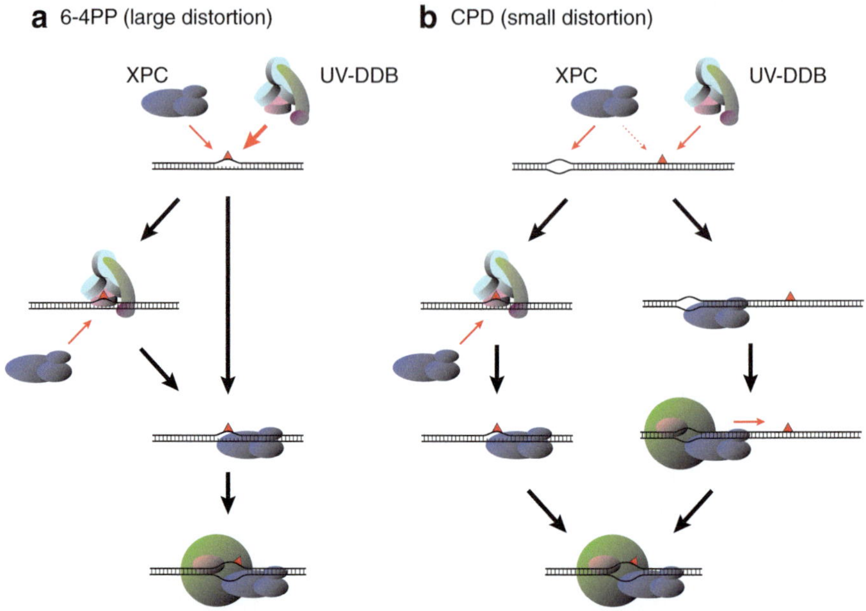

Fig. 1.2 Diverse mechanisms of recognition of UV-induced DNA photolesions. (**a**) 6-4PPs, associated with large helix distortions, are efficiently recognized by either XPC or UV-DDB, and the UV-DDB-dependent and UV-DDB-independent pathways can operate in parallel. (**b**) Because CPDs induce too small helix distortion to allow efficient recognition by XPC, the UV-DDB-dependent recruitment of XPC makes a major contribution to repair of CPDs. However, if XPC happens to bind to sites with mismatched bases or partially unwound duplex 5′ of a CPD, the NER machinery has the potential to find the lesion through a 5′–3′ scanning mechanism

ATPase/helicase subunit in TFIIH, also plays an essential role in NER [87, 88]. In contrast to XPD, however, the helicase, but not ATPase, activity of XPB is dispensable for dual incisions [89, 90]. The precise roles of the two ATPase/helicase subunits remain to be elucidated.

1.4.2 XPA Supports the Lesion Recognition Functions of TFIIH

Human XPA protein consists of 273 amino acids (calculated molecular mass: 31 kDa) and contains a zinc finger motif [91]. XPA is recruited to lesion sites after XPC and TFIIH, presumably through physical interactions with these factors [92–96]. In particular, the C-terminal part of XPA interacts with TFIIH, explaining why causative mutations resulting in even small C-terminal deletion compromise cellular NER functions [95]. XPA interacts also with RPA [28–31] and ERCC1-XPF [38–40] and is therefore thought to serve as an important scaffold for the NER pre-incision complex, depositing other NER factors around the lesion in the correct arrangement. Although XPA exhibits a DNA-binding activity with moderate specificity for damaged DNA [97, 98], this probably reflects its affinity for kinked DNA structures containing junctions of double- and single-stranded segments [99]. In the NER pre-incision complex, XPA may localize at the border of the unwound region [100].

Apart from this structural role in pre-incision complex assembly, XPA also participates in remodeling and activation of the TFIIH complex, which are crucial for lesion verification. Upon recruitment to a lesion site, XPA dissociates the CDK-activating kinase (CAK) module (CDK7, Cyclin H, and MAT-1 subunits) from TFIIH, thereby substantially stimulating its ATPase/helicase activities [101]. Both the helicase activities of the remaining seven-subunit TFIIH core complex (Core7) and inhibition of the helicases by bulky DNA lesions are stimulated by the presence of XPA [26]. On the other hand, Core7 exhibits a much stronger non-specific interaction with single-stranded DNA than the TFIIH holo-complex. Therefore, XPA may promote timely dissociation of CAK, which would be important for coordinating the specificity of XPC-dependent recruitment of TFIIH to lesion sites, ensuring the efficiency of the subsequent lesion verification and dual incision processes.

Our biochemical results suggest that when XPC binds to a site with mismatched bases but no damage, TFIIH (presumably together with XPA and/or XPC) can search nearby for a relevant lesion [76]. Although it remains to be determined whether such a mechanism indeed operates *in vivo*, this process would have some similarity to lesion recognition in prokaryotic NER. Although UvrA alone has some specific binding affinity for damaged DNA, the UvrA–UvrB complex has been implicated in lesion recognition, either directly or through sliding on DNA driven by UvrB helicase activity [102–105]. Further studies would shed light on the possible diversity of lesion recognition mechanisms in the mammalian GG-NER system.

1.5 UV-DDB as a Suppressor of UV-Induced Mutagenesis and Carcinogenesis

As befits a versatile lesion recognition factor, stable DNA binding by XPC is influenced primarily by the degree of disruption and/or destabilization of base pairs in the DNA duplex. Therefore, efficiencies of recognition by XPC and subsequent repair can vary tremendously depending on the type of lesion. The UV-induced CPD is a typical example of a difficult substrate for GG-NER; consequently, CPDs tend to persist for long periods in genomic DNA once cells are exposed to UV. Importantly, cytosines in CPDs are highly susceptible to deamination and conversion to uracils [106], which then induce misincorporation of adenines by TLS at high frequency. This could be the mechanism underlying the UV signature mutations (C to T, CC to TT) often associated with skin cancer [107, 108]. In mammalian cells, this problem is partially overcome by the presence of a unique lesion recognition factor, UV-DDB, dedicated to UV-induced DNA photolesions.

1.5.1 UV-DDB Assists XPC in Recognizing UV-Induced Photolesions

UV-DDB is a heterodimeric protein complex composed of the DDB1 and DDB2 subunits [109], and mutations in the *DDB2* gene have been implicated in XP genetic complementation group E [110, 111]. The human *DDB2* gene encodes a 427-amino acid protein (calculated molecular mass: 48 kDa) [112] that contains seven WD40 repeats adopting a β-propeller structure. The N-terminal helix–loop–helix motif in DDB2 serves as the binding interface for DDB1, which also contains three β-propeller domains (BPA, BPB, and BPC) [113].

In comparison with XPC, UV-DDB has much higher binding affinity and specificity for double-stranded DNA containing a UV-induced 6-4PP. Moreover, UV-DDB also exhibits significant binding to CPDs, albeit the affinity is lower than that for 6-4PPs [15–17]. Although it does not efficiently recognize other substrates of NER, such as bulky chemical adducts [114, 115], UV-DDB can bind to DNA containing an abasic site or mismatched bases with high affinity [15, 17]. When UV-DDB is bound to a target site with a photolesion, an evolutionarily conserved three-amino acid (FQH) hairpin on a plane of the DDB2 β-propeller is inserted into the minor groove, causing the two affected pyrimidine residues to flip out of the DNA duplex and directly interact with a binding pocket on DDB2 [113]. In this sense, UV-DDB is clearly distinct from XPC, which adopts an indirect lesion recognition strategy and thus averts direct contact with lesions.

Accumulating evidence indicates that the principal role of UV-DDB in GG-NER is to promote recruitment of XPC to sites with UV-induced photolesions, but not to substitute for XPC [116–119]. Consequently, in XPC-deficient cells, GG-NER is defective even though UV-DDB is functional. Because CPDs are poorly recognized by XPC alone, the presence of functional UV-DDB significantly increases the rate of CPD removal from the global genome [120, 121]. By contrast, repair of 6-4PPs occurs quite efficiently even in UV-DDB-deficient cells, most likely due to the ability of XPC to recognize these lesions. It should be noted that even in the presence of UV-DDB, CPDs are repaired much more slowly than 6-4PPs. As with other NER-deficient XP complementation groups, the hereditary defect in UV-DDB is associated with a predisposition to skin cancer, indicating that accumulation of unrepaired CPDs can cause skin cancer even when most 6-4PPs are removed.

1.5.2 UV-DDB and Cellular DNA Damage Responses

Transcription of the *DDB2* and *XPC* genes is under control of the p53 tumor suppressor [120, 122]. Consequently, inactivation of the *TP53* gene by mutations decreases cellular GG-NER capacity, leading to additional mutations that further drive the carcinogenic process. On the other hand, expression of DDB2 enhances normal p53 functions, thereby forming a positive-feedback loop [123]. DDB2 may regulate DNA damage-induced apoptosis either positively or negatively [124–127], although *Ddb2*-deficient mice exhibit marked resistance to UV-induced apoptosis of skin cells [128]. These model mice manifest a predisposition to skin cancer induced by exposure to UV, but not chemical carcinogens, consistent with the binding specificities of UV-DDB for various DNA lesions.

Several avenues of research indicate that post-translational protein modifications are involved in regulation of UV-DDB functions. Although DDB1 was originally identified as a component of UV-DDB, it turned out to play a much broader role as the adaptor molecule for the CUL4-RBX1 ubiquitin ligase (CRL4) module [129, 130]. DDB2 is a member of the DDB1- and CUL4-associated factors (DCAFs), and a number of other DCAFs can substitute for DDB2 and alter the substrate specificity of the ubiquitin ligase complex [131]. In the case of the CRL4 complex containing DDB2 ($CRL4^{DDB2}$), the interaction of DDB2 with a DNA lesion activates the ubiquitin ligase through release of the COP9 signalosome and conjugation of CUL4 to NEDD8 [132–134]. The activated $CRL4^{DDB2}$ then ubiquitinates various proteins in the vicinity of the lesion, including DDB2, XPC, and histones [16, 135–138]. Although the precise biological functions of this ubiquitination are not yet fully understood, it has been proposed that the degree of ubiquitination must be regulated at an appropriate level to ensure handover of the lesion from UV-DDB to XPC and a smooth transition to the subsequent repair process [139, 140]. DDB2 undergoes degradation by the UPS in response to UV irradiation [141, 142], a process that requires the N-terminal tail of DDB2 [133, 139, 143]. On the other hand, UV-induced degradation of XPC is not as pronounced, and the stability of DDB2 after UV irradiation correlates with the expression level of XPC [139]. We propose that if XPC is successfully recruited to a site where UV-DDB binds to a UV-induced photolesion, ubiquitination by activated $CRL4^{DDB2}$ preferentially targets XPC rather than DDB2, so that DDB2 is allowed to escape degradation and turn over to recognize multiple lesions. SUMOylation of XPC may be involved in such functional interactions between DDB2 and XPC [144, 145].

1.6 Roles of Chromatin Structures in Regulation of Lesion Recognition

Chromatin structures play pivotal roles in the regulation of various nuclear functions. For instance, interactions of transcription factors with their corresponding DNA elements are profoundly affected by the state of chromatin condensation, as well as nucleosome positioning around their target sites. Such functional modulation of chromatin structures is mediated by epigenetic marks, such as DNA methylation and post-translational modifications of histones, as well as by histone variants, histone chaperones, and chromatin remodeling factors. Similarly, lesion recognition for GG-NER may be under the control of chromatin structure.

UV-induced CPDs tend to be generated preferentially at the outer surface of the nucleosome core, whereas 6-4PPs show more random distribution [146–149]. The positioning of lesions within the nucleosome array can profoundly affect repair efficiency. Biochemical studies of cell-free NER and nucleosome assembly revealed that lesions within the nucleosome core tend to evade interaction with XPC and subsequent repair reactions, suggesting that histone octamers must be evicted from lesion sites prior to repair [150, 151]. By contrast, UV-DDB is capable of binding to nucleosome cores containing 6-4PPs or abasic sites [152]. Using a strong nucleosome positioning sequence, orientation of the lesions within the nucleosome can be precisely regulated, either at the farthest or closest site from the histone surface. Intriguingly, UV-DDB can bind to the damaged nucleosomes with comparable affinities, independent of the lesion orientation. Crystal structures of these nucleosomes indicate that a flexible DNA backbone is exposed on the outer surface of the nucleosome at the lesion site, regardless of whether the lesion itself is oriented outward or inward. These findings suggest that UV-DDB may have the potential to sense structural abnormalities of DNA within the nucleosome, even though NER has no canonical lesion. Subsequently, UV-DDB may induce remodeling of the nucleosome such that the lesion *per se* can be inspected and made accessible to XPC and other NER factors [153–155].

Decondensation of chromatin is thought to be a prerequisite for transcriptional competence, whereas highly condensed heterochromatin prevents access by transcription factors and thus suppresses gene expression. Decondensation is often associated with acetylation of histones H3 and H4 and condensation with histone methylation, *e.g.*, trimethylated lysine 9 of histone H3 (H3K9me3), as well as DNA methylation at CpG sequences. The involvement of these characteristics of chromatin structures in GG-NER remains somewhat controversial. By analogy with transcriptional regulation, it is reasonable to assume that histone acetylation may increase the accessibility of lesions to UV-DDB and/or XPC, thereby promoting initiation of the repair reaction. Indeed, UV-DDB associates with some histone acetyltransferases [142, 156–158] and can induce decondensation of chromatin [159, 160]. On the other hand, we recently reported that global chromatin decondensation induced by treatment with histone deacetylase inhibitors has negative effects on recruitment of XPC to lesion sites [161]. In the same study, we demonstrated that XPC physically interacts with histones H3 and H1. Intriguingly, the N-terminal tail of histone H3 is involved in the interaction with XPC, and this interaction is markedly attenuated by acetylation of histone H3. Furthermore, when local UV irradiation is applied to cell nuclei through isopore membrane filters, certain types of acetylated histones, such as H3K27ac, tend to be underrepresented in the damaged areas. Taken together, these results suggest that transient formation of condensed heterochromatin-like structures may promote recruitment of XPC to lesion sites. Notably in this regard, heterochromatin protein-1 is recruited to local UV damage [162], as is H3K9ac, a typical mark of euchromatin [163]. Because UV-induced photolesions can occur at various genomic loci, the roles of such epigenetic chromatin modifications may be heterogeneous, depending on the original chromatin state, but distinct from their functions in transcriptional regulation. We speculate that global decondensation of chromatin not only increases lesion accessibility but also allows non-specific interaction of XPC with DNA, which could interfere with efficient search for lesions within the huge genome. Further studies are necessary to understand the precise roles of chromatin structures in regulation of GG-NER.

1.7 Conclusions

To prevent UV-induced carcinogenesis, multiple XP-related factors must cooperate to ensure efficient recognition and removal of DNA photolesions. XPC recognizes both 6-4PPs and bulky chemical base adducts, and its indirect mode of damage recognition underlies the exceptionally broad substrate specificity of GG-NER. However, binding of XPC does not in itself guarantee the presence of a lesion. To avoid adverse incisions at lesion-free sites, damage must be verified by TFIIH and XPA, a process in which DNA strands are minutely inspected by a scanning mechanism. In addition, UV-DDB directly recognizes UV-induced photolesions and mediates efficient recruitment of XPC, which is particularly relevant for repair of CPDs. By using these different strategies to probe for structural abnormalities of DNA, the GG-NER system simultaneously attains high levels of efficiency, accuracy, and versatility [86]. Further studies would provide insight into the *in vivo* regulation of the lesion recognition process, including the roles of post-translational protein modifications and chromatin structures.

Acknowledgments The author's work was supported by Grants-in-Aid from the Japan Society for the Promotion of Science (JSPS KAKENHI Grant Numbers JP23116008, JP16H06307, and JP16H01311) to K.S.

References

1. Kraemer KH, Lee MM, Andrews AD, Lambert WC. The role of sunlight and DNA repair in melanoma and nonmelanoma skin cancer. The xeroderma pigmentosum paradigm. Arch Dermatol. 1994;130:1018–21. https://doi.org/10.1001/archderm.1994.01690080084012.
2. Friedberg EC, Walker GC, Siede W, Wood RD, Schultz RA, Ellenberger T, editors. DNA repair and mutagenesis. 2nd ed. Washington: ASM Press; 2006.
3. Gillet LCJ, Schärer OD. Molecular mechanisms of mammalian global genome nucleotide excision repair. Chem Rev. 2006;106:253–76. https://doi.org/10.1021/cr040483f.
4. Cleaver JE, Crowley E. UV damage, DNA repair and skin carcinogenesis. Front Biosci. 2002;7:d1024–43. https://doi.org/10.2741/A829.
5. Friedberg EC. How nucleotide excision repair protects against cancer. Nat Rev Cancer. 2001;1:22–33. https://doi.org/10.1038/35094000.
6. Schärer OD. Nucleotide excision repair in eukaryotes. Cold Spring Harb Perspect Biol. 2013;5:a012609. https://doi.org/10.1101/cshperspect.a012609.
7. Brooks PJ, Wise DS, Berry DA, Kosmoski JV, Smerdon MJ, Somers RL, Mackie H, Spoonde AY, Ackerman EJ, Coleman K, Tarone RE, Robbins JH. The oxidative DNA lesion 8,5′-(*S*)-cyclo-2′-deoxyadenosine is repaired by the nucleotide excision repair pathway and

blocks gene expression in mammalian cells. J Biol Chem. 2000;275:22355–62. https://doi.org/10.1074/jbc.M002259200.
8. Kuraoka I, Bender C, Romieu A, Cadet J, Wood RD, Lindahl T. Removal of oxygen free-radical-induced 5′,8-purine cyclodeoxynucleosides from DNA by the nucleotide excision-repair pathway in human cells. Proc Natl Acad Sci U S A. 2000;97:3832–7. https://doi.org/10.1073/pnas.070471597.
9. Wood RD. Mammalian nucleotide excision repair proteins and interstrand crosslink repair. Environ Mol Mutagen. 2010;51:520–6. https://doi.org/10.1002/em.20569.
10. Langevin F, Crossan GP, Rosado IV, Arends MJ, Patel KJ. Fancd2 counteracts the toxic effects of naturally produced aldehydes in mice. Nature. 2011;475:53–8. https://doi.org/10.1038/nature10192.
11. Noll DM, Mason TM, Miller PS. Formation and repair of interstrand cross-links in DNA. Chem Rev. 2006;106:277–301. https://doi.org/10.1021/cr040478b.
12. Riedl T, Hanaoka F, Egly JM. The comings and goings of nucleotide excision repair factors on damaged DNA. EMBO J. 2003;22:5293–303. https://doi.org/10.1093/emboj/cdg489.
13. Sugasawa K, Ng JMY, Masutani C, Iwai S, van der Spek PJ, Eker APM, Hanaoka F, Bootsma D, Hoeijmakers JHJ. Xeroderma pigmentosum group C protein complex is the initiator of global genome nucleotide excision repair. Mol Cell. 1998;2:223–32. https://doi.org/10.1016/S1097-2765(00)80132-X.
14. Volker M, Moné MJ, Karmakar P, van Hoffen A, Schul W, Vermeulen W, Hoeijmakers JHJ, van Driel R, van Zeeland AA, Mullenders LHF. Sequential assembly of the nucleotide excision repair factors in vivo. Mol Cell. 2001;8:213–24. https://doi.org/10.1016/s1097-2765(01)00281-7.
15. Fujiwara Y, Masutani C, Mizukoshi T, Kondo J, Hanaoka F, Iwai S. Characterization of DNA recognition by the human UV-damaged DNA-binding protein. J Biol Chem. 1999;274:20027–33. https://doi.org/10.1074/jbc.274.28.20027.
16. Sugasawa K, Okuda Y, Saijo M, Nishi R, Matsuda N, Chu G, Mori T, Iwai S, Tanaka K, Tanaka K, Hanaoka F. UV-induced ubiquitylation of XPC protein mediated by UV-DDB-ubiquitin ligase complex. Cell. 2005;121:387–400. https://doi.org/10.1016/j.cell.2005.02.035.
17. Wittschieben BØ, Iwai S, Wood RD. DDB1-DDB2 (xeroderma pigmentosum group E) protein complex recognizes a cyclobutane pyrimidine dimer, mismatches, apurinic/apyrimidinic sites, and compound lesions in DNA. J Biol Chem. 2005;280:39982–9. https://doi.org/10.1074/jbc.M507854200.
18. Hanawalt PC, Spivak G. Transcription-coupled DNA repair: two decades of progress and surprises. Nat Rev Mol Cell Biol. 2008;9:958–70. https://doi.org/10.1038/nrm2549.
19. Compe E, Egly JM. TFIIH: when transcription met DNA repair. Nat Rev Mol Cell Biol. 2012;13:343–54. https://doi.org/10.1038/nrm3350.
20. Araújo SJ, Nigg EA, Wood RD. Strong functional interactions of TFIIH with XPC and XPG in human DNA nucleotide excision repair, without a preassembled repairosome. Mol Cell Biol. 2001;21:2281–91. https://doi.org/10.1128/MCB.21.7.2281-2291.2001.
21. Li RY, Calsou P, Jones CJ, Salles B. Interactions of the transcription/DNA repair factor TFIIH and XP repair proteins with DNA lesions in a cell-free repair assay. J Mol Biol. 1998;281:211–8. https://doi.org/10.1006/jmbi.1998.1949.
22. Yokoi M, Masutani C, Maekawa T, Sugasawa K, Ohkuma Y, Hanaoka F. The xeroderma pigmentosum group C protein complex XPC-HR23B plays an important role in the recruitment of transcription factor IIH to damaged DNA. J Biol Chem. 2000;275:9870–5.
23. Evans E, Moggs JG, Hwang JR, Egly JM, Wood RD. Mechanism of open complex and dual incision formation by human nucleotide excision repair factors. EMBO J. 1997;16:6559–73. https://doi.org/10.1093/emboj/16.21.6559.
24. Mu D, Wakasugi M, Hsu DS, Sancar A. Characterization of reaction intermediates of human excision repair nuclease. J Biol Chem. 1997;272:28971–9. https://doi.org/10.1074/jbc.272.46.28971.

25. Tapias A, Auriol J, Forget D, Enzlin JH, Schärer OD, Coin F, Coulombe B, Egly JM. Ordered conformational changes in damaged DNA induced by nucleotide excision repair factors. J Biol Chem. 2004;279:19074–83. https://doi.org/10.1074/jbc.M312611200.
26. Li C-L, Golebiowski FM, Onishi Y, Samara NL, Sugasawa K, Yang W. Tripartite DNA lesion recognition and verification by XPC, TFIIH, and XPA in nucleotide excision repair. Mol Cell. 2015;59:1025–34. https://doi.org/10.1016/j.molcel.2015.08.012.
27. de Laat WL, Appeldoorn E, Sugasawa K, Weterings E, Jaspers NGJ, Hoeijmakers JHJ. DNA-binding polarity of human replication protein A positions nucleases in nucleotide excision repair. Genes Dev. 1998;12:2598–609. https://doi.org/10.1101/gad.12.16.2598.
28. He Z, Henricksen LA, Wold MS, Ingles CJ. RPA involvement in the damage-recognition and incision steps of nucleotide excision repair. Nature. 1995;374:566–9. https://doi.org/10.1038/374566a0.
29. Li L, Lu X, Peterson CA, Legerski RJ. An interaction between the DNA repair factor XPA and replication protein A appears essential for nucleotide excision repair. Mol Cell Biol. 1995;15:5396–402. https://doi.org/10.1128/MCB.15.10.5396.
30. Matsuda T, Saijo M, Kuraoka I, Kobayashi T, Nakatsu Y, Nagai A, Enjoji T, Masutani C, Sugasawa K, Hanaoka F. DNA repair protein XPA binds replication protein A (RPA). J Biol Chem. 1995;270:4152–7. https://doi.org/10.1074/jbc.270.8.4152.
31. Patrick SM, Turchi JJ. Xeroderma pigmentosum complementation group a protein (XPA) modulates RPA-DNA interactions via enhanced complex stability and inhibition of strand separation activity. J Biol Chem. 2002;277:16096–101. https://doi.org/10.1074/jbc.M200816200.
32. Matsunaga T, Mu D, Park CH, Reardon JT, Sancar A. Human DNA repair excision nuclease. Analysis of the roles of the subunits involved in dual incisions by using anti-XPG and anti-ERCC1 antibodies. J Biol Chem. 1995;270:20862–9. https://doi.org/10.1074/jbc.270.35.20862.
33. Moggs JG, Yarema KJ, Essigmann JM, Wood RD. Analysis of incision sites produced by human cell extracts and purified proteins during nucleotide excision repair of a 1,3-intrastrand d(GpTpG)-cisplatin adduct. J Biol Chem. 1996;271:7177–86. https://doi.org/10.1074/jbc.271.12.7177.
34. O'Donovan A, Davies AA, Moggs JG, West SC, Wood RD. XPG endonuclease makes the 3′ incision in human DNA nucleotide excision repair. Nature. 1994;371:432–5. https://doi.org/10.1038/371432a0.
35. Sijbers AM, de Laat WL, Ariza RR, Biggerstaff M, Wei YF, Moggs JG, Carter KC, Shell BK, Evans E, de Jong MC, Rademakers S, de Rooij J, Jaspers NGJ, Hoeijmakers JHJ, Wood RD. Xeroderma pigmentosum group F caused by a defect in a structure-specific DNA repair endonuclease. Cell. 1996;86:811–22. https://doi.org/10.1016/S0092-8674(00)80155-5.
36. Ito S, Kuraoka I, Chymkowitch P, Compe E, Takedachi A, Ishigami C, Coin F, Egly JM, Tanaka K. XPG stabilizes TFIIH, allowing transactivation of nuclear receptors: implications for Cockayne syndrome in XP-G/CS patients. Mol Cell. 2007;26:231–43. https://doi.org/10.1016/j.molcel.2007.03.013.
37. Zotter A, Luijsterburg MS, Warmerdam DO, Ibrahim S, Nigg A, van Cappellen WA, Hoeijmakers JHJ, van Driel R, Vermeulen W, Houtsmuller AB. Recruitment of the nucleotide excision repair endonuclease XPG to sites of UV-induced DNA damage depends on functional TFIIH. Mol Cell Biol. 2006;26:8868–79. https://doi.org/10.1128/MCB.00695-06.
38. Li L, Elledge SJ, Peterson CA, Bales ES, Legerski RJ. Specific association between the human DNA repair proteins XPA and ERCC1. Proc Natl Acad Sci U S A. 1994;91:5012–6. https://doi.org/10.1073/pnas.91.11.5012.
39. Li L, Peterson CA, Lu X, Legerski RJ. Mutations in XPA that prevent association with ERCC1 are defective in nucleotide excision repair. Mol Cell Biol. 1995;15:1993–8. https://doi.org/10.1128/MCB.15.4.1993.

40. Park CH, Sancar A. Formation of a ternary complex by human XPA, ERCC1, and ERCC4(XPF) excision repair proteins. Proc Natl Acad Sci U S A. 1994;91:5017–21. https://doi.org/10.1073/pnas.91.11.5017.
41. Staresincic L, Fagbemi AF, Enzlin JH, Gourdin AM, Wijgers N, Dunand-Sauthier I, Giglia-Mari G, Clarkson SG, Vermeulen W, Schärer OD. Coordination of dual incision and repair synthesis in human nucleotide excision repair. EMBO J. 2009;28:1111–20. https://doi.org/10.1038/emboj.2009.49.
42. Constantinou A, Gunz D, Evans E, Lalle P, Bates PA, Wood RD, Clarkson SG. Conserved residues of human XPG protein important for nuclease activity and function in nucleotide excision repair. J Biol Chem. 1999;274:5637–48. https://doi.org/10.1074/jbc.274.9.5637.
43. Wakasugi M, Reardon JT, Sancar A. The non-catalytic function of XPG protein during dual incision in human nucleotide excision repair. J Biol Chem. 1997;272:16030–4. https://doi.org/10.1074/jbc.272.25.16030.
44. Aboussekhra A, Biggerstaff M, Shivji MKK, Vilpo JA, Moncollin V, Podust VN, Protić M, Hübscher U, Egly JM, Wood RD. Mammalian DNA nucleotide excision repair reconstituted with purified protein components. Cell. 1995;80:859–68. https://doi.org/10.1016/0092-8674(95)90289-9.
45. Araújo SJ, Tirode F, Coin F, Pospiech H, Syväoja JE, Stucki M, Hübscher U, Egly JM, Wood RD. Nucleotide excision repair of DNA with recombinant human proteins: definition of the minimal set of factors, active forms of TFIIH, and modulation by CAK. Genes Dev. 2000;14:349–59. https://doi.org/10.1101/gad.14.3.349.
46. Shivji MK, Podust VN, Hübscher U, Wood RD. Nucleotide excision repair DNA synthesis by DNA polymerase epsilon in the presence of PCNA, RFC, and RPA. Biochemistry. 1995;34:5011–7. https://doi.org/10.1021/bi00015a012.
47. Gary R, Ludwig DL, Cornelius HL, MacInnes MA, Park MS. The DNA repair endonuclease XPG binds to proliferating cell nuclear antigen (PCNA) and shares sequence elements with the PCNA-binding regions of FEN-1 and cyclin-dependent kinase inhibitor p21. J Biol Chem. 1997;272:24522–9. https://doi.org/10.1074/jbc.272.39.24522.
48. Ogi T, Lehmann AR. The Y-family DNA polymerase k (pol k) functions in mammalian nucleotide-excision repair. Nat Cell Biol. 2006;8:640–2. https://doi.org/10.1038/ncb1417.
49. Ogi T, Limsirichaikul S, Overmeer RM, Volker M, Takenaka K, Cloney R, Nakazawa Y, Niimi A, Miki Y, Jaspers NG, Mullenders LHF, Yamashita S, Fousteri MI, Lehmann AR. Three DNA polymerases, recruited by different mechanisms, carry out NER repair synthesis in human cells. Mol Cell. 2010;37:714–27. https://doi.org/10.1016/j.molcel.2010.02.009.
50. Moser J, Kool H, Giakzidis I, Caldecott K, Mullenders LHF, Fousteri MI. Sealing of chromosomal DNA nicks during nucleotide excision repair requires XRCC1 and DNA ligase III alpha in a cell-cycle-specific manner. Mol Cell. 2007;27:311–23. https://doi.org/10.1016/j.molcel.2007.06.014.
51. Legerski R, Peterson C. Expression cloning of a human DNA repair gene involved in xeroderma pigmentosum group C. Nature. 1992;359:70–3. https://doi.org/10.1038/359070a0.
52. Masutani C, Sugasawa K, Yanagisawa J, Sonoyama T, Ui M, Enomoto T, Takio K, Tanaka K, van der Spek PJ, Bootsma D, Hoeijmakers JHJ, Hanaoka F. Purification and cloning of a nucleotide excision repair complex involving the xeroderma pigmentosum group C protein and a human homologue of yeast RAD23. EMBO J. 1994;13:1831–43.
53. Shivji MKK, Eker APM, Wood RD. DNA repair defect in xeroderma pigmentosum group C and complementing factor from HeLa cells. J Biol Chem. 1994;269:22749–57.
54. Min J-H, Pavletich NP. Recognition of DNA damage by the Rad4 nucleotide excision repair protein. Nature. 2007;449:570–5. https://doi.org/10.1038/nature06155.
55. Anantharaman V. Peptide-N-glycanases and DNA repair proteins, Xp-C/Rad4, are, respectively, active and inactivated enzymes sharing a common transglutaminase fold. Hum Mol Genet. 2001;10:1627–30. https://doi.org/10.1093/hmg/10.16.1627.
56. Lee J-H, Choi JM, Lee C, Yi KJ, Cho Y. Structure of a peptide:N-glycanase-Rad23 complex: insight into the deglycosylation for denatured glycoproteins. Proc Natl Acad Sci U S A. 2005;102:9144–9. https://doi.org/10.1073/pnas.0502082102.

57. Miller RD, Prakash L, Prakash S. Defective excision of pyrimidine dimers and interstrand DNA crosslinks in rad7 and rad23 mutants of *Saccharomyces cerevisiae*. Mol Gen Genet. 1982;188:235–9. https://doi.org/10.1007/BF00332681.
58. Sugasawa K, Ng JMY, Masutani C, Maekawa T, Uchida A, van der Spek PJ, Eker APM, Rademakers S, Visser C, Aboussekhra A, Wood RD, Hanaoka F, Bootsma D, Hoeijmakers JHJ. Two human homologs of Rad23 are functionally interchangeable in complex formation and stimulation of XPC repair activity. Mol Cell Biol. 1997;17:6924–31. https://doi.org/10.1128/MCB.17.12.6924.
59. Ng JMY, Vermeulen W, van der Horst GTJ, Bergink S, Sugasawa K, Vrieling H, Hoeijmakers JHJ. A novel regulation mechanism of DNA repair by damage-induced and RAD23-dependent stabilization of xeroderma pigmentosum group C protein. Genes Dev. 2003;17:1630–45. https://doi.org/10.1101/gad.260003.
60. Okuda Y, Nishi R, Ng JMY, Vermeulen W, van der Horst GTJ, Mori T, Hoeijmakers JHJ, Hanaoka F, Sugasawa K. Relative levels of the two mammalian Rad23 homologs determine composition and stability of the xeroderma pigmentosum group C protein complex. DNA Repair (Amst). 2004;3:1285–95. https://doi.org/10.1016/j.dnarep.2004.06.010.
61. van der Spek PJ, Visser CE, Hanaoka F, Smit B, Hagemeijer A, Bootsma D, Hoeijmakers JHJ. Cloning, comparative mapping, and RNA expression of the mouse homologues of the *Saccharomyces cerevisiae* nucleotide excision repair gene *RAD23*. Genomics. 1996;31:20–7. https://doi.org/10.1006/geno.1996.0004.
62. Watkins JF, Sung P, Prakash L, Prakash S. The *Saccharomyces cerevisiae* DNA repair gene *RAD23* encodes a nuclear protein containing a ubiquitin-like domain required for biological function. Mol Cell Biol. 1993;13:7757–65. https://doi.org/10.1128/MCB.13.12.7757.
63. Sugasawa K, Masutani C, Uchida A, Maekawa T, van der Spek PJ, Bootsma D, Hoeijmakers JHJ, Hanaoka F. HHR23B, a human Rad23 homolog, stimulates XPC protein in nucleotide excision repair in vitro. Mol Cell Biol. 1996;16:4852–61. https://doi.org/10.1128/MCB.16.9.4852.
64. Masutani C, Araki M, Sugasawa K, van der Spek PJ, Yamada A, Uchida A, Maekawa T, Bootsma D, Hoeijmakers JHJ, Hanaoka F. Identification and characterization of XPC-binding domain of hHR23B. Mol Cell Biol. 1997;17:6915–23. https://doi.org/10.1128/MCB.17.12.6915.
65. Araki M, Masutani C, Takemura M, Uchida A, Sugasawa K, Kondoh J, Ohkuma Y, Hanaoka F. Centrosome protein centrin 2/caltractin 1 is part of the xeroderma pigmentosum group C complex that initiates global genome nucleotide excision repair. J Biol Chem. 2001;276:18665–72. https://doi.org/10.1074/jbc.M100855200.
66. Nishi R, Okuda Y, Watanabe E, Mori T, Iwai S, Masutani C, Sugasawa K, Hanaoka F. Centrin 2 stimulates nucleotide excision repair by interacting with xeroderma pigmentosum group C protein. Mol Cell Biol. 2005;25:5664–74. https://doi.org/10.1128/MCB.25.13.5664-5674.2005.
67. Mu D, Park CH, Matsunaga T, Hsu DS, Reardon JT, Sancar A. Reconstitution of human DNA repair excision nuclease in a highly defined system. J Biol Chem. 1995;270:2415–8. https://doi.org/10.1074/jbc.270.6.2415.
68. Errabolu R, Sanders MA, Salisbury JL. Cloning of a cDNA encoding human centrin, an EF-hand protein of centrosomes and mitotic spindle poles. J Cell Sci. 1994;107(Pt 1):9–16.
69. Salisbury JL, Suino KM, Busby R, Springett M. Centrin-2 is required for centriole duplication in mammalian cells. Curr Biol. 2002;12:1287–92. https://doi.org/10.1016/S0960-9822(02)01019-9.
70. Batty D, Rapic'-Otrin V, Levine AS, Wood RD. Stable binding of human XPC complex to irradiated DNA confers strong discrimination for damaged sites. J Mol Biol. 2000;300:275–90. https://doi.org/10.1006/jmbi.2000.3857.
71. Hey T, Lipps G, Sugasawa K, Iwai S, Hanaoka F, Krauss G. The XPC-HR23B complex displays high affinity and specificity for damaged DNA in a true-equilibrium fluorescence assay. Biochemistry. 2002;41:6583–7. https://doi.org/10.1021/bi012202t.

72. Kusumoto R, Masutani C, Sugasawa K, Iwai S, Araki M, Uchida A, Mizukoshi T, Hanaoka F. Diversity of the damage recognition step in the global genomic nucleotide excision repair in vitro. Mutat Res. 2001;485:219–27. https://doi.org/10.1016/S0921-8777(00)00082-3.
73. Sugasawa K, Okamoto T, Shimizu Y, Masutani C, Iwai S, Hanaoka F. A multistep damage recognition mechanism for global genomic nucleotide excision repair. Genes Dev. 2001;15:507–21. https://doi.org/10.1101/gad.866301.
74. Sugasawa K, Shimizu Y, Iwai S, Hanaoka F. A molecular mechanism for DNA damage recognition by the xeroderma pigmentosum group C protein complex. DNA Repair (Amst). 2002;1:95–107. https://doi.org/10.1016/S1568-7864(01)00008-8.
75. Buterin T, Meyer C, Giese B, Naegeli H. DNA quality control by conformational readout on the undamaged strand of the double helix. Chem Biol. 2005;12:913–22. https://doi.org/10.1016/j.chembiol.2005.06.011.
76. Sugasawa K, Akagi J-I, Nishi R, Iwai S, Hanaoka F. Two-step recognition of DNA damage for mammalian nucleotide excision repair: directional binding of the XPC complex and DNA strand scanning. Mol Cell. 2009;36:642–53. https://doi.org/10.1016/j.molcel.2009.09.035.
77. Kong M, Liu L, Chen X, Driscoll KI, Mao P, Böhm S, Kad NM, Watkins SC, Bernstein KA, Wyrick JJ, Min J-H, Van Houten B. Single-molecule imaging reveals that Rad4 employs a dynamic DNA damage recognition process. Mol Cell. 2016;64:376–87. https://doi.org/10.1016/j.molcel.2016.09.005.
78. Camenisch U, Träutlein D, Clement FC, Fei J, Leitenstorfer A, Ferrando-May E, Naegeli H. Two-stage dynamic DNA quality check by xeroderma pigmentosum group C protein. EMBO J. 2009;28:2387–99. https://doi.org/10.1038/emboj.2009.187.
79. Chen X, Velmurugu Y, Zheng G, Park B, Shim Y, Kim Y, Liu L, Van Houten B, He C, Ansari A, Min J-H. Kinetic gating mechanism of DNA damage recognition by Rad4/XPC. Nat Commun. 2015;6:5849. https://doi.org/10.1038/ncomms6849.
80. Hess MT, Schwitter U, Petretta M, Giese B, Naegeli H. Bipartite substrate discrimination by human nucleotide excision repair. Proc Natl Acad Sci U S A. 1997;94:6664–9. https://doi.org/10.1016/0092-8674(95)90289-9.
81. McAteer K, Jing Y, Kao J, Taylor JS, Kennedy MA. Solution-state structure of a DNA dodecamer duplex containing a Cis-syn thymine cyclobutane dimer, the major UV photoproduct of DNA. J Mol Biol. 1998;282:1013–32. https://doi.org/10.1006/jmbi.1998.2062.
82. Coin F, Marinoni JC, Rodolfo C, Fribourg S, Pedrini AM, Egly JM. Mutations in the XPD helicase gene result in XP and TTD phenotypes, preventing interaction between XPD and the p44 subunit of TFIIH. Nat Genet. 1998;20:184–8. https://doi.org/10.1038/2491.
83. Schaeffer L, Moncollin V, Roy R, Staub A, Mezzina M, Sarasin A, Weeda G, Hoeijmakers JHJ, Egly JM. The ERCC2/DNA repair protein is associated with the class II BTF2/TFIIH transcription factor. EMBO J. 1994;13:2388–92.
84. Mathieu N, Kaczmarek N, Naegeli H. Strand- and site-specific DNA lesion demarcation by the xeroderma pigmentosum group D helicase. Proc Natl Acad Sci U S A. 2010;107:17545–50. https://doi.org/10.1073/pnas.1004339107.
85. Naegeli H, Bardwell L, Friedberg EC. The DNA helicase and adenosine triphosphatase activities of yeast Rad3 protein are inhibited by DNA damage. A potential mechanism for damage-specific recognition. J Biol Chem. 1992;267:392–8.
86. Naegeli H, Sugasawa K. The xeroderma pigmentosum pathway: decision tree analysis of DNA quality. DNA Repair (Amst). 2011;10:673–83. https://doi.org/10.1016/j.dnarep.2011.04.019.
87. Hwang JR, Moncollin V, Vermeulen W, Seroz T, van Vuuren H, Hoeijmakers JHJ, Egly JM. A 3′→5′ XPB helicase defect in repair/transcription factor TFIIH of xeroderma pigmentosum group B affects both DNA repair and transcription. J Biol Chem. 1996;271:15898–904. https://doi.org/10.1074/jbc.271.27.15898.
88. Tirode F, Busso D, Coin F, Egly JM. Reconstitution of the transcription factor TFIIH: assignment of functions for the three enzymatic subunits, XPB, XPD, and cdk7. Mol Cell. 1999;3:87–95. https://doi.org/10.1016/S1097-2765(00)80177-X.

89. Coin F, Oksenych V, Egly JM. Distinct roles for the XPB/p52 and XPD/p44 subcomplexes of TFIIH in damaged DNA opening during nucleotide excision repair. Mol Cell. 2007;26:245–56. https://doi.org/10.1016/j.molcel.2007.03.009.
90. Oksenych V, Bernardes de Jesus B, Zhovmer A, Egly JM, Coin F. Molecular insights into the recruitment of TFIIH to sites of DNA damage. EMBO J. 2009;28:2971–80. https://doi.org/10.1038/emboj.2009.230.
91. Tanaka K, Miura N, Satokata I, Miyamoto I, Yoshida MC, Satoh Y, Kondo S, Yasui A, Okayama H, Okada Y. Analysis of a human DNA excision repair gene involved in group A xeroderma pigmentosum and containing a zinc-finger domain. Nature. 1990;348:73–6. https://doi.org/10.1038/348073a0.
92. Bunick CG, Miller MR, Fuller BE, Fanning E, Chazin WJ. Biochemical and structural domain analysis of xeroderma pigmentosum complementation group C protein. Biochemistry. 2006;45:14965–79. https://doi.org/10.1021/bi061370o.
93. Nishi R, Sakai W, Tone D, Hanaoka F, Sugasawa K. Structure-function analysis of the EF-hand protein centrin-2 for its intracellular localization and nucleotide excision repair. Nucleic Acids Res. 2013;41:6917–29. https://doi.org/10.1093/nar/gkt434.
94. Nocentini S, Coin F, Saijo M, Tanaka K, Egly JM. DNA damage recognition by XPA protein promotes efficient recruitment of transcription factor II H. J Biol Chem. 1997;272:22991–4. https://doi.org/10.1074/jbc.272.37.22991.
95. Park CH, Mu D, Reardon JT, Sancar A. The general transcription-repair factor TFIIH is recruited to the excision repair complex by the XPA protein independent of the TFIIE transcription factor. J Biol Chem. 1995;270:4896–902. https://doi.org/10.1074/jbc.270.9.4896.
96. You J-S, Wang M, Lee S-H. Biochemical analysis of the damage recognition process in nucleotide excision repair. J Biol Chem. 2003;278:7476–85. https://doi.org/10.1074/jbc.M210603200.
97. Asahina H, Kuraoka I, Shirakawa M, Morita EH, Miura N, Miyamoto I, Ohtsuka E, Okada Y, Tanaka K. The XPA protein is a zinc metalloprotein with an ability to recognize various kinds of DNA damage. Mutat Res. 1994;315:229–37. https://doi.org/10.1016/0921-8777(94)90034-5.
98. Jones CJ, Wood RD. Preferential binding of the xeroderma pigmentosum group A complementing protein to damaged DNA. Biochemistry. 1993;32:12096–104. https://doi.org/10.1021/bi00096a021.
99. Missura M, Buterin T, Hindges R, Hübscher U, Kaspárková J, Brabec V, Naegeli H. Double-check probing of DNA bending and unwinding by XPA-RPA: an architectural function in DNA repair. EMBO J. 2001;20:3554–64. https://doi.org/10.1093/emboj/20.13.3554.
100. Krasikova YS, Rechkunova NI, Maltseva EA, Petruseva IO, Lavrik OI. Localization of xeroderma pigmentosum group A protein and replication protein A on damaged DNA in nucleotide excision repair. Nucleic Acids Res. 2010;38:8083–94. https://doi.org/10.1093/nar/gkq649.
101. Coin F, Oksenych V, Mocquet V, Groh S, Blattner C, Egly JM. Nucleotide excision repair driven by the dissociation of CAK from TFIIH. Mol Cell. 2008;31:9–20. https://doi.org/10.1016/j.molcel.2008.04.024.
102. Goosen N, Moolenaar GF. Role of ATP hydrolysis by UvrA and UvrB during nucleotide excision repair. Res Microbiol. 2001;152:401–9. https://doi.org/10.1016/S0923-2508(01)01211-6.
103. Kad NM, Wang H, Kennedy GG, Warshaw DM, Van Houten B. Collaborative dynamic DNA scanning by nucleotide excision repair proteins investigated by single- molecule imaging of quantum-dot-labeled proteins. Mol Cell. 2010;37:702–13. https://doi.org/10.1016/j.molcel.2010.02.003.
104. Selby CP, Sancar A. Structure and function of the (A) BC excinuclease of *Escherichia coli*. Mutat Res. 1990;236:203–11. https://doi.org/10.1016/0921-8777(90)90005-P.
105. Van Houten B, Croteau DL, DellaVecchia MJ, Wang H, Kisker C. "Close-fitting sleeves": DNA damage recognition by the UvrABC nuclease system. Mutat Res. 2005;577:92–117. https://doi.org/10.1016/j.mrfmmm.2005.03.013.

106. Peng W, Shaw BR. Accelerated deamination of cytosine residues in UV-induced cyclobutane pyrimidine dimers leads to CC→TT transitions. Biochemistry. 1996;35:10172–81. https://doi.org/10.1021/bi960001x.
107. Stary A, Sarasin A. Molecular mechanisms of UV-induced mutations as revealed by the study of DNA polymerase h in human cells. Res Microbiol. 2002;153:441–5. https://doi.org/10.1016/S0923-2508(02)01343-8.
108. You YH, Lee DH, Yoon JH, Nakajima S, Yasui A, Pfeifer GP. Cyclobutane pyrimidine dimers are responsible for the vast majority of mutations induced by UVB irradiation in mammalian cells. J Biol Chem. 2001;276:44688–94. https://doi.org/10.1074/jbc.M107696200.
109. Keeney S, Chang GJ, Linn S. Characterization of a human DNA damage binding protein implicated in xeroderma pigmentosum E. J Biol Chem. 1993;268:21293–300.
110. Nichols AF, Itoh T, Graham JA, Liu W, Yamaizumi M, Linn S. Human damage-specific DNA-binding protein p48. Characterization of XPE mutations and regulation following UV irradiation. J Biol Chem. 2000;275:21422–8. https://doi.org/10.1074/jbc.M000960200.
111. Rapić-Otrin V, Navazza V, Nardo T, Botta E, McLenigan M, Bisi DC, Levine AS, Stefanini M. True XP group E patients have a defective UV-damaged DNA binding protein complex and mutations in DDB2 which reveal the functional domains of its p48 product. Hum Mol Genet. 2003;12:1507–22. https://doi.org/10.1093/hmg/ddg174.
112. Dualan R, Brody T, Keeney S, Nichols AF, Admon A, Linn S. Chromosomal localization and cDNA cloning of the genes (DDB1 and DDB2) for the p127 and p48 subunits of a human damage-specific DNA binding protein. Genomics. 1995;29:62–9. https://doi.org/10.1006/geno.1995.1215.
113. Scrima A, Konícková R, Czyzewski BK, Kawasaki Y, Jeffrey PD, Groisman R, Nakatani Y, Iwai S, Pavletich NP, Thomä NH. Structural basis of UV DNA-damage recognition by the DDB1-DDB2 complex. Cell. 2008;135:1213–23. https://doi.org/10.1016/j.cell.2008.10.045.
114. Payne A, Chu G. Xeroderma pigmentosum group E binding factor recognizes a broad spectrum of DNA damage. Mutat Res. 1994;310:89–102. https://doi.org/10.1016/0027-5107(94)90012-4.
115. Reardon JT, Nichols AF, Keeney S, Smith CA, Taylor JS, Linn S, Sancar A. Comparative analysis of binding of human damaged DNA-binding protein (XPE) and *Escherichia coli* damage recognition protein (UvrA) to the major ultraviolet photoproducts: T[c,s]T, T[t,s]T, T[6-4]T, and T[Dewar]T. J Biol Chem. 1993;268:21301–8.
116. Fitch ME, Nakajima S, Yasui A, Ford JM. In vivo recruitment of XPC to UV-induced cyclobutane pyrimidine dimers by the DDB2 gene product. J Biol Chem. 2003;278:46906–10. https://doi.org/10.1074/jbc.M307254200.
117. Moser J, Volker M, Kool H, Alekseev S, Vrieling H, Yasui A, Van Zeeland AA, Mullenders LHF. The UV-damaged DNA binding protein mediates efficient targeting of the nucleotide excision repair complex to UV-induced photo lesions. DNA Repair (Amst). 2005;4:571–82. https://doi.org/10.1016/j.dnarep.2005.01.001.
118. Nishi R, Alekseev S, Dinant C, Hoogstraten D, Houtsmuller AB, Hoeijmakers JHJ, Vermeulen W, Hanaoka F, Sugasawa K. UV-DDB-dependent regulation of nucleotide excision repair kinetics in living cells. DNA Repair (Amst). 2009;8:767–76. https://doi.org/10.1016/j.dnarep.2009.02.004.
119. Wang Q-E, Zhu Q, Wani G, Chen J, Wani AA. UV radiation-induced XPC translocation within chromatin is mediated by damaged-DNA binding protein, DDB2. Carcinogenesis. 2004;25:1033–43. https://doi.org/10.1093/carcin/bgh085.
120. Hwang BJ, Ford JM, Hanawalt PC, Chu G. Expression of the p48 xeroderma pigmentosum gene is p53-dependent and is involved in global genomic repair. Proc Natl Acad Sci U S A. 1999;96:424–8. https://doi.org/10.1073/pnas.96.2.424.
121. Tang JY, Hwang BJ, Ford JM, Hanawalt PC, Chu G. Xeroderma pigmentosum p48 gene enhances global genomic repair and suppresses UV-induced mutagenesis. Mol Cell. 2000;5:737–44. https://doi.org/10.1016/S1097-2765(00)80252-X.
122. Adimoolam S, Ford JM. p53 and DNA damage-inducible expression of the xeroderma pigmentosum group C gene. Proc Natl Acad Sci U S A. 2002;99:12985–90. https://doi.org/10.1073/pnas.202485699.

123. Itoh T, O'Shea C, Linn S. Impaired regulation of tumor suppressor p53 caused by mutations in the xeroderma pigmentosum DDB2 gene: mutual regulatory interactions between p48(DDB2) and p53. Mol Cell Biol. 2003;23:7540–53. https://doi.org/10.1128/MCB.23.21.7540-7553.2003.
124. Bagchi S, Raychaudhuri P. Damaged-DNA binding protein-2 drives apoptosis following DNA damage. Cell Div. 2010;5:3. https://doi.org/10.1186/1747-1028-5-3.
125. Itoh T, Iwashita S, Cohen MB, Meyerholz DK, Linn S. Ddb2 is a haploinsufficient tumor suppressor and controls spontaneous germ cell apoptosis. Hum Mol Genet. 2007;16:1578–86. https://doi.org/10.1093/hmg/ddm107.
126. Stoyanova T, Roy N, Kopanja D, Bagchi S, Raychaudhuri P. DDB2 decides cell fate following DNA damage. Proc Natl Acad Sci U S A. 2009;106:10690–5. https://doi.org/10.1073/pnas.0812254106.
127. Stubbert LJ, Smith JM, Hamill JD, Arcand TL, McKay BC. The anti-apoptotic role for p53 following exposure to ultraviolet light does not involve DDB2. Mutat Res. 2009;663:69–76. https://doi.org/10.1016/j.mrfmmm.2009.01.010.
128. Itoh T, Cado D, Kamide R, Linn S. DDB2 gene disruption leads to skin tumors and resistance to apoptosis after exposure to ultraviolet light but not a chemical carcinogen. Proc Natl Acad Sci U S A. 2004;101:2052–7. https://doi.org/10.1073/pnas.0306551101.
129. Angers S, Li T, Yi X, MacCoss MJ, Moon RT, Zheng N. Molecular architecture and assembly of the DDB1-CUL4A ubiquitin ligase machinery. Nature. 2006;443:590–3. https://doi.org/10.1038/nature05175.
130. Higa LA, Wu M, Ye T, Kobayashi R, Sun H, Zhang H. CUL4-DDB1 ubiquitin ligase interacts with multiple WD40-repeat proteins and regulates histone methylation. Nat Cell Biol. 2006;8:1277–83. https://doi.org/10.1038/ncb1490.
131. Lee J, Zhou P. DCAFs, the missing link of the CUL4-DDB1 ubiquitin ligase. Mol Cell. 2007;26:775–80. https://doi.org/10.1016/j.molcel.2007.06.001.
132. Cavadini S, Fischer ES, Bunker RD, Potenza A, Lingaraju GM, Goldie KN, Mohamed WI, Faty M, Petzold G, Beckwith REJ, Tichkule RB, Hassiepen U, Abdulrahman W, Pantelic RS, Matsumoto S, Sugasawa K, Stahlberg H, Thomä NH. Cullin-RING ubiquitin E3 ligase regulation by the COP9 signalosome. Nature. 2016;531:598–603. https://doi.org/10.1038/nature17416.
133. Fischer ES, Scrima A, Böhm K, Matsumoto S, Lingaraju GM, Faty M, Yasuda T, Cavadini S, Wakasugi M, Hanaoka F, Iwai S, Gut H, Sugasawa K, Thomä NH. The molecular basis of CRL4DDB2/CSA ubiquitin ligase architecture, targeting, and activation. Cell. 2011;147:1024–39. https://doi.org/10.1016/j.cell.2011.10.035.
134. Groisman R, Polanowska J, Kuraoka I, Sawada J-I, Saijo M, Drapkin R, Kisselev AF, Tanaka K, Nakatani Y. The ubiquitin ligase activity in the DDB2 and CSA complexes is differentially regulated by the COP9 signalosome in response to DNA damage. Cell. 2003;113:357–67. https://doi.org/10.1016/S0092-8674(03)00316-7.
135. Chen X, Zhang Y, Douglas L, Zhou P. UV-damaged DNA-binding proteins are targets of CUL-4A-mediated ubiquitination and degradation. J Biol Chem. 2001;276:48175–82. https://doi.org/10.1074/jbc.M106808200.
136. Kapetanaki MG, Guerrero-Santoro J, Bisi DC, Hsieh CL, Rapić-Otrin V, Levine AS. The DDB1-CUL4ADDB2 ubiquitin ligase is deficient in xeroderma pigmentosum group E and targets histone H2A at UV-damaged DNA sites. Proc Natl Acad Sci U S A. 2006;103:2588–93. https://doi.org/10.1073/pnas.0511160103.
137. Matsuda N, Azuma K, Saijo M, Iemura S-I, Hioki Y, Natsume T, Chiba T, Tanaka K, Tanaka K. DDB2, the xeroderma pigmentosum group E gene product, is directly ubiquitylated by Cullin 4A-based ubiquitin ligase complex. DNA Repair (Amst). 2005;4:537–45. https://doi.org/10.1016/j.dnarep.2004.12.012.
138. Wang H, Zhai L, Xu J, Joo H-Y, Jackson S, Erdjument-Bromage H, Tempst P, Xiong Y, Zhang Y. Histone H3 and H4 ubiquitylation by the CUL4-DDB-ROC1 ubiquitin ligase facilitates cellular response to DNA damage. Mol Cell. 2006;22:383–94. https://doi.org/10.1016/j.molcel.2006.03.035.

139. Matsumoto S, Fischer ES, Yasuda T, Dohmae N, Iwai S, Mori T, Nishi R, Yoshino K-I, Sakai W, Hanaoka F, Thomä NH, Sugasawa K. Functional regulation of the DNA damage-recognition factor DDB2 by ubiquitination and interaction with xeroderma pigmentosum group C protein. Nucleic Acids Res. 2015;43:1700–13. https://doi.org/10.1093/nar/gkv038.
140. Puumalainen M-R, Lessel D, Rüthemann P, Kaczmarek N, Bachmann K, Ramadan K, Naegeli H. Chromatin retention of DNA damage sensors DDB2 and XPC through loss of p97 segregase causes genotoxicity. Nat Commun. 2014;5:3695. https://doi.org/10.1038/ncomms4695.
141. Fitch ME, Cross IV, Turner SJ, Adimoolam S, Lin CX, Williams KG, Ford JM. The DDB2 nucleotide excision repair gene product p48 enhances global genomic repair in p53 deficient human fibroblasts. DNA Repair (Amst). 2003;2:819–26. https://doi.org/10.1016/S1568-7864(03)00066-1.
142. Rapić-Otrin V, McLenigan MP, Bisi DC, Gonzalez M, Levine AS. Sequential binding of UV DNA damage binding factor and degradation of the p48 subunit as early events after UV irradiation. Nucleic Acids Res. 2002;30:2588–98. https://doi.org/10.1093/nar/30.11.2588.
143. Pines A, Vrouwe MG, Marteijn JA, Typas D, Luijsterburg MS, Cansoy M, Hensbergen P, Deelder A, de Groot A, Matsumoto S, Sugasawa K, Thoma N, Vermeulen W, Vrieling H, Mullenders L. PARP1 promotes nucleotide excision repair through DDB2 stabilization and recruitment of ALC1. J Cell Biol. 2012;199:235–49. https://doi.org/10.1083/jcb.201112132.
144. Akita M, Tak Y-S, Shimura T, Matsumoto S, Okuda-Shimizu Y, Shimizu Y, Nishi R, Saitoh H, Iwai S, Mori T, Ikura T, Sakai W, Hanaoka F, Sugasawa K. SUMOylation of xeroderma pigmentosum group C protein regulates DNA damage recognition during nucleotide excision repair. Sci Rep. 2015;5:10984. https://doi.org/10.1038/srep10984.
145. van Cuijk L, van Belle GJ, Turkyilmaz Y, Poulsen SL, Janssens RC, Theil AF, Sabatella M, Lans H, Mailand N, Houtsmuller AB, Vermeulen W, Marteijn JA. SUMO and ubiquitin-dependent XPC exchange drives nucleotide excision repair. Nat Commun. 2015;6:7499. https://doi.org/10.1038/ncomms8499.
146. Gale JM, Nissen KA, Smerdon MJ. UV-induced formation of pyrimidine dimers in nucleosome core DNA is strongly modulated with a period of 10.3 bases. Proc Natl Acad Sci U S A. 1987;84:6644–8.
147. Gale JM, Smerdon MJ. UV induced (6-4) photoproducts are distributed differently than cyclobutane dimers in nucleosomes. Photochem Photobiol. 1990;51:411–7. https://doi.org/10.1111/j.1751-1097.1990.tb01732.x.
148. Pehrson JR. Thymine dimer formation as a probe of the path of DNA in and between nucleosomes in intact chromatin. Proc Natl Acad Sci U S A. 1989;86:9149–53.
149. Pehrson JR. Probing the conformation of nucleosome linker DNA in situ with pyrimidine dimer formation. J Biol Chem. 1995;270:22440–4. https://doi.org/10.1038/311532a0.
150. Hara R, Mo J, Sancar A. DNA damage in the nucleosome core is refractory to repair by human excision nuclease. Mol Cell Biol. 2000;20:9173–81. https://doi.org/10.1128/MCB.20.24.9173-9181.2000.
151. Yasuda T, Sugasawa K, Shimizu Y, Iwai S, Shiomi T, Hanaoka F. Nucleosomal structure of undamaged DNA regions suppresses the non-specific DNA binding of the XPC complex. DNA Repair (Amst). 2005;4:389–95. https://doi.org/10.1016/j.dnarep.2004.10.008.
152. Osakabe A, Tachiwana H, Kagawa W, Horikoshi N, Matsumoto S, Hasegawa M, Matsumoto N, Toga T, Yamamoto J, Hanaoka F, Thomä NH, Sugasawa K, Iwai S, Kurumizaka H. Structural basis of pyrimidine-pyrimidone (6-4) photoproduct recognition by UV-DDB in the nucleosome. Sci Rep. 2015;5:16330. https://doi.org/10.1038/srep16330.
153. Hara R, Sancar A. The SWI/SNF chromatin-remodeling factor stimulates repair by human excision nuclease in the mononucleosome core particle. Mol Cell Biol. 2002;22:6779–87. https://doi.org/10.1128/MCB.22.19.6779-6787.2002.
154. Hara R, Sancar A. Effect of damage type on stimulation of human excision nuclease by SWI/SNF chromatin remodeling factor. Mol Cell Biol. 2003;23:4121–5. https://doi.org/10.1128/MCB.23.12.4121-4125.2003.

155. Ura K, Araki M, Saeki H, Masutani C, Ito T, Iwai S, Mizukoshi T, Kaneda Y, Hanaoka F. ATP-dependent chromatin remodeling facilitates nucleotide excision repair of UV-induced DNA lesions in synthetic dinucleosomes. EMBO J. 2001;20:2004–14. https://doi.org/10.1093/emboj/20.8.2004.
156. Datta A, Bagchi S, Nag A, Shiyanov P, Adami GR, Yoon T, Raychaudhuri P. The p48 subunit of the damaged-DNA binding protein DDB associates with the CBP/p300 family of histone acetyltransferase. Mutat Res. 2001;486:89–97. https://doi.org/10.1016/S0921-8777(01)00082-9.
157. Martinez E, Palhan VB, Tjernberg A, Lymar ES, Gamper AM, Kundu TK, Chait BT, Roeder RG. Human STAGA complex is a chromatin-acetylating transcription coactivator that interacts with pre-mRNA splicing and DNA damage-binding factors in vivo. Mol Cell Biol. 2001;21:6782–95. https://doi.org/10.1128/MCB.21.20.6782-6795.2001.
158. Matsunuma R, Niida H, Ohhata T, Kitagawa K, Sakai S, Uchida C, Shiotani B, Matsumoto M, Nakayama KI, Ogura H, Shiiya N, Kitagawa M. UV damage-induced phosphorylation of HBO1 triggers $CRL4^{DDB2}$-mediated degradation to regulate cell proliferation. Mol Cell Biol. 2015;36:394–406. https://doi.org/10.1128/MCB.00809-15.
159. Adam S, Dabin J, Chevallier O, Leroy O, Baldeyron C, Corpet A, Lomonte P, Renaud O, Almouzni G, Polo SE. Real-time tracking of parental histones reveals their contribution to chromatin integrity following DNA damage. Mol Cell. 2016;64:65–78. https://doi.org/10.1016/j.molcel.2016.08.019.
160. Luijsterburg MS, Lindh M, Acs K, Vrouwe MG, Pines A, van Attikum H, Mullenders LH, Dantuma NP. DDB2 promotes chromatin decondensation at UV-induced DNA damage. J Cell Biol. 2012;197:267–81. https://doi.org/10.1083/jcb.201106074.
161. Kakumu E, Nakanishi S, Shiratori HM, Kato A, Kobayashi W, Machida S, Yasuda T, Adachi N, Saito N, Ikura T, Kurumizaka H, Kimura H, Yokoi M, Sakai W, Sugasawa K. Xeroderma pigmentosum group C protein interacts with histones: regulation by acetylated states of histone H3. Genes Cells. 2017;22:310–27. https://doi.org/10.1111/gtc.12479.
162. Luijsterburg MS, Dinant C, Lans H, Stap J, Wiernasz E, Lagerwerf S, Warmerdam DO, Lindh M, Brink MC, Dobrucki JW, Aten JA, Fousteri MI, Jansen G, Dantuma NP, Vermeulen W, Mullenders LHF, Houtsmuller AB, Verschure PJ, van Driel R. Heterochromatin protein 1 is recruited to various types of DNA damage. J Cell Biol. 2009;185:577–86. https://doi.org/10.1083/jcb.200810035.
163. Guo R, Chen J, Mitchell DL, Johnson DG. GCN5 and E2F1 stimulate nucleotide excision repair by promoting H3K9 acetylation at sites of damage. Nucleic Acids Res. 2011;39:1390–7. https://doi.org/10.1093/nar/gkq983.

Chapter 2
Disorders with Deficiency in TC-NER: Molecular Pathogenesis of Cockayne Syndrome and UV-Sensitive Syndrome

Chaowan Guo and Tomoo Ogi

Abstract Nucleotide excision repair (NER) is one of the most important DNA repair systems involved in removing a wide range of DNA damage from the genome. NER consists of two sub-pathways: the global genome nucleotide excision repair (GG-NER) pathway, which removes DNA lesions generated in the whole genome (as described in Chap. 1 of this book), and the transcription-coupled nucleotide excision repair (TC-NER) pathway, which removes lesions specifically from the transcribed strands of actively transcribed genes. At least 20 factors are involved in the TC-NER process, and mutations in the genes responsible for coding these factors may mainly result in two human genetic disorders: Cockayne syndrome (CS) and UV-sensitive syndrome (UV^SS). Despite similar molecular defects in TC-NER, CS and UV^SS show distinct clinical phenotypes. CS patients display severe developmental and neurological abnormalities as well as premature ageing, whereas UV^SS individuals only show milder cutaneous abnormalities, such as hypersensitivity to UV light. The molecular basis for the difference in the clinical features remains unclear. In this chapter, we will specifically describe the historical progress and recent findings of TC-NER and summarize the current understanding of the molecular pathogenesis of CS and UV^SS.

2.1 Introduction

The maintenance of genomic integrity and the accurate replication of the genome are critical processes for life. Genome integrity is constantly threatened by the by-products of normal cellular metabolic processes and environmental agents, such as ultraviolet (UV) exposure, ionizing radiation, and numerous genotoxic chemicals

C. Guo · T. Ogi (✉)
Department of Genetics, Research Institute of Environmental Medicine (RIEM), Nagoya University, Nagoya, Japan

Department of Human Genetics and Molecular Biology, Graduate School of Medicine, Nagoya University, Nagoya, Japan
e-mail: togi@riem.nagoya-u.ac.jp

C. Nishigori, K. Sugasawa (eds.), *DNA Repair Disorders*,
https://doi.org/10.1007/978-981-10-6722-8_2

[1, 2]. Unrepaired DNA lesions in the genome lead to diverse outcomes. At the cellular level, DNA lesions hamper DNA replication, transcription, and chromosome segregation, and thus, they result in cell cycle arrest, apoptosis, or necrosis. At the organismal level, DNA lesions are implicated in the development of cancer, ageing, and several genetic diseases [1, 3]. To prevent the deleterious consequences, living organisms across the evolutionary scale employ a sophisticated network of DNA repair systems. The DNA repair system is generally divided into five major sub-pathways, and each sub-pathway deals with structurally different types of DNA lesions, including direct damage removal, base excision repair (BER), nucleotide excision repair (NER), mismatch repair, and DNA double-strand break (DSB) repair [4]. Malfunctions in these repair mechanisms are associated with a variety of human disorders. Here, we mainly focus on NER, which is a distinct excision repair machinery that excises and removes a diversity of DNA lesions, including UV-induced photolesions, some forms of oxidative DNA lesions, and bulky base adducts induced by chemicals from the environment or from metabolic products [5]. Several other dedicated DNA repair pathways and their associated human disorders are addressed in other chapters of this book. The entire NER process consists of several sequential steps. It begins with the recognition of a DNA lesion, followed by the unwinding of the double-stranded DNA around the lesions and dual incisions of the damaged DNA strand on both sides of the lesions, and after the lesion containing the oligonucleotide is removed, a single-strand DNA "patch" is synthesized and is then ligated with the parental strand to complete the repair [5]. NER can be divided into two sub-pathways: global genome nucleotide excision repair (GG-NER) [6] and transcription-coupled nucleotide excision repair (TC-NER) [7, 8]. GG-NER operates in the damage removal from both expressed and silent genomic regions. Several genetic diseases, such as xeroderma pigmentosum (XP) and trichothiodystrophy (TTD), are associated with mutations in genes involved in GG-NER [9] (detailed information is described in Chap. 1 and Chaps. 3, 4, 5, 6, 7, 8, and 9 of this book). In this chapter, we focus on another sub-pathway of NER, the TC-NER, which is a versatile mechanism for the recognition and repair of DNA lesions from actively transcribed genes.

2.2 Transcription-Coupled Nucleotide Excision Repair

2.2.1 Molecular Mechanism of TC-NER

TC-NER was first described as a sub-pathway of NER in the early 1980s [10], and primary researchers noticed that NER removed lesions from the entire genome with varying efficiencies and that certain lesions, such as UV-induced cyclobutane pyrimidine dimers (CPDs), were removed more rapidly from the transcribed strands than the strands opposite the actively transcribed genes [11, 12]. This repair process was thereby named transcription-coupled nucleotide excision repair for its involvement in gene transcription. TC-NER was subsequently shown to operate for other lesions, the so-called bulky DNA adducts, including pyrimidine-pyrimidones (6-4) photoproducts (6-4PPs), cisplatin-induced intrastrand-crosslinks, BPDE (benzo(a)pyrene diolepoxide), and other polycyclic aromatic amines, such as acetyl-aminofluorene and aflatoxin, and some nitrosamines, such as MNNG (*N*-methyl-*N*′-nitro-*N*-nitrosoguanidine) and 4-NQO (4-nitroquinoline oxide) [13]. Transcription-coupled NER for such DNA lesions is shown to widely operate in *E. coli* [14], *S. cerevisiae* [15, 16], and rodents [11], as well as in human cells [15, 17, 18].

GG-NER and TC-NER mainly differ in the damage recognition steps. In GG-NER, the detection of helix-distorting lesions or aberrant chromatin structures is mediated by the indirect binding of the XPC complex (XPC-RAD23B-Centrin2) to the undamaged strand opposite the actual lesion. However, certain DNA lesions, such as UV-induced CPDs, which only cause small helix distortions and thus are not recognized by the XPC complex, require additional factors for detection. In this case, the UV-damaged DNA-binding protein (UV-DDB) complex, which comprises DDB1 and DDB2 (also known as XPE), binds to the lesions and promotes the subsequent binding of the XPC complex. Arrival of the XPC complex at the site of DNA damage then triggers the recruitment of the multifactor complex TFIIH [19].

Unlike GG-NER, in the TC-NER sub-pathway, the DNA lesion is recognized through a stalled elongating form of RNA polymerase II (RNA pol IIo) at the DNA damage site [7]. Cockayne syndrome B protein (CSB) interacts with the stalled RNA pol IIo, and it changes the DNA conformation around the damage site, altering the interface between RNA pol IIo and DNA [20]. CSB also recruits the CSA, Cullin-RING ubiquitin ligases (CRLs), as well as other TC-NER factors, including the recently identified UVSSA and its binding partner USP7 [21–23], to the stalled RNA pol IIo for the subsequent damage processing (the details are described later in this chapter).

After damage recognition, further damage processing in both GG-NER and TC-NER occurs via a common pathway [6, 24]. The general transcription factor TFIIH is recruited to the damaged site after lesion recognition, and its helicase subunits, XPB and XPD, are implicated in the unwinding of the damaged DNA. XPA, together with RPA, is responsible for the lesion-verification process and the stabilization of the pre-incision complex. The lesion is then incised by the ERCC1-XPF endonuclease complex from the 5′ side and by another endonuclease, XPG, from the 3′ side. The final gap-filling and ligation step is carried out by a series of proteins, including the proliferating cell nuclear antigen (PCNA), replication factor C (RFC), several DNA polymerases, and ligase-I or the ligase-III-XRCC1 complex, depending on the cell cycle and cellular circumstances (see Chap. 1 for detailed information).

2.2.2 *TC-NER-Deficient Disorders*

In humans, defects in genes responsible for the TC-NER processes may cause genodermatosis. Two representative disorders discussed here are Cockayne syndrome and UV-sensitive syndrome. Patients affected by these disorders are characterized by UV-photosensitivity, but there are some variations in the clinical features.

Cockayne syndrome is a rare autosomal recessive disorder with a multitude of clinical symptoms and was first reported by Edward Alfred Cockayne in 1936 [25]. CS patients have a characteristic face with deep-set eyes, prominent ears, and a wizened facial appearance. CS individuals exhibit three major clinical symptoms: profound growth failure (generally beginning in infancy but, in some cases, present at birth), premature ageing, and neurodevelopmental and later neurological dysfunctions, such as mental retardation and microcephaly. CS patients also display several minor clinical symptoms, including ocular abnormalities, dental caries, progressive sensorineural deafness, and cutaneous photosensitivity of the skin, but in contrast to XP patients, they do not have pigmentary changes and are not found to have susceptibility to sunlight-induced skin cancer [26]. CS is a progressive disorder, and the severity of CS is quite variable. Based on the severity of symptoms, CS patients can be generally classified into three subtypes: Type I (classical or moderate form of CS), Type II (early-onset or severe form of CS), and Type III (late-onset or milder form of CS) [27, 28]. Since cells from CS patients show hypersensitivity to UV irradiation, cell complementation analyses have defined two basic genetic complementation groups in CS, namely, CS-A and CS-B [20], which are caused by mutations in two genes essential for TC-NER, *CSA/ERCC8* and *CSB/ERCC6*. In addition to CS-A and CS-B, a few patients belong to xeroderma pigmentosum (XP) complementation groups -B, -D, -F, or -G (XP-B, XP-D, XP-F, or XP-G), and they exhibit overlapping features of CS, XP, and/or Fanconi anaemia (XP-B/CS, XP-D/CS, XP-G/CS, and XP-F/CS/FA) [29–31]. These XP/CS and XP/CS/FA patients generally suffer from very severe CS with major neurological symptoms, which are mostly lethal in infancy. The specified clinical aspects of CS will be further addressed in Chap. 10 of this book.

UV-sensitive syndrome is another human genetic disease with defects in TC-NER [32]. In contrast to other dedicated disorders with photosensitivity (*e.g.* XP and CS), $UV^{S}S$ patients only display relatively mild clinical manifestations limited to the skin, including acute sunburn and freckles, but no increased skin cancer susceptibility (distinct from XP patients, who develop malignant tumours in sun-exposed areas at an early age) or developmental or neurological abnormalities (in contrast to CS patients; see above).

The history of UV-sensitive syndrome began in the 1980s. Fujiwara *et al.* first described a Japanese mildly photosensitive patient, with complete absence of recovery of RNA synthesis after UV irradiation [33]. In 1994, Itoh *et al.* further reported two normally developed Japanese siblings, Kps2 and Kps3, who only showed slight cutaneous photosensitivity and cutaneous pigmentation [34]; the authors implied that a subset of photosensitive individuals are clearly distinguished from XP or CS. In 1995, Dr. Yamaizumi, at Kumamoto University, Japan, proposed a general category of UV-sensitive syndrome to include Kps2, Kps3, and the previously reported patient, UVS1KO [27], assigning all three individuals to the same complementation group by a cell fusion complementation test after UV irradiation [35] (note that UVS1KO turned out to be in the CSB complementation group and another patient, XP24KO, who had been previously assigned to XP-E, was reassigned to $UV^{S}S$). To date, very few $UV^{S}S$ patients have been reported, possibly because the

Table 2.1 UV^{S}S cases reported to date and their characteristics

Complementation groups	UV^{S}S/CS-A	UV^{S}S/CS-B	UV^{S}S-A
Causative gene	*CSA* (*ERCC8*)	*CSB* (*ERCC6*)	*UVSSA* (*KIAA1530*)
HGNC/ID	3439	3438	29304
Locations	5q12.1	10q11.23	4p16.3
Protein (MW)	396 aa (45 kDa)	1493 aa (170 kDa)	709 aa (80 kDa)
Domain(s)	WD repeat domain	Helicase motif Acidic domain Nuclear localization signal	VHS domain Nuclear localization signal
Pathogenic mutations of UV^{S}S	p.Trp361Cys (hom)	p.Arg77* (hom)	p.Lys123* (hom) p.Ile31Phefs*9 (hom) p.Cys32Arg (hom)

clinical symptoms of UV^{S}S are very mild and individuals with UV^{S}S may seek for medical care only when they are affected by an acute sunburn. The diagnosis is also difficult because they do not display any other apparent devastating clinical symptoms such as CS or XP patients do. In total, seven UV^{S}S patients have been reported to date, and they belong to three complementation groups, which are defined by specific mutations in *CSA* [36], *CSB* [37], and the recently identified gene *UVSSA/KIAA1530* [21–23]. Table 2.1 lists the UV^{S}S cases reported to date and their characteristics.

2.2.3 Diagnostic Methods for TC-NER-Deficient Disorders

The clinical features of the CS and UV^{S}S patients have some similarities but also marked differences. However, the acute cellular responses and sensitivities to UV light in CS- and UV^{S}S-derived cells are basically identical. Fibroblasts derived from both CS and UV^{S}S patients exhibit severe sensitivity to UV irradiation and are also defective in the recovery of RNA synthesis. However, the global genome repair of UV-induced CPDs and 6-4PPs (GG-NER is mostly defective in XP cells) is normal [21–23]. These characteristics suggest that the cells from UV^{S}S and CS are specifically defective in TC-NER. Two assays, unscheduled DNA synthesis (UDS) and recovery of RNA synthesis (RRS) after UV damage, are commonly used for measuring NER activities in patient-derived cells. These are widely applied for basic research and the clinical diagnosis of NER-deficient disorders. UDS represents damage-induced, non-S phase DNA repair synthesis, and it reflects the entire NER activities (GG-NER and TC-NER), whereas RRS is applied for the measurement of TC-NER activity only as the unrepaired DNA damage elicits a transient inhibition of messenger RNA (mRNA) transcription [38]; note that the ribosomal RNA (rRNA) synthesis also contributes to the recovery of total RNA synthesis after UV damage but that this recovery rate is much slower for rRNA synthesis than it is for mRNA synthesis [39]. It has been shown that the repair of CPDs in the rRNA genes is less efficient than that in the genome overall suggesting the inefficient repair of rRNA genes may resulting the slow recovery of rRNA synthesis [40]. In principle, UDS accounts for both GG-NER and TC-NER activities, but the contribution of TC-NER to UDS is usually ignorable because of its limited target. Generally, the RRS and UDS activities in fibroblasts derived from undiagnosed patients and healthy individuals are compared after UV-C irradiation. The cells from CS or UV^{S}S typically display a low level of RRS activity because of the TC-NER deficiency, but they retain nearly normal UDS activity, while the cells from XP patients usually exhibit reduced activities in both UDS and RRS, except for the cells derived from the XP-C and XP-E patients. For the rapid and accurate diagnosis of NER-deficient disorders, our laboratory developed a semi-automated assay system for non-radioactive UDS and RRS measurements by the incorporation of the alkyne-conjugated nucleoside analogues, 5-ethynyl-2′-deoxyuridine (EdU) and 5-ethynyluridine (EU), followed by the fluorescent-azide coupling reaction, click chemistry [41–44]. These fluorescence-based UDS/RRS assays combined with lentivirus-based complementation tests enable us to systematically determine the pathogenic genes of various NER-deficient individuals within 1–2 weeks.

2.3 Molecular Pathogenesis of Cockayne Syndrome and UV-Sensitive Syndrome

As mentioned above, TC-NER deficiency in humans is associated with devastating CS and very mild UV^{S}S. The molecular basis underlying these clinical differences is still unclear. In the last 10 years, various factors involved in the early step of TC-NER have been identified (*e.g.* UVSSA and its partner USP7), and more studies have focused on the molecular mechanism of the initiation process of TC-NER, including post-translational modifications of TC-NER factors as well as chromatin-remodelling proteins [45]. In this section, we will review the roles of the core TC-NER proteins that are involved in facilitating TC-NER and provide recent understandings of the molecular pathogenesis of CS and UV^{S}S.

2.3.1 Proteins Involved in the Initiation Step of TC-NER

2.3.1.1 CSA and CSB

The human *CSA* gene is located on chromosome 5q12.1 and consists of 12 exons. The *CSA* gene was identified by the complementation of UV sensitivity of a CSA cell line [46], and it is also known as *ERCC8* (excision repair cross-complementing rodent repair deficiency, complementation group 8) because the gene also complements UV sensitivity in a rodent cell line with mutations in the corresponding gene [47]. CSA is a 45-kDa protein that belongs to the WD-repeat protein family, various members of which play regulatory roles in many cellular processes, including cell division, signal transduction, mRNA modification, and transcription [46, 48]. The CSA protein contains seven WD40 repeat motifs, which form a beta-propeller architecture that serves as a scaffold for protein-protein interactions [49]. In TC-NER, the CSA protein has been reported to interact with CSB, DDB1, UVSSA, XAB2, and the p44 subunit of the TFIIH complex [21, 23, 46, 50–52].

The human *CSB* (also known as *ERCC6*) gene is located on chromosome 10q11.23 and is organized into 21 exons. The *CSB* gene encodes a 170-kDa protein, which contains a region of helicase-like ATPase motifs characteristic of the expanding and diverse SWI2/SNF2 protein family [53], whose members are implicated in chromatin remodelling during transcription. The CSB protein has seven helicase motifs: helicase motifs I, IA, II, and III, which are located in the inner surface of domain 1 and are mainly involved in ATP-binding and energy transduction, and helicase motifs IV, V, and VI, which are located in domain 2 and may be involved in DNA binding. Domain 1 and domain 2 are separated by a linker region between motif III and motif IV. CSB is a DNA-dependent ATPase, but it does not have DNA helicase activity [53, 54]. CSB binds directly to the core histones and has recently been shown to modulate the conformation of double-stranded DNA and to have ATP-dependent chromatin-remodelling activity. In addition to its critical roles in transcription and TC-NER, CSB also functions in oxidative DNA damage repair [20, 55] and in mitochondrial DNA (mtDNA) repair [56].

The majority of CS cases identified world-wide (approximately 75%) have mutations in the *CSB/ERCC6* gene, whereas mutations in the *CSA/ERCC8* genes have been found in the remaining 25% of CS patients [26, 28, 57]. Note that in Japan, *CSA* mutations are dominant in the CS cases (~70%). Genetic assays, along with the clinical data, suggest that mutations in *CSB* are distributed along the gene, and almost all types of mutations are included, while most mutations in *CSA* are involved in the WD40 repeat motifs, except for the null mutations. As mentioned above, WD40 repeat motifs are important for the construction of a beta-propeller structure as well as protein-protein interactions; thus, *CSA* mutations within the WD repeat motifs may alter the domain structure and affect its ability to associate with other proteins, *e.g.* DDB1 [58]. See Chap. 10 for further information on the genotype-phenotype correlations of CS.

2.3.1.2 UVSSA and USP7

The human *UVSSA* (also known as *KIAA1530*) gene is located on chromosome 4p16.3, and it consists of 13 exons. The *UVSSA* gene was recently identified as a causative gene for UV-sensitive syndrome complementation group A by three research groups in Japan and in the Netherlands [21–23]. The *UVSSA* gene encodes an 80-kDa protein, which contains a conserved domain with homology to the Vps27, Hrs, and STAM (VHS) domain [59] at its N-terminus and a conserved but poorly characterized DUF2043 (domain of unknown function) domain near its C-terminus, as well as a conventional nuclear localization signal. The VHS domain was previously shown to be involved in the binding of ubiquitin or ubiquitinated proteins [60]. By whole-exome sequencing analyses, three different mutations in the *UVSSA* gene were identified from four UV^{S}S-A patients. In addition to two null mutations, one missense mutation (p.Cys32Arg) was located in the VHS domain, resulting in a complete lack of TC-NER activity and implying that mutational changes in the VHS domain may alter the interactions between the UVSSA and other ubiquitinated TC-NER proteins. Furthermore, UV^{S}S-A cells expressing UVSSA deletion mutants without either the VHS domain or the DUF2043 domain failed to complement the TC-NER deficiency, indicating that both domains are required for the UVSSA function. The UVSSA protein is recruited to the TC-NER machinery after UV damage by interacting with the core conventional TC-NER factors, such as stalled RNA Pol IIo, CSA, CSB, and TFIIH, suggesting that UVSSA is a novel TC-NER factor [23]. In addition to these known TC-NER factors, UVSSA also interacts with USP7, which is a member of the ubiquitin-specific proteases (USPs) that recognizes and removes the ubiquitin chain from proteins [21, 22]. UVSSA binds to the TRAF (tumour necrosis factor receptor-associated factor) domain of USP7 and suppresses its deubiquitinating activity; on the contrary, USP7 also stabilizes UVSSA from rapid degradation mediated by the proteasome [61]. The UVSSA-USP7 complex binds directly with CSA regardless of the UV treatment after UV irradiation; it is recruited to the stalled RNA pol IIo on UV-damaged chromatin by CSB and functions in protecting CSB from UV-induced degradation [21, 50]. Another model for the recruitment of UVSSA-USP7 to the TC-NER multifactor machinery is also proposed. In this model, the UVSSA protein directly interacts with the elongating form of RNA pol II in a UV-independent manner, and it subsequently recruits USP7 to join the core TC-NER reaction [22].

2.3.1.3 General Model of the TC-NER Initiation Process

The core GG-NER reaction in mammals has been successfully reconstituted *in vitro* with purified proteins [6, 62, 63]. In contrast, the molecular mechanism of TC-NER has been fully resolved only in *Escherichia coli*. After DNA damage (*e.g.* UV irradiation), an arrested RNA polymerase at the DNA damage site is displaced by the transcription-repair coupling factor (TRCF), which is encoded by the *E. coli mfd* gene. TRCF mediates the removal of DNA lesions from the transcribed strands by recruiting the UvrABC multi-endonuclease complex [64, 65]. In mammalian cells, genetic and cell biological evidence indicates that the CSA, CSB, and UVSSA proteins play critical roles in the initiation step of TC-NER, but the exact molecular mechanisms are not completely understood. A well-considered model of the TC-NER initiation process suggests that the CSA and CSB protein complex is important for the recruitment of other TC-NER factors to the stalled RNA pol IIo at a lesion on the transcribed strand of active genes and then it regulates the assembly of the core reaction machinery [8]. In this model, CSB and XPG interact loosely with the hyper-phosphorylated, elongating form of RNA pol IIo during transcription. Upon UV damage, the interaction between CSB and RNA pol IIo becomes tight [66, 67]. The stalled RNA pol IIo provides a signal for chromatin remodelling and stimulates the recruitment of the chromatin-remodelling factors histone acetyltransferase p300 and HMGN1 [68, 69] via CSB and CSA [70]. The loose chromatin structure around the stalled RNA pol IIo enables access to the core TC-NER factors, and this is important for the later resumption of transcription after the removal of the DNA lesion [71]. CSA is recruited to the stalled RNA pol IIo by CSB in the nuclear matrix [72]. CSA is a part of a ubiquitin E3 ligase complex, which consists of DDB1, Cul4A (a member of the cullin family of ubiquitin ligase E3 subunits), and a well-known ubiquitin ligase component, Roc1 [51]. The active ubiquitin E3 ligase mediates the UV-dependent ubiquitination and consequent degradation of CSB [73] and RNA pol IIo [74]. Meanwhile, UVSSA-USP7 is also recruited to the TC-NER complex in a CSA- or RNA pol IIo-dependent manner. UVSSA cooperates with USP7 to stabilize the CSB protein [21, 22]. The UVSSA-USP7 complex is also considered to be involved in the regulation of ubiquitination around or directly on stalled RNA pol IIo, which may be important for the precise coordination of the repair factors and the modulation of RNA pol IIo removal or backtracking (details will be addressed below). This mechanism is considered a key step that enables the binding of TFIIH and the following repair processes [75]. However, the details remain largely unknown.

2.3.2 Molecular Pathogenesis of CS and UV^{S}S

Several studies on rare "progeroid" syndromes have highlighted that the DNA transcription arrest, abnormal replication control, and the impaired repair of the genome DNA might be particularly relevant to the ageing process [76]. Although CS and UV^{S}S cells display similar deficiencies in TC-NER, which is required for repairing DNA damage on transcribed genes and facilitating transcription resumption, it is still unlikely that only this single cellular defect fully explains the strikingly distinct symptoms between these two disorders. Importantly, XP patients are defective in both the GG-NER and TC-NER sub-pathways. However, aside from hyper-photosensitivity, most XP patients do not display developmental abnormality like CS patients (see elsewhere in this book). Moreover, the complete lack of NER activity in *Xpa*$^{-/-}$ mice only results in a very mild ageing phenotype [77]. These observations imply that the causative genes for CS or UV^{S}S probably also participate in several other biological pathways outside of TC-NER and that different mutations in these genes may impair their functions in one or several pathways. Several models have been proposed for explaining the underlying molecular basis of the clinical differences observed among patients with CS and UV^{S}S, and most of these models focus on additional functions of CS proteins in basal transcription regulation as well as in the repair of oxidative DNA damage, which is usually repaired by base excision repair. Since UVSSA is assumed to be not involved in these processes, the much milder clinical features of UV^{S}S-A patients can be explained.

2.3.2.1 Model 1: The Role of CS Proteins in Oxidative DNA Damage Repair

One well-defined model supported by numerous genetic and cellular studies postulates that the accelerated ageing and neurological degeneration phenotypes observed only in CS patients might be related to their impaired repair of the oxidative DNA damage produced by endogenous reactive oxygen species (ROS) [78]. Fibroblasts derived from CS-A and CS-B patients have been reported to elicit an increased steady-state level of endogenous ROS and to display a deficiency in the repair of oxidative DNA damage in these cells [55, 79, 80]. Menoni *et al.* further showed that CSB binds to oxidative DNA lesions in living cells, suggesting the role of CSB in the oxidative DNA damage response [81]. It is known that oxidative metabolism during neural development may result in a substantial level of oxidative DNA damage, and thus, the accumulation of unrepaired oxidative DNA lesions and the following transcription arrest may result in apoptosis and consequential neurological degeneration, which is a typical clinical feature observed in CS patients [82]. Indeed, cells from CS patients show a higher sensitivity to reagents that induce oxidative DNA damage (*e.g.* hydrogen peroxide) than normal or $UV^{S}S$ cells [83]. Moreover, a homozygous missense mutation in the *CSA* gene (p.Trp361Cys) was found from a $UV^{S}S$/CS-A patient [36]; although the patient-derived cells showed hypersensitivity to UV irradiation, they were not sensitive to H_2O_2 treatment, implying a separation of function of CSA in response to UV- and ROS-induced DNA damage. Fei *et al.* further demonstrated that this CSA mutant protein had greatly reduced binding ability to UVSSA, leading to the failure of UVSSA translocation to the TC-NER complex upon UV treatment [50]. This could ultimately cause $UV^{S}S$ in this CSA-hypomorphic patient, while an unaffected role of CSA in the oxidative DNA damage repair prevents the CS phenotype. Taken together, these findings support the dependence of the severe CS phenotype on the repair deficiency of oxidative DNA damage.

2.3.2.2 Model 2: The Role of CSB in Basal Transcription

Another hypothesis focuses on the role of the CSB protein in basal transcription. The regulation of gene expression is crucial for the maintenance of cellular homeostasis. The CSB protein is a chromatin-remodelling factor, and it loosely binds with RNA Pol IIo during transcription elongation [84, 85]. This implies that malfunction of the CSB protein may result in transcription defects, even in the absence of DNA damage. Genetic alterations in transcription factors are associated with several human disorders, most of which share overlapping clinical features (*e.g.* congenital defects) with CS [86]. This suggests that the severe phenotype of CS may be somewhat caused by a defect in the transcription of genes related to development and ageing, while the photosensitivity features observed in the CS and $UV^{S}S$ patients are commonly caused by defects in TC-NER.

It is also worth mentioning that the hypotheses described above may be challenged by the identification of a near-5′ stop-gain mutation in the *CSB* gene from two unrelated UV^SS patients (UV^SS/CS-B). The mutation (p.Arg77*) results in the undetectable expression of the CSB protein, indicating that the absence of the CSB protein alone does not fully explain the striking phenotype variability observed in CS and UV^SS [37]. In spite of this, several reports further indicate that an evolutionarily conserved CSB-PGBD3 fusion protein, which is derived from a 2.5-kb piggybac transposon (PGBD3) insertion into the intron 5 of the *CSB* gene, may be involved in the CS severe phenotype, although the detailed mechanism needs to be further elucidated by clinical and experimental evidence.

2.3.2.3 Model 3: The Role of TC-NER in the Processing of Stalled RNA Pol IIo

The last model involves a cooperation of TC-NER factors in the processing of stalled RNA pol IIo to backtrack and remove RNA pol IIo, which primarily makes the DNA lesions accessible for TC-NER or GG-NER. Upon DNA damage (*e.g.* UV irradiation or H_2O_2 treatment), the elongating form of RNA pol II stalls at DNA lesions and results in transcription arrest. The stalled RNA pol IIo masks the DNA lesion within a 35-nucleotide "footprint," and it may thereby prevent efficient DNA repair from occurring [87, 88]. In normal cells, TC-NER factors are recruited to a DNA damage site and facilitate RNA pol IIo backtracking for enabling the lesion to be further recognized and repaired by the TFIIH complex followed by repair replication, after which transcription is restored. The precise backtracking molecular mechanism of the damage arrested RNA polymerase in prokaryotes was studied [64, 65]. However, the molecular basis of this process in eukaryotes still remains largely unknown [89]. It is suggested that the critical step (transcript cleavage [90]) in the backtracking of stalled RNA pol IIo from the DNA lesion may be mediated by the elongation factor TFIIS [91]. Some chromatin remodellers, including HMGN1 and histone acetyl transferases (HAT), which are recruited by the CS protein complex upon DNA damage, form an open chromatin behind the stalled RNA pol IIo (R-loop formation and spliceosome displacement), and then they facilitate backtracking [89]. These alterations of the chromatin structure are also necessary for the activation of the TFIIH complex before the dual incision of the DNA lesions. Damage-induced ubiquitination also plays a regulatory role in TC-NER, especially in the RNA pol IIo processing [92]. Several ubiquitin E3 ligases, including BRCA1-BARD1 [93], CSA-DDB1-Cul4A-Roc1 ($CRL4^{CSA}$) [74], and NEDD4 [94], have been proposed to be involved in the K48-linked ubiquitination and the subsequent degradation of RNA pol IIo by the 26S proteasome when the lesion-stalled RNA pol IIo cannot be properly processed [95]. Recently, Nakazawa *et al.* further identified the UVSSA-dependent ubiquitination of stalled RNA pol IIo, which is not subject to proteasomal degradation. After UV irradiation, but not H_2O_2 treatment, K63-ubiquitination on RBP1, the largest subunit of RNA pol II, is observed, which may be involved in backtracking regulation. However, the precise function of this ubiquitination needs further investigation [23]. In cell lines derived from CS patients, CSA- or UVSSA-dependent ubiquitination of RNA pol IIo is significantly reduced, which thereby induces a prolonged stalling of RNA pol IIo at the DNA damage site, resulting no transcription resumption. This may subsequently activate cellular stress response signalling and eventually leads to apoptosis in the CS cells. It is conceivable that these cellular features may contribute to the severe neurological abnormities and premature ageing of the patients. Conversely, in the UV^{S}S-A cells, although the processing of stalled RNA pol IIo (backtracking) is compromised, the UV-dependent degradation of elongating RNA pol IIo is still operating [23]. The lack of UVSSA protein in the UV^{S}S cells may not interfere the CSA- and UV-induced ubiquitination of RNA pol IIo, which results in the rapid removal of stalled RNA pol IIo from the DNA damaged sites, thereby making the lesions accessible by an alternative repair pathway, such as GG-NER. Importantly, the stress response signalling stimulated by persistently stalled RNA pol II is also avoided, and this may explain the relatively milder clinical phenotype observed in UV^{S}S patients.

2.4 Concluding Remarks and Future Prospect

In this chapter, we provided a general review on the molecular mechanisms of TC-NER and recent findings on the molecular pathogenesis of TC-NER-deficient disorders. TC-NER has been studied for more than 30 years, and the growing evidence provided by numerous clinical and experimental studies enabled us to roughly depict how this versatile repair system correlates with ageing and human diseases. However, the precise molecular mechanism of TC-NER, especially the initiation step, remains largely unknown, and the exact genotype-phenotype correlation of CS and UV^{S}S is also unclear. The recent discovery of the UVSSA-USP7 complex allowed us to test the abovementioned models in detail, which helped us to understand the processing of stalled RNA pol IIo at the DNA lesion. Further identification of new TC-NER factors, which may also be involved in other biological processes, will help us to achieve full understanding of the TC-NER in future. Post-translation modifications, such as ubiquitination, are shown to play a crucial role in TC-NER. Further studies should focus on how these different modification events effectively coordinate with each other to ensure proper DNA damage repair. Another interesting area for further investigation is the differential regulation of the TC-NER factors in repairing UV-induced damage and oxidative DNA lesions, and this may provide new insights in the phenotypical difference between CS and UV^{S}S.

References

1. Friedberg EC, et al. DNA repair: from molecular mechanism to human disease. DNA Repair (Amst). 2006;5(8):986–96.
2. Gates KS. An overview of chemical processes that damage cellular DNA: spontaneous hydrolysis, alkylation, and reactions with radicals. Chem Res Toxicol. 2009;22:1747–60.
3. Hoeijmakers JH. DNA damage, aging, and cancer. N Engl J Med. 2009;361:1475–85.

4. Hoeijmakers JH. Genome maintenance mechanisms for preventing cancer. Nature. 2001;411:366–74.
5. Scharer OD. Nucleotide excision repair in eukaryotes. Cold Spring Harb Perspect Biol. 2013;5:a012609.
6. Gillet LC, Scharer OD. Molecular mechanisms of mammalian global genome nucleotide excision repair. Chem Rev. 2006;106(2):253–76.
7. Hanawalt PC, Spivak G. Transcription-coupled DNA repair: two decades of progress and surprises. Nat Rev Mol Cell Biol. 2008;9:958–70.
8. Vermeulen W, Fousteri M. Mammalian transcription-coupled excision repair. Cold Spring Harb Perspect Biol. 2013;5:a012625.
9. Lehmann AR. DNA repair-deficient diseases, xeroderma pigmentosum, Cockayne syndrome and trichothiodystrophy. Biochimie. 2003;85(11):1101–11.
10. Mayne LV, Lehmann AR. Failure of RNA synthesis to recover after UV irradiation: an early defect in cells from individuals with Cockayne's syndrome and xeroderma pigmentosum. Cancer Res. 1982;42:1473–8.
11. Bohr VA, et al. DNA repair in an active gene: removal of pyrimidine dimers from the DHFR gene of CHO cells is much more efficient than in the genome overall. Cell. 1985;40(2):359–69.
12. Bohr VA, Phillips DH, Hanawalt PC. Heterogeneous DNA damage and repair in the mammalian genome. Cancer Res. 1987;47(24 Pt 1):6426–36.
13. Ljungman M, Lane DP. Transcription [mdash] guarding the genome by sensing DNA damage. Nat Rev Cancer. 2004;4:727–37.
14. Mellon I, Hanawalt PC. Induction of the *Escherichia coli* lactose operon selectively increases repair of its transcribed DNA strand. Nature. 1989;342(6245):95–8.
15. Leadon SA, Lawrence DA. Strand-selective repair of DNA damage in the yeast GAL7 gene requires RNA polymerase II. J Biol Chem. 1992;267(32):23175–82.
16. Sweder KS, Hanawalt PC. Preferential repair of cyclobutane pyrimidine dimers in the transcribed strand of a gene in yeast chromosomes and plasmids is dependent on transcription. Proc Natl Acad Sci U S A. 1992;89(22):10696–700.
17. Mellon I, Spivak G, Hanawalt PC. Selective removal of transcription-blocking DNA damage from the transcribed strand of the mammalian DHFR gene. Cell. 1987;51(2):241–9.
18. Leadon SA, Lawrence DA. Preferential repair of DNA damage on the transcribed strand of the human metallothionein genes requires RNA polymerase II. Mutat Res. 1991;255(1):67–78.
19. Sugasawa K. A multistep damage recognition mechanism for global genomic nucleotide excision repair. Genes Dev. 2001;15:507–21.
20. Stevnsner T, et al. The role of Cockayne syndrome group B (CSB) protein in base excision repair and aging. Mech Ageing Dev. 2008;129:441–8.
21. Zhang X. Mutations in UVSSA cause UV sensitive syndrome and destabilize ERCC6 in transcription-coupled DNA repair. Nat Genet. 2012;44:593–7.
22. Schwertman P. UV sensitive syndrome protein UVSSA recruits USP7 to regulate transcription-coupled repair. Nat Genet. 2012;44:598–602.
23. Nakazawa Y. Mutations in UVSSA cause UV sensitive syndrome and impair RNA polymerase IIo processing in transcription-coupled nucleotide-excision repair. Nat Genet. 2012;44:586–92.
24. Fagbemi AF, Orelli B, Scharer OD. Regulation of endonuclease activity in human nucleotide excision repair. DNA Repair. 2011;10:722–9.
25. Cockayne EA. Dwarfism with retinal atrophy and deafness. Arch Dis Child. 1936;11(61):1–8.
26. Laugel V. Cockayne syndrome: the expanding clinical and mutational spectrum. Mech Ageing Dev. 2013;134:161–70.
27. Cleaver JE, Thomas GH. Clinical syndromes associated with DNA repair deficiency and enhanced sun sensitivity. Arch Dermatol. 1993;129:348–50.
28. Nance MA, Berry SA. Cockayne syndrome: review of 140 cases. Am J Med Genet. 1992;42(1):68–84.
29. Berneburg M, Lehmann AR. Xeroderma pigmentosum and related disorders: defects in DNA repair and transcription. Adv Genet. 2001;43:71–102.

30. Kraemer KH, et al. Xeroderma pigmentosum, trichothiodystrophy and Cockayne syndrome: a complex genotype-phenotype relationship. Neuroscience. 2007;145(4):1388–96.
31. Kashiyama K, et al. Malfunction of nuclease ERCC1-XPF results in diverse clinical manifestations and causes Cockayne syndrome, xeroderma pigmentosum, and Fanconi anemia. Am J Hum Genet. 2013;92(5):807–19.
32. Spivak G. UV sensitive syndrome. Mutat Res. 2005;577:162–9.
33. Fujiwara Y, et al. A new human photosensitive subject with a defect in the recovery of DNA synthesis after ultraviolet-light irradiation. J Investig Dermatol. 1981;77:256–63.
34. Itoh T, Ono T, Yamaizumi M. A new UV-sensitive syndrome not belonging to any complementation groups of xeroderma pigmentosum or Cockayne syndrome: siblings showing biochemical characteristics of Cockayne syndrome without typical clinical manifestations. Mutat Res. 1994;314:233–48.
35. Itoh T, et al. UVs syndrome, a new general category of photosensitive disorder with defective DNA repair, is distinct from xeroderma pigmentosum variant and rodent complementation group I. Am J Hum Genet. 1995;56:1267–76.
36. Nardo T. A UV sensitive syndrome patient with a specific CSA mutation reveals separable roles for CSA in response to UV and oxidative DNA damage. Proc Natl Acad Sci U S A. 2009;106:6209–14.
37. Horibata K. Complete absence of Cockayne syndrome group B gene product gives rise to UV sensitive syndrome but not Cockayne syndrome. Proc Natl Acad Sci U S A. 2004;101:15410–5.
38. Lehmann AR, Stevens S. A rapid procedure for measurement of DNA repair in human fibroblasts and for complementation analysis of xeroderma pigmentosum cells. Mutat Res. 1980;69(1):177–90.
39. Ayaki H, Hara R, Ikenaga M. Recovery from ultraviolet tight-induced depression of ribosomal RNA synthesis in normal human, xeroderma pigmentosum and Cockayne syndrome cells. J Radiat Res. 1996;37(2):107–16.
40. Christians FC, Hanawalt PC. Lack of transcription-coupled repair in mammalian ribosomal RNA genes. Biochemistry. 1993;32(39):10512–8.
41. Kolb HC, Finn MG, Sharpless KB. Click chemistry: diverse chemical function from a few good reactions. Angew Chem Int Ed Engl. 2001;40(11):2004–21.
42. Jia N, et al. A rapid, comprehensive system for assaying DNA repair activity and cytotoxic effects of DNA-damaging reagents. Nat Protoc. 2015;10(1):12–24.
43. Limsirichaikul S, et al. A rapid non-radioactive technique for measurement of repair synthesis in primary human fibroblasts by incorporation of ethynyl deoxyuridine (EdU). Nucleic Acids Res. 2009;37(4):e31.
44. Nakazawa Y, et al. A semi-automated non-radioactive system for measuring recovery of RNA synthesis and unscheduled DNA synthesis using ethynyluracil derivatives. DNA Repair (Amst). 2010;9:506–16.
45. Li S. Implication of posttranslational histone modifications in nucleotide excision repair. Int J Mol Sci. 2012;13(10):12461–86.
46. Henning KA, et al. The Cockayne syndrome group A gene encodes a WD repeat protein that interacts with CSB protein and a subunit of RNA polymerase II TFIIH. Cell. 1995;82(4):555–64.
47. Itoh T, et al. Rodent complementation group 8 (ERCC8) corresponds to Cockayne syndrome complementation group A. Mutat Res. 1996;362(2):167–74.
48. Zhou HX, Wang G. Predicted structures of two proteins involved in human diseases. Cell Biochem Biophys. 2001;35(1):35–47.
49. Xu C, Min J. Structure and function of WD40 domain proteins. Protein Cell. 2011;2(3):202–14.
50. Fei J, Chen J. KIAA1530 protein is recruited by Cockayne syndrome complementation group protein A (CSA) to participate in transcription-coupled repair (TCR). J Biol Chem. 2012;287:35118–26.
51. Groisman R. The ubiquitin ligase activity in the DDB2 and CSA complexes is differentially regulated by the COP9 signalosome in response to DNA damage. Cell. 2003;113:357–67.

52. Nakatsu Y, et al. XAB2, a novel tetratricopeptide repeat protein involved in transcription-coupled DNA repair and transcription. J Biol Chem. 2000;275(45):34931–7.
53. Troelstra C, et al. ERCC6, a member of a subfamily of putative helicases, is involved in Cockayne's syndrome and preferential repair of active genes. Cell. 1992;71(6):939–53.
54. Christiansen M, et al. Functional consequences of mutations in the conserved SF2 motifs and post-translational phosphorylation of the CSB protein. Nucleic Acids Res. 2003;31(3):963–73.
55. de Waard H, et al. Different effects of CSA and CSB deficiency on sensitivity to oxidative DNA damage. Mol Cell Biol. 2004;24:7941–8.
56. Aamann MD, et al. Cockayne syndrome group B protein promotes mitochondrial DNA stability by supporting the DNA repair association with the mitochondrial membrane. FASEB J. 2010;24(7):2334–46.
57. Laugel V, et al. Mutation update for the CSB/ERCC6 and CSA/ERCC8 genes involved in Cockayne syndrome. Hum Mutat. 2010;31(2):113–26.
58. Jin J, et al. A family of diverse Cul4-Ddb1-interacting proteins includes Cdt2, which is required for S phase destruction of the replication factor Cdt1. Mol Cell. 2006;23(5):709–21.
59. Lohi O, et al. VHS domain -- a longshoreman of vesicle lines. FEBS Lett. 2002;513(1):19–23.
60. Dikic I, Wakatsuki S, Walters KJ. Ubiquitin-binding domains - from structures to functions. Nat Rev Mol Cell Biol. 2009;10(10):659–71.
61. Higa M, et al. Stabilization of ultraviolet (UV)-stimulated scaffold protein a by interaction with ubiquitin-specific peptidase 7 is essential for transcription-coupled nucleotide excision repair. J Biol Chem. 2016;291(26):13771–9.
62. de Laat WL, Jaspers NG, Hoeijmakers JH. Molecular mechanism of nucleotide excision repair. Genes Dev. 1999;13(7):768–85.
63. Petruseva IO, Evdokimov AN, Lavrik OI. Molecular mechanism of global genome nucleotide excision repair. Acta Nat. 2014;6(1):23–34.
64. Epshtein V, et al. UvrD facilitates DNA repair by pulling RNA polymerase backwards. Nature. 2014;505(7483):372–7.
65. Selby CP, Sancar A. Molecular mechanism of transcription-repair coupling. Science. 1993;260(5104):53–8.
66. Laine JP, Egly JM. When transcription and repair meet: a complex system. Trends Genet. 2006;22(8):430–6.
67. Fousteri M, Mullenders LH. Transcription-coupled nucleotide excision repair in mammalian cells: molecular mechanisms and biological effects. Cell Res. 2008;18(1):73–84.
68. Ramanathan B, Smerdon MJ. Enhanced DNA repair synthesis in hyperacetylated nucleosomes. J Biol Chem. 1989;264(19):11026–34.
69. Lim JH, et al. Chromosomal protein HMGN1 enhances the acetylation of lysine 14 in histone H3. EMBO J. 2005;24(17):3038–48.
70. Fousteri M, et al. Cockayne syndrome A and B proteins differentially regulate recruitment of chromatin remodeling and repair factors to stalled RNA polymerase II in vivo. Mol Cell. 2006;23(4):471–82.
71. Lukas J, Lukas C, Bartek J. More than just a focus: the chromatin response to DNA damage and its role in genome integrity maintenance. Nat. Cell Biol. 2011;13:1161–9.
72. Kamiuchi S, et al. Translocation of Cockayne syndrome group A protein to the nuclear matrix: possible relevance to transcription-coupled DNA repair. Proc Natl Acad Sci U S A. 2002;99(1):201–6.
73. Groisman R. CSA-dependent degradation of CSB by the ubiquitin-proteasome pathway establishes a link between complementation factors of the Cockayne syndrome. Genes Dev. 2006;20:1429–34.
74. Bregman DB. UV-induced ubiquitination of RNA polymerase II: a novel modification deficient in Cockayne syndrome cells. Proc Natl Acad Sci U S A. 1996;93:11586–90.
75. Mullenders L. DNA damage mediated transcription arrest: step back to go forward. DNA Repair (Amst). 2015;36:28–35.

76. Capell BC, Tlougan BE, Orlow SJ. From the rarest to the most common: insights from progeroid syndromes into skin cancer and aging. J Investig Dermatol. 2009;129(10):2340–50.
77. Andressoo JO. An Xpb mouse model for combined xeroderma pigmentosum and Cockayne syndrome reveals progeroid features upon further attenuation of DNA repair. Mol Cell Biol. 2009;29:1276–90.
78. Tsutakawa SE, Cooper PK. Transcription-coupled repair of oxidative DNA damage in human cells: mechanisms and consequences. Cold Spring Harb Symp Quant Biol. 2000;65:201–15.
79. Trapp C, et al. Deficiency of the Cockayne syndrome B (CSB) gene aggravates the genomic instability caused by endogenous oxidative DNA base damage in mice. Oncogene. 2007;26(27):4044–8.
80. D'Errico M, et al. The role of CSA in the response to oxidative DNA damage in human cells. Oncogene. 2007;26(30):4336–43.
81. Menoni H, Hoeijmakers JH, Vermeulen W. Nucleotide excision repair-initiating proteins bind to oxidative DNA lesions in vivo. J Cell Biol. 2012;199:1037–46.
82. Hoeijmakers JH. Genome maintenance mechanisms are critical for preventing cancer as well as other aging-associated diseases. Mech Ageing Dev. 2007;128(7–8):460–2.
83. Spivak G, Hanawalt PC. Host cell reactivation of plasmids containing oxidative DNA lesions is defective in Cockayne syndrome but normal in UV-sensitive syndrome fibroblasts. DNA Repair (Amst). 2006;5:13–22.
84. Citterio E. ATP-dependent chromatin remodeling by the Cockayne syndrome B DNA repair-transcription-coupling factor. Mol Cell Biol. 2000;20:7643–53.
85. van den Boom V, et al. DNA damage stabilizes interaction of CSB with the transcription elongation machinery. J Cell Biol. 2004;166(1):27–36.
86. Villard J. Transcription regulation and human diseases. Swiss Med Wkly. 2004;134(39–40):571–9.
87. Svejstrup JQ. Mechanisms of transcription-coupled DNA repair. Nat Rev Mol Cell Biol. 2002;3(1):21–9.
88. Tornaletti S, Reines D, Hanawalt PC. Structural characterization of RNA polymerase II complexes arrested by a cyclobutane pyrimidine dimer in the transcribed strand of template DNA. J Biol Chem. 1999;274(34):24124–30.
89. Sahoo M, Klumpp S. Backtracking dynamics of RNA polymerase: pausing and error correction. J Phys Condens Matter. 2013;25(37):374104.
90. Sigurdsson S, Dirac-Svejstrup AB, Svejstrup JQ. Evidence that transcript cleavage is essential for RNA polymerase II transcription and cell viability. Mol Cell. 2010;38(2):202–10.
91. Jensen A, Mullenders LH. Transcription factor IIS impacts UV-inhibited transcription. DNA Repair (Amst). 2010;9(11):1142–50.
92. van Cuijk L, Vermeulen W, Marteijn JA. Ubiquitin at work: the ubiquitous regulation of the damage recognition step of NER. Exp Cell Res. 2014;329(1):101–9.
93. Starita LM. BRCA1/BARD1 ubiquitinate phosphorylated RNA polymerase II. J Biol Chem. 2005;280:24498–505.
94. Anindya R, Aygun O, Svejstrup JQ. Damage-induced ubiquitylation of human RNA polymerase II by the ubiquitin ligase Nedd4, but not Cockayne syndrome proteins or BRCA1. Mol Cell. 2007;28:386–97.
95. Wilson MD, Harreman M, Svejstrup JQ. Ubiquitylation and degradation of elongating RNA polymerase II: the last resort. Biochim Biophys Acta. 2013;1829:151–7.

Chapter 3
Neurological Symptoms in Xeroderma Pigmentosum

Fumio Kanda, Takehiro Ueda, and Chikako Nishigori

Some xeroderma pigmentosum (XP) patients display progressive neurological manifestations, including cognitive deterioration. XP consists of eight different clinical subtypes: complementation groups A through G of nucleotide excision repair-deficient type and variant type. Among these groups A, B, D, F, and G are accompanied by neurological symptoms with various frequencies [1, 2]. In particular, serious neurological complications are frequently observed in patients with XP genetic complementation group A (XP-A).

In Japan, most of XP-A patients harbor the identical founder mutation, which fail to yield XP-A protein, and they show severe neurological symptoms and become bedridden by the age of 20 [3]. The precise mechanisms underlying neurological deterioration in XP remain unclear, and no effective treatments are available. Currently, while advances in prevention and treatment for skin cancer have improved the dermatological prognosis, neurological complications have become the most serious problem for daily activity and life expectancy [4].

This chapter focuses on the most severe neurological complications of XP-A, whereby deterioration extends to both the central and peripheral nervous systems. In another groups of XP, patients may suffer milder and slower development of neurological problems than those in the XP-A patients.

F. Kanda (✉)
Division of Neurology, Kobe University Hospital, Kobe, Japan
e-mail: kanda@med.kobe-u.ac.jp

T. Ueda
Division of Neurology, Kobe University Graduate School of Medicine, Kobe, Japan

C. Nishigori
Department of Dermatology, Graduate School of Medicine, Kobe University, Kobe, Hyogo, Japan

C. Nishigori, K. Sugasawa (eds.), *DNA Repair Disorders*,
https://doi.org/10.1007/978-981-10-6722-8_3

3.1 The Neurological Manifestations and Natural History of X-PA

Generally, neurological abnormalities are not the first symptoms identified in patients with XP, and diagnosis of XP is never made based only on neurological problems. Guardians of XP patients do not identify any problems at birth or delays in subsequent developmental milestones. However, the age at which patients start walking is slightly delayed, with an average onset at 1.5 years. In general, neurological symptoms are not noticeable until around 5 years of age, even when the skin symptoms are obvious. Families who observe carefully may notice that the patient tends to fall more often than other children of a similar age. A majority of XP-A patients show some degree of mental retardation by school age.

After the age of 5, various kinds of neurological symptoms become apparent in XP-A patients. Walking becomes unstable, and as a result, they fall frequently. In addition, dysarthria becomes gradually more noticeable. Neurological examinations reveal reduction or loss of tendon reflexes with a consistently positive Babinski sign. In audiometry, sensorineural hearing impairment is detected in many patients, but is not so severe at this stage as to interfere with their daily life.

Around the age of 10, gait abnormalities and intellectual regression become obvious in all cases. Following this, neurological deterioration progresses unremittingly. In addition to intellectual deterioration, worsening of dysarthria begins to interfere patient social adaptation. Tendon reflexes are almost lost in all extremities following progression of peripheral neuropathy; thus it is necessary to continue physical therapy to prevent joint contracture. Gradual reduction of facial expressions is also observed as an extrapyramidal symptom. In some patients, involuntary movements such as myoclonus and/or chorea may be observed, predominantly in the upper extremities. Deafness progresses steadily, resulting in all patients requiring hearing aids. In addition, autonomic failure, including dysuria and constipation, also becomes apparent.

At the age of 15 or over, most patients are unable to walk independently and require wheelchairs. Mental and physical activities are progressively reduced. The patients entirely lose their facial expressions, such as "masked face," and always open their mouths with drooling. In addition to dysarthria, dysphagia is also a problem, especially when accompanied with choking and aspiration pneumonia. Percutaneous endoscopic gastrostomy is recommended in some patients, and urinary catheterization and diapers are necessary for bladder and bowel problems. By the end of the second decade, cognitive impairment becomes so severe that patients need help in all activities of daily living.

In the third decade, almost all patients are bedridden. Spontaneous speech and voluntary movement become extremely reduced, and daily activity is almost lost. Placement of tracheotomy may be needed for frequent aspiration pneumonia, and laryngeal spasm is life-threatening. In addition, there are risks of sudden death from unknown causes.

To semiquantitatively evaluate the severity of neurological deterioration, we have proposed a severity scale for neurological conditions in XP-A [5]. The scale consists of three sections, section 1, activity of daily life; section 2, motor functions; and section 3,

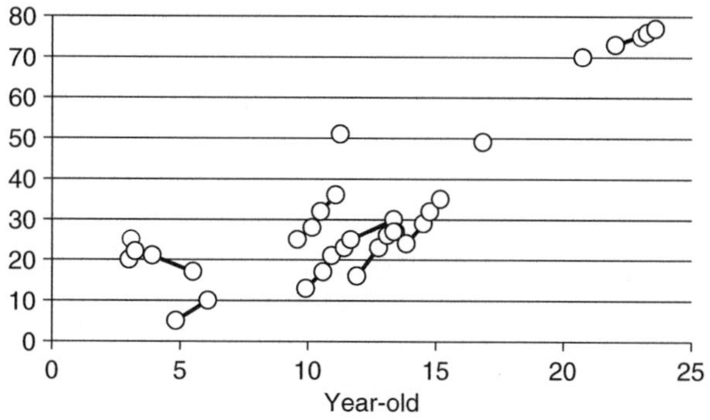

Fig. 3.1 Total scores of neurological disabilities for patients with XP-A. Markers connected with solid lines reflect the scores of a single patient

cognition, as well as the inclusion of a global disability score. The number of items in sections 1, 2, and 3 is 12, 6, and 2, respectively. Each item is graded on a scale from 0 to 4 based on the severity of the defect. For the total score, a minimum of zero reflects no neurological deficit, and the maximum score is 80. Figure 3.1 shows the correlation between patient age and total score for XP-A patients. Scores are the lowest at 5 years old and then increase with age. In patients over 20 years old, the total score is close to 80.

3.2 MRI Findings

Progressive brain atrophy is a hallmark of MRI findings in XP-A patients [6]. All brain tissue, including the skull, shows progressive reduction in volume as neurological problems become clinically apparent. In Fig. 3.2, brain MRI of a 21-year-old XP-A patient showing typical microcephaly was exhibited. No part of the brain, including the cerebral cortex, white matter, brainstem, and cerebellum, avoided progressive atrophy. In T2WI, there was no signal abnormality indicating cerebrovascular diseases, focal gliosis, or calcification. We measured the brain volume of XP-A patients. As shown in Fig. 3.3, brain volume was largest for patients younger than 5 years. Thereafter, brain volume continuously decreased as age increased.

3.3 Peripheral Neuropathy

Mild muscle weakness and diffuse muscular atrophy are observed in all extremities in XP-A patients. Surprisingly, patients do not complain of sensory symptoms such as numbness or hypoesthesia, but reduced tendon reflexes are commonly revealed by neurological examinations. Nerve conduction studies show subclinical peripheral neuropathy from an early age. As shown in Fig. 3.4, sensory dominant axonal neuropathy progresses with age, especially in the lower extremities. Beyond 10 years of age, sensory nerve action potentials cannot be elicited in the sural nerves.

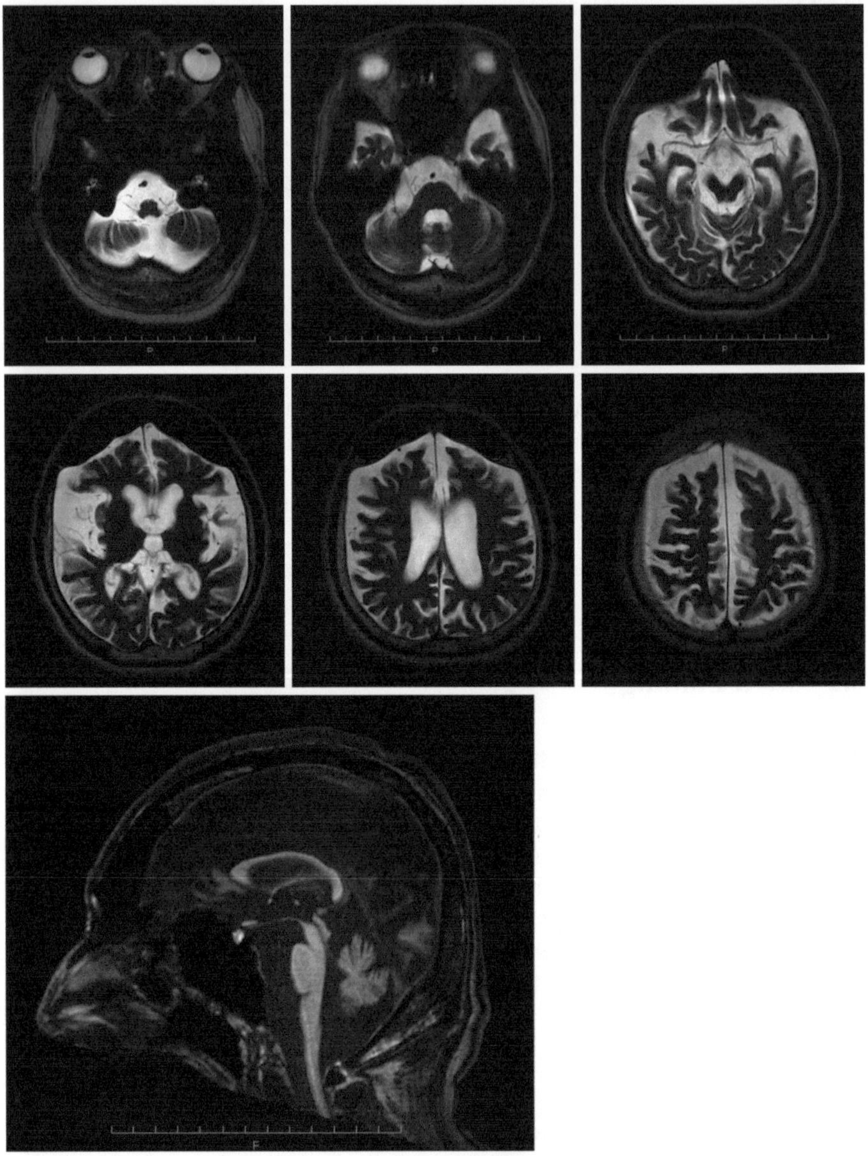

Fig. 3.2 Axial views of T2WI and a sagittal view of T1WI of the brain MRI in 21-year-old male patient with XP-A. There was severe atrophy of entire brain structures with enlargement of frontal sinus

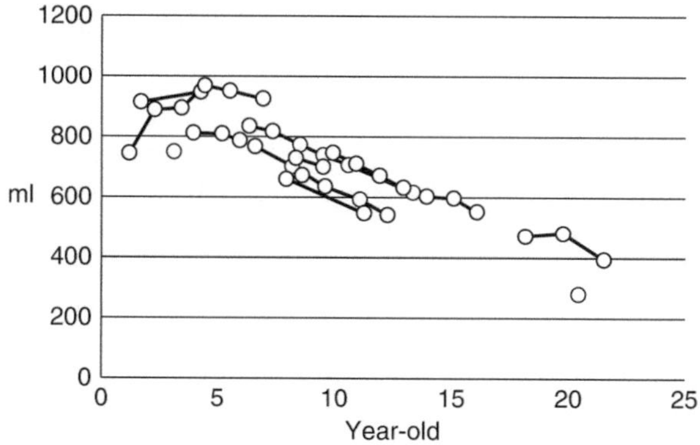

Fig. 3.3 Volumetric analysis of total brain volume in patients with XP-A. Markers connected with solid lines reflect the scores of the same patient. Brain atrophy starts around the age of 5 and continuously progresses with age

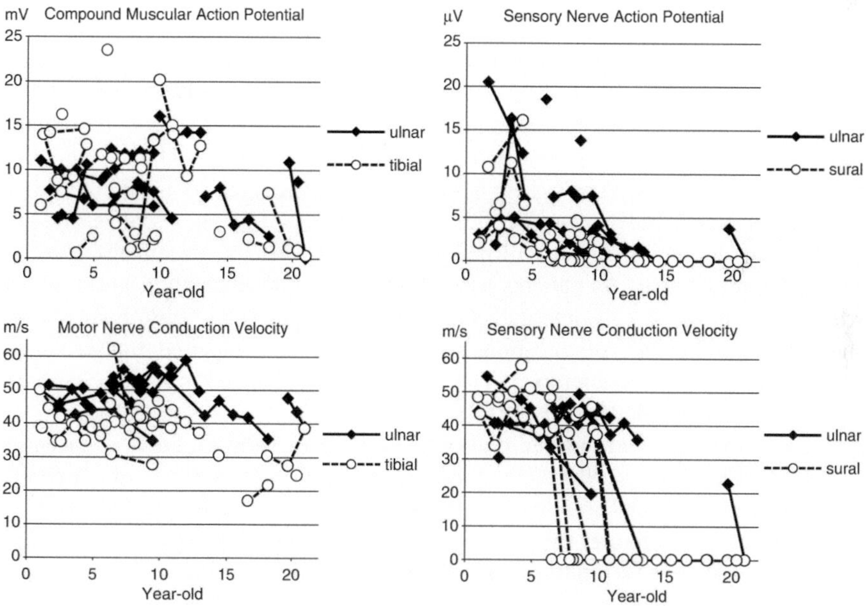

Fig. 3.4 Nerve conduction studies in patients with XP-A. Closed diamonds and open circles represent upper extremities (ulnar nerve) and lower extremities (tibial or sural nerve), respectively. Markers connected with solid or dashed lines reflect the scores of the same patient. In the second decade, no sensory action potentials could be evoked in the lower extremities

3.4 Neuropathological Findings in Autopsy Cases

With regard to neuropathological findings, there are only a small number of reports on advanced adult cases [7–10]. Histopathological studies have identified a wide range of neuronal cell loss and gliosis throughout the cerebrum but no characteristic findings specific to XP-A. Reduction of Purkinje cells and increase of torpedo and Bergman glia are seen in the cerebellar cortex. Immunohistochemical studies have not detected neurofibrillary tangles or senile plaques, thus neurodegeneration in XP-A brain is not consistent with mere accelerated aging. In the peripheral nerves, extreme loss of large myelinated fibers was observed. Again, there were no specific findings in peripheral nerves in XP-A. Further intensive neuropathological studies are required.

3.5 The Cause of Neurological Complications in XP-A

Because neurons do not divide or proliferate postnatally, it is easy to imagine that serious neurological problems may occur in situations where DNA repair fails. Cumulative DNA damage has been implicated in the functional deterioration and degeneration of long-living post-mitotic cells, such as neurons. Unlike the skin, however, the central nervous system is surrounded by the skull, so it is unlikely that DNA in nerve cells are damaged by ultraviolet radiation. Though oxidative stress is emphasized as a possible cause of neuronal damage [10, 11], the precise mechanisms remain unclear. As patients assigned to the complementation group C, in which only the global genome nucleotide excision repair (GG-NER) is impaired, show no neurological complications, it is believed that impaired transcription-coupled nucleotide excision repair (TC-NER) may be essential for the development of neurological disorders in XP [12].

Using patient-derived induced pluripotent stem cells (iPSCs), Fu et al. recently found that compromised NER activity not only influences the survival of nondividing neurons but also promotes the apoptosis of dividing neural stem cells [13]. It is expected that in the future, there will be further research on this pathogenesis and the development of new treatments for neurodegeneration in XP-A.

References

1. Robbins JH, Brumback RA, Mendiones M, et al. Neurological disease in xeroderma pigmentosum. Documentation of a late onset type of the juvenile onset form. Brain. 1991;114:1335–61.
2. Moriwaki S, Nishigori C, Imamura S, et al. A case of xeroderma pigmentosum complementation group F with neurological abnormalities. Br J Dermatol. 1993;128:91–4.
3. Nishigori C, Moriwaki S, Takebe H, et al. Gene alterations and clinical characteristics of xeroderma pigmentosum group A patients in Japan. Arch Dermatol. 1994;130:191–7.

4. Anttinen A, Koulu L, Nikoskelainen E, et al. Neurological symptoms and natural course of xeroderma pigmentosum. Brain. 2008;131:1979–89.
5. Nakano E, Masaki T, Kanda F, et al. The present status of xeroderma pigmentosum in Japan and a tentative severity classification scale. Exp Dermatol. 2016;25(Suppl 3):28–33.
6. Ueda T, Kanda F, Aoyama N, et al. Neuroimaging features of xeroderma pigmentosum group A. Brain Behav. 2012;2:1–5.
7. Kanda T, Oda M, Yonezawa M, et al. Peripheral neuropathy in xeroderma pigmentosum. Brain. 1990;113:1025–44.
8. Itoh M, Hayashi M, Shioda K, et al. Neurodegeneration in hereditary nucleotide repair disorders. Brain Dev. 1999;21:326–33.
9. Hayashi M, Araki S, Kohyama J, et al. Brainstem and basal ganglia lesions in xeroderma pigmentosum group A. J Neuropathol Exp Neurol. 2004;63:1048–57.
10. Lai JP, Liu YC, Alimchandani M, et al. The influence of DNA repair on neurological degeneration, cachexia, skin cancer and internal neoplasms: autopsy report of four xeroderma pigmentosum patients (XP-A, XP-C and XP-D). Acta Neuropathol Commun. 2013;1:4.
11. Weissman L, de Souza-Pinto NC, Stevnsner T, et al. DNA repair, mitochondria, and neurodegeneration. Neuroscience. 2007;145:1318–29.
12. Jaarsma D, van der Pluijm I, de Waard MC, et al. Age-related neuronal degeneration: complementary roles of nucleotide excision repair and transcription-coupled repair in preventing neuropathology. PLoS Genet. 2011;7:e1002405.
13. Fu L, Xu X, Ren R, et al. Modeling xeroderma pigmentosum associated neurological pathologies with patients-derived iPSCs. Protein Cell. 2016;7:210–21.

Chapter 4
Hearing Impairment in Xeroderma Pigmentosum: Animal Models and Human Studies

Takeshi Fujita and Daisuke Yamashita

Abstract Progressive neurological symptoms, including hearing loss, occur in some patients with xeroderma pigmentosum (XP). Patients with neurodegeneration commonly have mutations in XP-A, XP-B, XP-D, XP-F, or XP-G. Typically, audiograms of patients with XP who have sensorineural hearing loss are downsloping. The degree of hearing loss is directly correlated with neurological involvement, including cognitive impairment. Thus, for audiometric assessment, auditory brainstem response (ABR) or other objective tests are sometimes required instead of pure-tone audiometry. In the human temporal bone, XP-mediated pathology includes atrophy of the organ of Corti, stria vascularis, and spiral ganglion neurons. *Xpa*-deficient mice also showed significant loss of spiral ganglion neurons in the cochlea. Several studies show that the cochlea and nervous system in patients with XP are susceptible to persistent genomic stress, such as reactive oxygen species (ROS), which leads to early onset of sensorineural hearing loss. Regular audiometric monitoring of the hearing status of patients with XP to identify the need for auditory interventions, such as hearing aids, is important for maintaining their quality of life.

Keywords Hearing impairment · Xeroderma pigmentosum · Sensorineural hearing loss · Spiral ganglion neurons · ROS · Hearing aid · Pure-tone audiometry · ABR

T. Fujita
Department of Otolaryngology, Kindai University Faculty of Medicine, Osakasayama, Osaka, Japan

D. Yamashita (✉)
Department of Otolaryngology-Head and Neck Surgery, Kobe University Graduate School of Medicine, Kobe, Hyogo, Japan

C. Nishigori, K. Sugasawa (eds.), *DNA Repair Disorders*,
https://doi.org/10.1007/978-981-10-6722-8_4

4.1 Introduction

Xeroderma pigmentosum (XP) typically starts with skin symptoms. Progressive neurological symptoms, including hearing loss, occur in some patients with XP (described in Chap. 3). Exchanging information with others is an important aspect of everyday life, which can be seriously impaired in patients with hearing loss. Taking care of patients' hearing is important for maintaining their quality of life.

In this chapter, we review the clinical features, pathophysiology, and animal research about hearing in patients with XP.

4.1.1 Clinical Features of Hearing in Patients with XP

Approximately 25% of patients with XP demonstrate progressive neurologic abnormalities in the United States (more often in Japan, 50%) including microcephaly, diminished or absent deep tendon stretch reflexes, progressive sensorineural hearing loss (see Sect. 4.2), and progressive cognitive impairment [1, 2]. The earliest clinical neurologic abnormalities are frequently absence of deep tendon reflexes and high-frequency hearing loss [3].

XP is classified into seven genetic complementation groups deficient in nucleotide excision repair (A through G) and an XP variant type. The relative frequency and severity of cutaneous and neurological symptoms differ depending on the subtype. Patients with neurodegeneration commonly have mutations in XP-A, XP-B, XP-D, XP-F, or XP-G. However, not all patients with mutations in these genes develop neurodegeneration. Thus, it is clinically difficult to predict neurological disease [4].

There are few publications describing the audiologic characterization of hearing loss in patients with XP. Longridge described two patients with XP, aged 29 and 24 years, who had progressive bilateral sensorineural hearing loss and reduced speech discrimination [5]. The 29-year-old man showed neurological symptoms, such as fidgety hand movements, facial grimacing, and slurred speech between 9 and 10 years of age. His first audiometric test at 20 years of age showed approximately 55 dB of sensorineural hearing loss. The 24-year-old woman had mild sensorineural hearing loss of approximately 20 dB at 15 years of age. The hearing loss progressed to approximately 40 dB, and she required a hearing aid by age 19. Both audiograms were downsloping. The author hypothesized that the deafness was because of degeneration of central auditory pathways. Kenyon et al. reported three cases of XP with progressive downsloping SNHL, preserved speech discrimination, and impaired visual suppression of the vestibuloocular reflex. Audiometric assessment suggested that the hearing loss was cochlear in origin [6]. Totonchy et al. reported the status of hearing and neurological function in 77 patients with XP [4]. Overall, 71% ($n = 56$) of patients had normal hearing, and 23% ($n = 18$) had sensorineural hearing loss. They evaluated neurological involvement in patients with XP using the scale of XP-type neurological degeneration [7]. XP-type neurological degeneration was evaluated by the degree of mental retardation and the ability to walk and speak. Patients with neurological involvement atypical of XP were classified as non-XP-type neurological involvement. Of patients with XP without neurological involvement, 89% ($n = 49$) had normal hearing, and only 11% ($n = 6$) had hearing loss. In contrast, 76% ($n = 13$) of patients in the XP-type neurological degeneration group had hearing loss, including nine with sensorineural hearing loss and four with an unknown type of hearing loss. Pure-tone hearing thresholds were similar within each XP group (XP-type neurological degeneration, no neurological involvement, and non-XP-type neurological involvement). However, there was a distinct difference in hearing thresholds between the neurological and the non-neurological XP groups. The authors concluded that the degree of hearing loss was directly correlated with neurological involvement [4].

4.1.2 Severe Hearing Impairment in Patients with XP

XP occurs in all races, although it occurs at a higher frequency in Japan (1:22,000) [8] than in the United States (1:250,000) [9]. Approximately 55% of all Japanese patients with XP are assigned to the XP complementation group A (XP-A), and this proportion is higher than in other countries (e.g., 40%) [7, 10]. Most patients with XP-A exhibit severe neurological manifestations. In XP-A patients, the onset of neurological symptoms is between 3 and 8 years of age and manifests as cognitive and cerebellar signs [3, 7]. They usually have normal size and weight at birth [11]. However, even 1-year-old patients show a decline of deep tendon reflexes [12]. Typical symptoms include sensorineural hearing loss; progressive intellectual impairment, which may progress to slurred speech in severe cases; loss of the ability to walk; difficulty in swallowing; and a need of feeding gastrostomy. Patients with XP who have neurological degeneration have a high rate of mortality [11]. After onset, neurological symptoms progress slowly, eventually leading to premature death and affecting the whole nervous system [3].

Hearing loss may appear in XP-A patients from the age of approximately 6 years through their 20s, eventually becoming severe in their late 30s–40s. In patients with XP-A harboring a homozygous IVS3-1G > C missplice mutation of *XPA*, hearing impairment typically manifests as bilateral severe to profound hearing loss; the audiogram is usually horizontal or worse at higher frequencies [4, 7]. Here we review the audiological characteristics of two XP-A patients.

Case No. 1 An 8-year-old boy was diagnosed with XP-A at 1 year of age. His 2-year-old brother was also diagnosed with XP-A. The 8-year-old boy's auditory brainstem response (ABR) (see Sect. 4.2.2) was almost normal at 2 years of age. He gradually developed cognitive impairment and joint contractures and then became wheelchair bound. At 8 years of age, he could use only three-word sentences. The parents thought the delay in language development was because of cognitive impairment. However, ABR showed profound sensorineural hearing loss. Immediately following diagnosis, he began wearing hearing aids and attending a rehabilitation center. After this, he showed good responses to sounds in his surroundings.

Case No. 2 A 9-year-old girl was diagnosed with XP-A at 2 years of age. At her first visit to our office, she was 9 years of age, and her hearing was almost normal. Her mental development was diagnosed as that of a 5-year-old girl. She had slight contractures of the hip joint and was able to respond to pure-tone audiometry (see Sect. 4.2.1). Only 6 months later, she showed severe hearing loss (Fig. 4.1) despite avoiding exposure to sunlight. She started to wear hearing aids and could respond to sounds in her surroundings. In this case, hearing impairment progressed suddenly and rapidly.

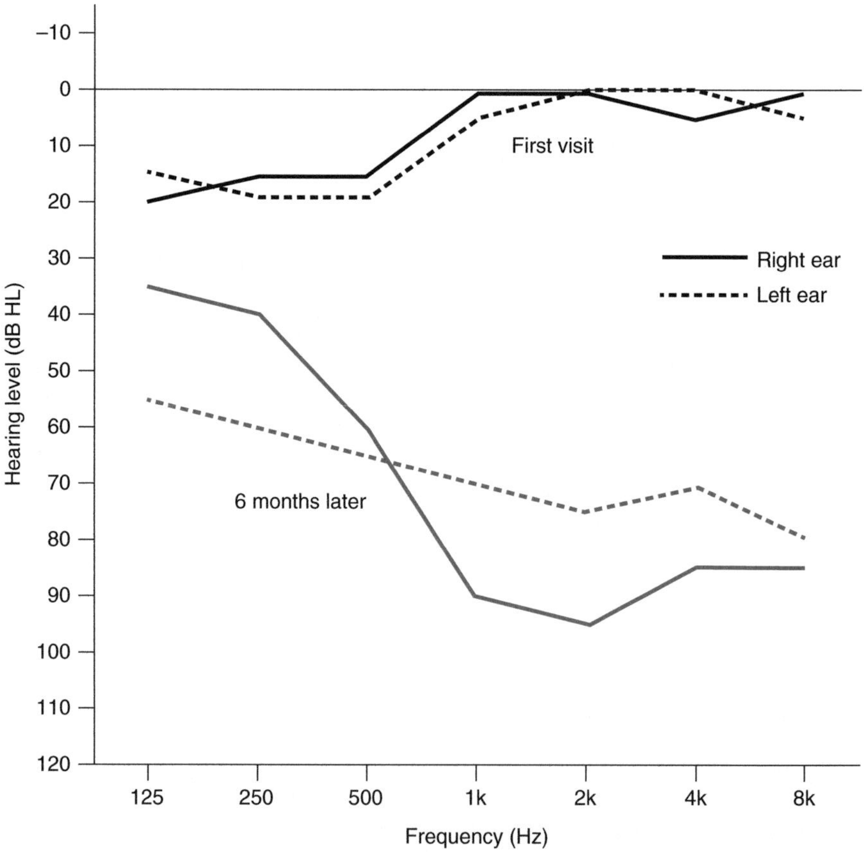

Fig. 4.1 Pure-tone audiometric data of case 2. The hearing level was almost normal at the first visit. At 6 months later, she showed severe hearing loss in both ears

Most of our XP-A patients had progressive bilateral sensorineural hearing loss. Typically, patients over 8 years of age showed severe to profound hearing loss, whereas hearing was preserved in patients up to approximately 2–4 years of age. Therefore, patients can acquire speech and language skills. However, in most cases, hearing loss and other neurologic complications, especially cognitive deterioration, progress concurrently as previously reported [4], making it difficult to precisely assess hearing and cognitive skills. For a patient that has difficulty responding to pure-tone audiometry, such as patients with cognitive deterioration, ABR with sedation is necessary.

4.2 Audiological Tests

The purpose of audiologic testing is to provide a quantitative assessment of hearing, and the resulting audiometric profile indicates the pattern and degree of hearing loss. Sound is perceived when the acoustic signal is conducted to the sensory organ of hearing (the cochlea) and then successfully transduced into neural impulses that travel to the brain. Hearing can be disrupted at any stage in this sequence. Therefore, hearing loss is broadly differentiated into two categories: sensorineural hearing loss and conductive hearing loss.

Sensorineural hearing loss is the result of dysfunction of the inner ear, the sensory organ (cochlea and associated structures), or pathophysiology affecting the nerve pathways from the inner ear to the brain. Therefore, sensorineural hearing loss can be subdivided into sensory loss and neural loss (or retrocochlear loss). It remains unclear whether sensorineural hearing loss in patients with XP is sensory, neural, or both based on the pathology described in human temporal bone studies (see Sect. 4.3). A recent study using an animal model of XP-A suggested that hearing loss in XP-A is mainly because of neural loss rather than cochlear damage (see Sect. 4.4).

Conductive hearing loss is the result of deficits in the transmission of sound through the outer and middle ear. This type of hearing loss is not typically seen in XP. Conductive hearing loss in patients with XP may be a secondary effect of fluid accumulating in the middle ear space of patients who have difficulty sitting and standing up because of neurodegeneration.

4.2.1 Pure-Tone Audiometry

Pure-tone audiometry is used to quantitatively measure sensitivity to sound as a function of frequency. This test generates a graph called an audiogram, which plots detection thresholds for pure tones in dB HL (decibels Hearing level) as a function of frequency. Thresholds are plotted separately for each ear and for each mode of stimulus delivery (air conduction and bone conduction).

This behavioral test for the assessment of auditory function depends on voluntary responses from the listener. For older children and adults, listeners can indicate their response by raising a hand or pressing a button. However, babies and younger children cannot respond voluntarily. Patients with XP who have cognitive impairment also have difficulties in this test. In these cases, an objective audiometric test needs to be applied.

4.2.2 *Auditory Brainstem Response (ABR)*

ABR is an auditory evoked potential that assesses the function of the peripheral auditory system, including the eighth cranial nerve and the brainstem. ABR is an objective test, and listeners do not need to respond to sounds. There is a strong association between the ABR threshold (the lowest stimulus intensity at which wave V is reliably identified) and the behavioral threshold. Therefore, ABR is often used to estimate hearing sensitivity in babies and other populations unable to provide behavioral responses. For patients with XP who are younger than 3 years of age or older patients with cognitive impairment, ABR is an appropriate tool for hearing assessment. Because ABR requires patients to stay calm or sleep to avoid noise interference, patients with XP often need to be sedated during the test.

4.2.3 *Otoacoustic Emissions (OAE)*

Otoacoustic emissions (OAE) are sounds recorded in the ear canal that are generated as a result of the electromotility of outer hair cells. They are used as an objective test of cochlear function. Although OAE are a reliable, easy, and minimally invasive method of screening for normal cochlear function, they have some important limitations. OAE can only be found if pure-tone hearing thresholds are better than 30–40 dB HL. They are also susceptible to human error. If OAE are absent or reduced, patients should be referred for an ABR.

4.3 Histopathology of the Inner Ear in Patients with XP

There are few pathological studies on cochleae from autopsies of patients with XP. In a recent autopsy study, researchers described temporal bone histopathology in two patients with XP who have neurologic degeneration: a 44-year-old woman with mutations in the *XPA* gene and a 45-year-old with mutations in the *XPD* gene. In both cases, patients had progressive sensorineural hearing loss with a downsloping audiometric pattern. The audiologic phenotype was more severe in the XP-D case than in the XP-A case. Findings were similar in both cases, including atrophy of the organ of Corti, stria vascularis, and spiral ganglia leading to severe or profound sensorineural hearing loss by the third decade of life. Spiral ganglion neurons are the dendrites of cochlear nerve neurons that innervate inner hair cells. Afferent spiral ganglion neurons depend upon the support of neurotrophic factors released from hair cells and supporting cells of the cochlear sensory epithelium. When hair cells die, the afferent dendrites of spiral ganglion neurons retract, leading to secondary loss of the somata of spiral ganglion neurons [13]. Although the authors were unable to conclude whether neuronal degeneration was primary or secondary to hair cell loss in both cases, in the XP-A case, the presence of remaining hair cells in the middle and apical turns of the cochlea combined with severe spiral ganglion degeneration suggested primary neuronal degeneration [4, 14]. Robbins et al. described temporal bone histopathology in the case of an adult XP-C patient without apparent neurologic degeneration that died at age 49. Her last audiometric test at age 48 showed mild hearing loss at higher frequencies but no abnormalities in the inner ear. However, there was neuronal loss in the dorsal root ganglia. The authors concluded that primary neuronal degeneration had begun in the peripheral nervous system [15].

4.4 Animal Studies

Knockout mice have been generated for many of the genes defective in patients with XP. In addition, there is a report describing the audiologic phenotype in a XP animal model [16]. *Xpa*-deficient mice in CBA (15), C57BL/6, and CD-1 chimeric backgrounds were generated by the insertion of neomycin cassettes into exon 4 of the *Xpa* gene using embryonic stem cell techniques [17]. This study, which characterized the mechanisms of neurological dysfunction and hearing impairment in patients with XP by evaluating hearing loss in *Xpa*-deficient mice, is described in Sect. 4.4.1.

4.4.1 Hearing in Xpa*–Deficient Mice*

Mouse hearing was assessed by ABR. In this study, *Xpa*-deficient mice had ABR thresholds similar to wild-type mice at 20 weeks of age (20–40 dB). *Xpa*-deficient mice displayed sensorineural hearing loss with significantly higher hearing thresholds at frequencies of 4, 8, and 16 kHz relative to wild-type mice at 38–40 weeks of age. At 60 weeks of age, ABR thresholds of *Xpa*-deficient and wild-type mice had nearly reached the upper limit of detection. This is consistent with studies showing that XP-A patients have normal hearing through the age of 2–4 years before they begin to manifest hearing impairment.

4.4.2 Morphological Features of Xpa-Deficient Mice Cochleae

In *Xpa*-deficient mice, there were significantly fewer spiral ganglion neurons in the apical and middle turns of the cochlea relative to wild-type mice. In contrast, there were no differences in the thickness of the stria vascularis or the percentage of remaining hair cells between *Xpa*-deficient and wild-type mice. These data suggest that hearing loss in XP is mainly because of primary degeneration of the nervous system rather than secondary degeneration following hair cell loss.

4.5 Molecular Mechanism of Hearing Loss in XP

The progression of neurological deterioration is relentless in XP-A patients, even when they avoid sun exposure. UV radiation only penetrates the skin. Thus, UV photoproducts that lead to disorders such as skin cancers and corneal damage cannot be the cause of neurological disease. However, the accumulation of unrepaired oxidative DNA lesions in the nervous system, which lead to progressive neuronal death, may be a cause of neurological problems.

One of the major DNA repair pathways is nucleotide excision repair (NER). The gene products responsible for each of the XP complementation groups are also involved in NER, including recognition of DNA damage, unwinding of double-stranded DNA, excision of damaged DNA and flanking oligonucleotides, repair of errors during synthesis of original and daughter strands of DNA during replication, and ligation of newly synthesized fragments [18]. NER-related proteins exist in the cochleae of Fischer 344 rats, and the NER pathway is involved in the repair of DNA damage caused by cisplatin. In addition, XPA protein in spiral ganglion neurons translocates from the cytoplasm to the nucleus during cisplatin treatment, which is ototoxic to the inner ear [19]. These results indicate that XPA protein recognizes DNA damage resulting from exogenous and/or endogenous hazards to spiral ganglion neurons. *XPA* and *XPC* mRNA are also expressed in Fischer 344 rats. *XP* mRNA levels in the cochleae were up to sixfold (*XPC*) and threefold (*XPA*) greater than in the kidney, which has the highest level of *XP* DNA repair of all the major organs, including the brain, heart, lung, spleen, and muscle [20].

The generation of reactive oxygen species (ROS) in the cochlea is triggered by exposure to loud sounds and ototoxic drugs. This is often followed by caspase-mediated apoptotic cell death. ROS causes chronic damage to ear structures [21]. Moreover, Brooks reported that 8,5′-cyclopurine-2′-deoxynucleosides, which are induced by oxidative stress, may induce neurodegenerative DNA lesions in patients with XP [22]. We speculate that *XP* genome products defend the cochlea and nervous system against persistent genomic stress (e.g., from endogenous ROS) during normal metabolism. In patients with XP, the cochlea and nervous system are deficient in genes responsible for XP and are susceptible to these stresses. Therefore, they show early onset sensorineural hearing loss.

4.6 Patient Hearing Care

Based on the clinical features and basic research described so far, avoidance of exposure to high levels of noise/music should be a routine part of counseling for patients who are at risk for XP-related hearing loss. Regular audiometric evaluations are important for identifying the need for auditory intervention, such as hearing aids, and also important for the early identification of XP-type neurological degeneration [4].

Hearing aids boost sound levels so that they become audible to the hearing-impaired. Healthcare professionals need to tune hearing aids based on audiometric tests and avoid exposing patients' ears to too much sound pressure. Hearing aids were effective for cases no. 1 and no. 2 described in Sect. 4.1.2. Indeed, hearing loss and cognitive impairment progress concurrently in some patients with XP. In these cases, it is difficult to fit hearing aids with settings comfortable for the patients.

Cochlear implants restore useful hearing in patients with severe to profound hearing loss. They bypass the outer, middle, and inner ear and provide information through direct electrical stimulation of the spiral ganglion. There is a report of cochlear implantation in two patients with Cockayne syndrome, which is a disorder related and overlapping with XP [23]. Cochlear implants worked well in one case and not in the other case. The latter patient also showed cognitive impairment. The loss of spiral ganglion neurons in the temporal bone of humans and animals with XP indicates that cochlear implants have limited benefit in these cases. Moreover, cognitive impairment and other forms of neurodegeneration make the mapping and adjustment of cochlear implants difficult. Indications for cochlear implantation in patients with XP should be carefully considered.

Currently, patients with XP are diagnosed early, carefully protected from sunlight to minimize skin problems, and living longer than in the past. Therefore, it is increasingly important that hearing assessment is sufficient to identify the need for auditory interventions, such as hearing aids, to maintain the quality of life for patients with XP.

References

1. Bradford PT, Goldstein AM, Tamura D, et al. Cancer and neurologic degeneration in xeroderma pigmentosum: long term follow-up characterises the role of DNA repair. J Med Genet. 2011;48:168–76.
2. Nakano E, Masaki T, Kanda F, et al. The present status of xeroderma pigmentosum in Japan and a tentative severity classification scale. Exp Dermatol. 2016;25(suppl. 3):28–33.
3. Anttinen A, Koulu L, Nikoskelainen E, et al. Neurological symptoms and natural course of xeroderma pigmentosum. Brain. 2008;131(8):1979–89.
4. Totonchy MB, Tamura D, Pantell MS, et al. Auditory analysis of xeroderma pigmentosum 1971-2012: hearing function, sun sensitivity and DNA repair predict neurological degeneration. Brain. 2013;136:194–208.
5. Longridge NS. Audiological assessment of deafness associated with xeroderma pigmentosa. J Laryngol Otol. 1976;90:539–51.

6. Kenyon GS, Booth JB, Prasher DK, et al. Neuro-otological abnormalities in xeroderma pigmentosum with particular reference to deafness. Brain. 1985;108:771–84.
7. Nishigori C, Moriwaki S, Takebe H, et al. Gene alterations and clinical characteristics of xeroderma pigmentosum group A patients in Japan. Arch Dermatol. 1994;130:191–7.
8. Hirai Y, Kodama Y, Moriwaki S, et al. Heterozygous individuals bearing a founder mutation in the XPA DNA repair gene comprise nearly 1% of the Japanese population. Mutat Res. 2006;601:171–8.
9. Robbins JH, Kraemer KH, Lutzner MA, et al. Xeroderma pigmentosum. An inherited diseases with sun sensitivity, multiple cutaneous neoplasms, and abnormal DNA repair. Ann Intern Med. 1974;80:221–48.
10. Moriwaki S, Kraemer KH. Xeroderma pigmentosum – bridging a gap between clinic and laboratory. Photodermatol Photoimmunol Photomed. 2001;17:47–54.
11. DiGiovanna JJ, Kraemer KH. Shining a light on xeroderma pigmentosum. J Invest Dermatol. 2012;132:785–96.
12. Ueda T, Kanda F, Aoyama N, Fujii M, Nishigori C, Toda T. Neuroimaging features of xeroderma pigmentosum group A. Brain Behav. 2012;2(1):1–5.
13. Wong AC, Ryan AF. Mechanisms of sensorineural cell damage, death and survival in the cochlea. Front Aging Neurosci. 2015;7:58.
14. Viana LM, Seyyedi M, Brewer CC, et al. Histopathology of the inner ear in patients with xeroderma pigmentosum and neurologic degeneration. Otol Neurotol. 2013;34:1230–6.
15. Robbins JH, Kraemer KH, Merchant SN, et al. Adult-onset xeroderma pigmentosum neurological disease – observations in an autopsy case. Clin Neuropathol. 2002;21:18–23.
16. Shinomiya H, Yamashita D, Fujita T, et al. Hearing dysfunction in Xpa-deficient mice. Front Aging Neurosci. 2017;10(9):19.
17. Nakane H, Takeuchi S, Yuba S, et al. High incidence of ultraviolet-B-or chemical-carcinogen-induced skin tumours in mice lacking the xeroderma pigmentosum group A gene. Nature. 1995;377:165–8.
18. Fousteri M, Mullenders LH. Transcription-coupled nucleotide excision repair in mammalian cells: molecular mechanisms and biological effects. Cell Res. 2008;18:73–84.
19. Guthrie OW, Li-Korotky HS, Durrant JD, Balaban C. Cisplatin induces cytoplasmic to nuclear translocation of nucleotide excision repair factors among spiral ganglion neurons. Hear Res. 2008;239(1–2):79–91.
20. Guthrie OW, Carrero-Martínez FA. Real-time quantification of Xeroderma pigmentosum mRNA from the mammalian cochlea. Ear Hear. 2010;31(5):714–21.
21. Staecker H, Zheng QY, Van De Water H, et al. Oxidative stress in aging in the C57B16/J mouse cochlea. Acta Otolaryngol. 2001;121:666–72.
22. Brooks PJ. The 8,5′-cyclopurine-2′-deoxynucleosides: candidate neurodegenerative DNA lesions in xeroderma pigmentosum, and unique probes of transcription and nucleotide excision repair. DNA Repair. 2008;7:1168–79.
23. Morris DP, Alian W, Maessen H, et al. Cochlear implantation in Cockayne syndrome: our experience of two cases with different outcomes. Laryngoscope. 2007;117(5):939–43.

Chapter 5
Epidemiological Study of Xeroderma Pigmentosum in Japan: Genotype-Phenotype Relationship

Chikako Nishigori and Eiji Nakano

Abstract Xeroderma pigmentosum (XP) is a rare autosomal recessive hereditary disease caused by the deficiency of repairing DNA damage caused by ultraviolet radiation and some other compounds. Patients with XP display pigmentary change and numerous skin cancers in sun-exposed body sites, and some patients show exaggerated severe sunburn upon minimum sun exposure and neurological symptoms. We have conducted the nationwide survey for XP since 1980 as a research project supported for the intractable disease initiated by the Japanese Ministry of Health, Labour and Welfare. The frequency of each complementation group in Japan is considerably different from that in Western countries; in Japan, XP complementation group A is the most frequent, followed by variant type. Regarding skin cancers in XP, basal cell carcinoma was the most frequent cancer that patients with XP developed, followed by squamous cell carcinoma and malignant melanoma. The frequency of these skin cancers in patients with XP-A has decreased in these 20 years, and ages of onset of developing skin cancers are much older than those previously observed, which is greatly attributed to the education of sun protection for the patients with XP and their parents and guardians for these 20 years. In order to encourage the patients and their parents to perform appropriate sun protection for the prevention from skin cancers, definite diagnosis but not possible diagnosis is crucial. In addition, diagnosing at younger ages is important. On the other hand, the effective therapy for neurologic XP has not been established yet, and this needs to be done urgently.

5.1 Introduction

Xeroderma pigmentosum (XP) is an autosomal recessive hereditary photosensitive disease, in which patients display extreme hypersensitivity to ultraviolet radiation (UV) because of congenital defect of repair ability for UV-induced DNA damage. If

C. Nishigori (✉) · E. Nakano
Department of Dermatology, Graduate School of Medicine,
Kobe University, Kobe, Hyogo, Japan
e-mail: chikako@med.kobe-u.ac.jp

C. Nishigori, K. Sugasawa (eds.), *DNA Repair Disorders*,
https://doi.org/10.1007/978-981-10-6722-8_5

these patients do not take appropriate protection from sunlight, they will develop serious photoaging skin symptoms, xerosis of the skin, progressive development of freckle-like pigmentation, and multiple skin cancers at sun-exposed area regardless of young age. XP is classified into eight subtypes; A–G genetic complementation groups of nucleotide excision repair deficient type and variant type. Responsible gene for each subtype has been identified, and each displays characteristic clinical features (Table 5.1). In Japan, the frequency of XP is higher than other countries, and furthermore genetic complementation group A (XP-A), of which patients develop progressive neurological symptoms and its severity impacts their prognosis, accounts for about half of XP patients. Therefore, XP has been assigned as an "intractable disease" which is supported by the *Research Grant Initiative on Overcoming Intractable Skin Diseases* by the Japanese Ministry of Health, Labour and Welfare (MHLW) since the 1980s. In July of 2015, XP has been assigned to the status of intractable diseases, making patients eligible for support from the government. In this chapter, we briefly discuss about the clinical aspects of XP, especially focusing on the status in the present Japan in comparison with that 25 years previously, and describe how we have coped with XP and what remains to be resolved in the future.

Table 5.1 Clinical and cytological characteristics of XP complementation groups

					Skin symptom		
	Responsible gene	Number of patients and frequency (%)[a] Japan	UDS (%)	UV sensitivity (D_0) (J/m^2)	Sunburn	Age of onset of BCC (year)[b]	Neurological symptom
A	*XPA* 9q34.1 (31kD)	63 (67.7)	<5	0.4	+++	9.3	++
B	*XPB/ERCC3* 2q21 (89kD)	0	3–7		++		– ~ ++
C	*XPC* 3q25 (106kD)	2 (2.2)	10–20	1.0	++	14.8	–
D	*XPD/ERCC2* 19q13.2 (87kD)	5 (5.4)	20–50	0.77	++	31	– ~ ++
E	*DDB2* 11q12-p11.2 (48kD)	Rare	40–60	2.2–2.4	+	43.5	–
F	*XPF* 16p13.13 (126kD)	1 (1.1)	10–20	1.7–2.2	+	45.5	–
G	*ERCC5* 13q33 (133kD)	Rare	<5	0.6	+	32	+
V	*POLH* 6p21.1-6p12 (83kD)	22 (23.7)	75–100	2.4–4.5	+	40.8	–

XP is subdivided into nucleotide excision repair (NER) deficient groups A through G and variant type. In Japan, half of the patients belong to the XP complementation group A, the severest type with the lowest DNA repair capacity, and 25% of the patients are assigned to the variant type, which has an almost normal level of nucleotide excision repair but is deficient in POLH, that is involved in translesional replication (TLR)

[a]Nakano et al [5]

[b]Sato and Nishigori [3]

5.2 Epidemiology

In Japan, the frequency of XP in newborn babies had been believed to be approximately 1/100,000 or less [1], and now it is estimated that XP occurs in 1 of 22,000 people [2]. This figure means that XP is not so extremely rare as compared with the incidence in Western countries, where the frequency is one of one million people, still it is a rare disease. In the old literature of the early twentieth century written in Japanese, we can find the description of siblings of XP which seems to be similar to cases of XP-A. In the late1980s, epidemiological survey for XP has been conducted. At that time, the DNA repair test had just become available for clinical use to diagnose XP in a certain laboratory, and big effort has been made to diagnose patients with possible or probable XP by DNA repair tests or genetic complementation tests, and these cytological data and clinical information from all patients who have once visited the hospitals were compiled by the nationwide survey in 1988 [3]. After a quarter of a century, nationwide XP survey was conducted again by the Research Committee for Intractable Diseases supported by the Japanese MHLW; in 2012, patients with probable XP who visited the medical institution during 2010–2012 were enrolled. For the first survey, questionnaires were distributed to 616 hospitals requesting for the number of patients who visited the hospital between 2010 and 2012. We received replies from 403 institutions and then the second survey questionnaires were sent to the dermatologists who attended to patients with XP and asked for more detailed clinical information [4]. In the survey_1988, as many patients with XP as possible were enrolled and analyzed in order to grasp the whole picture and natural history of Japanese patients with XP, who had hardly received any treatment or care for the disease, because at that time, clinical methodology on how to care and educate the patients with XP had not been established, while the results of the survey–2012 represent the status of Japanese patients with XP who needed medical care during 2010-2012, and it is useful to search for the present problem to be resolved. Table 5.2 shows the frequencies of each clinical form enrolled in the two surveys. In both surveys, XP-A, in which both cutaneous

Table 5.2 Number of patients with each genetic complementation groups of XP patients

	A	B	C	D	E	F	G	V	Unknown	Total
Survey _1988[a]	117 (43.0)[c]		5 (1.8)[c]	5 (1.8)[c]	6 (2.2)[c]	12 (4.4)[c]	1	89[b] (32.7)[c]	37[d]	272 (100)
Survey _2012[e]	63 (67.7)[f]		2 (2.2)[f]	5 (5.4)[f]		1 (1.1)[f]		22 (23.7)[f]	77[g]	170

[a]Sato and Nishigori [3]
[b]Patients with clinically definite XP with the UDS level over >60%
[c]Number in parentheses: frequency(%)of each complementation group; the number of patients were divided by the total number of the patients
[d]Patients with reduced UDS but complementation groups has not been assigned
[e]Modified from the data from Nakano [5]
[f]Number in parentheses indicate the frequency out of the total numbers of patients excluded "unknown"
[g]Clinically the patient satisfies the criteria of XP, but the complementation group was not identified yet

symptoms and neurological symptoms are the most severe, accounts for about half of XP patients in Japan, followed by XP-V in which patients develop only cutaneous symptoms, accounts for about 25%. In the 1988 survey, all patients who were suspected of XP were subjected for the DNA repair tests for UDS and cell survival assay, but the responsible gene for XP has not been identified yet, and the diagnosis of XP has been done by genetic complementation tests using fusion technique or by the combination of clinical information, where patients manifesting typical severe exaggerated sunburn with UDS lower than 5% were diagnosed as "probable XP-A" and patients manifesting typical pigmented freckles with multiple skin cancer at younger ages with UDS higher than 60% were diagnosed as "probable XP-V." However in the 2012 survey, responsible genes for all XP clinical subtypes have been identified, and genetic analysis-based diagnosis has been made for most cases, which apparently increased the frequency of XP-A higher and decreased the frequency of XP-V slightly lower, since, in Japan, diagnosis of XP-A is very feasible because of its founder mutation, whereas diagnosis of XP-V needs several cumbersome examinations including UDS, POLH immunoblotting, and *POLH* gene sequencing, and not all patients cannot be genetically diagnosed. All the same, these frequency patterns are similar between the two surveys and differ substantially from that observed in Western countries, where XP-C and XP-D are the most common clinical subtypes. Patients with XP-E are rare, and there is no case report of patients with XP-B in Japan.

5.3 Cutaneous Symptoms of XP

XP Displaying Exaggerated Sunburn Reaction Followed by Pigmentary Change In patients with XP-A, XP-B, XP-D, XP-F, and XP-G, severe and exaggerated sunburn reaction occurs at the sun-exposed area upon a minimum sun exposure (e.g., face, nape, ear auricle, dorsum of the hand, and upper and lower limbs) (Fig. 5.1). Unlike normal sunburn, this exaggerated sunburn reaction is often associated with remarkable erythema, swelling, blister, and erosion and is exacerbated for 3–4 days after exposure and persists for at least 1 week. After having repeated such severe sunburn-like reaction, freckle-like small pigmented maculae are found at the sun-exposed area. Freckle-like pigmented maculae increase whenever sun exposure episode is repeated. In comparison with normal freckle, the sizes of freckle-like pigmented maculae in XP are various, and its color tone is heterogeneous, from pale-brown to brown or black-brown color. Small pigmented maculae can be found not only in the face but also in the nape, the dorsum of the hand, and the upper chest (Fig. 5.2). Sun-exposed area of the skin tends to be xerotic easily, and multiple malignant skin tumors (actinic keratosis, basal cell carcinoma, squamous cell carcinoma, malignant melanoma, etc.) will be found in the face and others at young age. If such patients do not perform strict protection from sunlight, malignant skin tumors may occur at ages of 30–60 years younger than healthy people, and the frequency is considered as 1000 times or more as much as healthy generation.

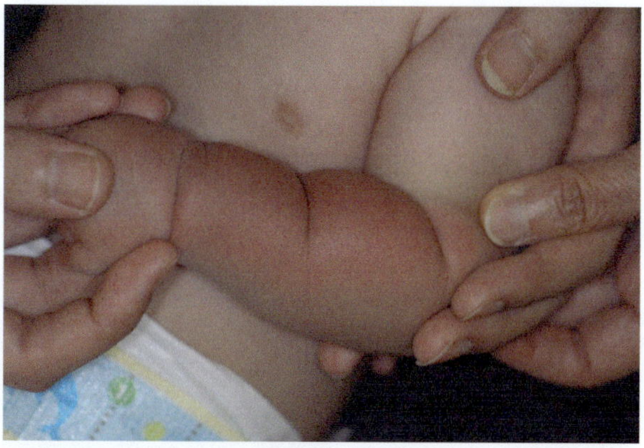

Fig. 5.1 XP-A patients (2 months old). Clearly demarcated edematous erythema was observed limited on the sun-exposed body site. This was her first sun exposure after birth

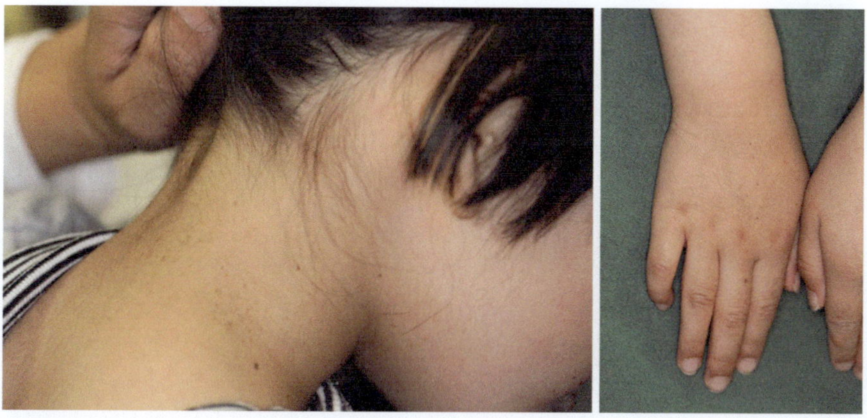

Fig. 5.2 XP-A patients (3 years old). Freckle-like pigmented maculae limited on the sun-exposed area. Whenever the patient experiences repeated severe exaggerated sunburn, freckle-like pigmented maculae develop in patients with XP

XP Displaying Abnormal Pigmentary Change Without Exaggerated Sunburn In patients with XP-C, XP-E, and XP-V, freckle-like pigmented maculae gradually progresses without the history of exaggerated sunburn at the sun-exposed skin sites. This pigmented maculae varies in size from miliary to rice grain size, and its color tone is heterogeneous, and its border is indistinct. Patients present with photoaging skin which is unsuitable for their age and symptoms are progressive. In some cases, depigmented maculae are also observed. Multiple skin malignant tumors occur at the sun-exposed body sites at younger age [6]. In patients with XP presenting with only pigmentary change, lifelong cumulative UV dose to which patients exposed, develop their skin symptoms. It often happens that patients become to know that they have XP for the first time when they visit the dermatologist to refer the skin cancers (Fig. 5.3). At that time, they have already received a substantial amount of UV. Recently, parents who experienced their children's exag-

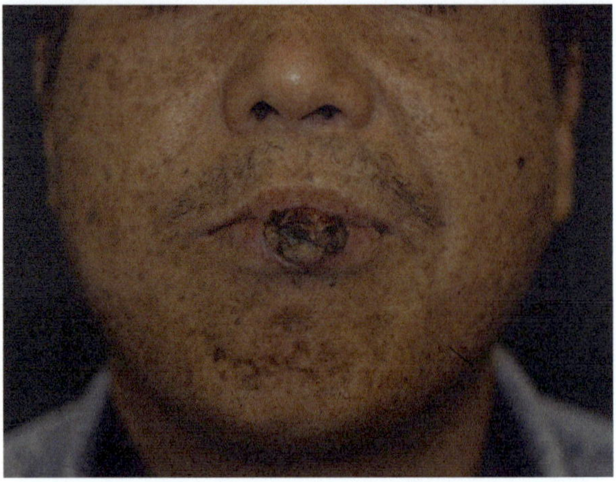

Fig. 5.3 XP-V (55 years old). He visited a hospital consulting the tumor on the lower lip and was referred to us to examine the genetic diagnosis for XP. He has grown up in the island, and he never protected himself from sunlight

gerated sunburn tend to protect their children from sun exposure, while parents whose children display only pigmentary change, but not exaggerated sunburn, easily overlooked the presence of the disease. Therefore, frequency of skin cancer tends to be greater in XP patients with only pigmented freckles, especially XP-C and XP-V, rather than the patients with XP who present with exaggerated sunburn, especially XP-A and XP-D. We will discuss this matter in the latter paragraph.

5.4 Neurological Symptom of XP

In Japan, progressive central and peripheral neurodegeneration occurs in approximately 100% of XP-A patients. For typical development in children patients with XP-A in which the most severe symptoms are found, in their childhood, they can acquire approximately age-appropriate functions, although they show slight delay in their development. The head is held up at an average of 3.5-month-old; rolling over, sitting position, pulling up to standing, and walking are achieved at an average of 6-month-old, 7-month-old, 12-month-old, and 15-month-old, respectively. The peak of physical performance is achieved at about 6 years old, gait disturbance occurs at about 12 years old, and wheelchairs are required at about 15 years old [7]. Deformity in the foot such as contracture in pes equinovarus and pes cavus can be found approximately around 6 years of age [8]. In some cases, it may be complicated with callosity, skin ulcer, contact dermatitis, and tinea pedis due to foot deformity and the use of prostheses. Regarding auditory function, hearing loss occurs at mostly around 4–7 years old [9], and wearing of hearing aid devices is required in the second half of school age. At about 15 years old, auditory function almost nonexistent. Regarding speech function, the peak is achieved at 5–6 years old. The language that they acquired once is maintained in spite of the progressive deafness during the elementary school period. However, they show dysarthria with decline of intellectual ability and advanced deafness, and their speech function disappears at about 15 years old. Involuntary movement, such as tremor and myoclonus, may be also found in older children. Deep tendon reflexes in the extremities gradually disappear. Progressive sensory-dominant axonal neuropathy is found by peripheral nerve conduction studies. This finding goes along well with our experience that XP patients with progressive neurodegenerative symptoms scarcely express their pain sensation during biopsy procedures. Brain CT and MRI reveal atrophy of all the cerebrum, brainstem, and cerebellum with ventricular dilatation [10, 11]. Currently, we are following up 35 patients with XP-A (0–48 years old), and among them, choke occurs in 18–20 years old, dysphagia may occur between 15 and 19 years old, and they frequently cause aspiration pneumonia. In some cases, tracheotomy may be required because of vocal cord paralysis and larynx dystonia at about 20 years (17–24 years old). Afterward, their general status becomes deteriorated and dies because of pneumonia or sudden death. Figure 5.4 shows the age distribution of patients with XP-A, indicating that in the 1987 survey, patients older than 20 years old were extremely rare, meaning that most patients died before 20 years old. In fact, previously we reported that patients with XP-A mostly die around the age of 20 years, because of aspiration pneumonia and sudden death [9]. However now, life expectancy of patient with XP-A became strikingly longer than that of 25 years

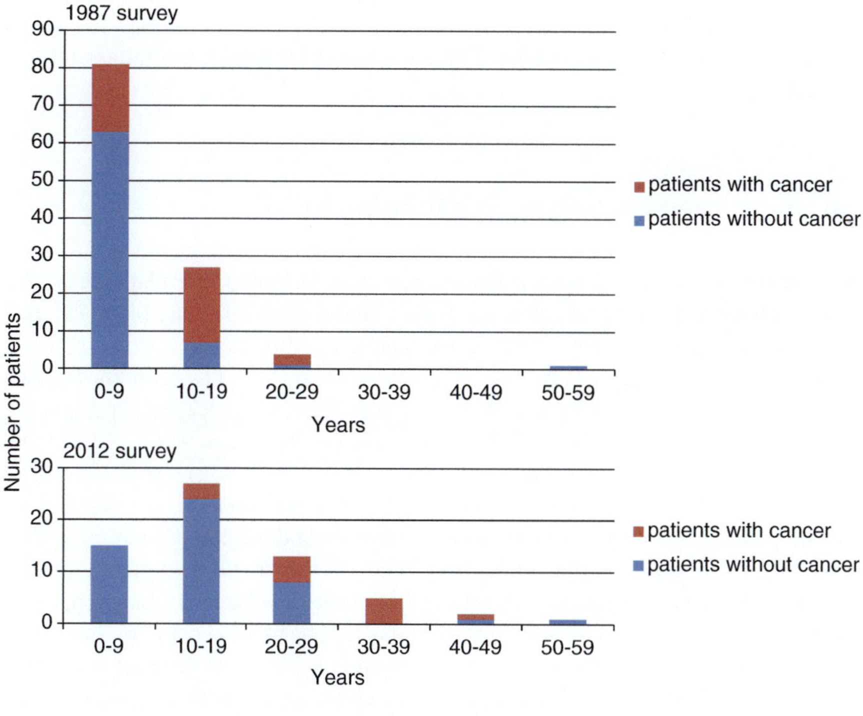

Fig. 5.4 Difference of age distribution of patients with XP-A between 1987 survey and 2012 survey. Life expectancy of patients with XP-A prolonged strikingly

before, and the number of patients with XP-A older than 20 years remarkably increased. The reason for this striking difference is not known. Advances in medical practice may be one reason. Whether strict sun protection contributes this elongation of life expectancy remains to be elucidated. XP-D patients in Western countries develop neurological symptoms frequently. Meanwhile XP-D patients in Japan commonly do not develop neurological symptom, and most of them can do normal work, if any. We speculated this difference because of the difference in the ability of ATP binding due to the difference in the mutation sites [4]. Rarely, some XP-F patients showed neurological symptoms [12].

5.5 Eye Manifestation of XP

In XP patients, the eye tissues exposed to UV are also involved. Therefore, they have lesions in the anterior ocular segment such as conjunctival xerosis and corneal drying, conjunctivitis, keratitis, evagination, corneal ulcer, and decrease of lacrimation. Since most of UVB spectrum does not reach the retina, no morbid change due to direct exposure to UV occurs, and conjunctivitis, corneal neovascularization (pannus), corneal drying, corneal cicatrization, ectropion, blepharitis, pigmentation of the conjunctiva, and cataract may occur. Abnormalities in the optic nerve as neurological symptom of XP may occur, including visual field disturbance and optic neuropathy. There are some reports of malignant tumors as well.

5.6 Genotype-Phenotype Relationship in XP

At least in some complementation groups, genotype-phenotype correlation has been noticed depending on the mutated site in the same responsible gene [4, 9, 13], and it explains the heterogeneity of clinical symptoms within the same complementation groups. In Japanese XP-A, three frequent mutation sites has been known, two nonsense mutation, c. 348 T>A, p.Y116X, and c. 682 C>T, p. R228X, and splicing mutation, IVS3-1G>C. This IVS3-1G>C mutation of *XPA* gene was reported to be the founder mutation among Japanese patients with XP-A, where 86% (25/29) of the patients harbored the homozygous IVS3-1G>C mutation and 14% (4/29) were the compound heterozygote of the founder mutation and the other mutation in *XPA* gene [9]. Consequently, the allele frequency of the founder mutation among Japanese XP-A patients was calculated as 93.1% (25 × 2 + 4/29 × 2). Later on, using the haplotype analysis, this founder mutation has been shown to originate 120 generation previously, 2400 years before, assuming a 20-year generation interval [14]. In the XP survey in 2012 as well, we could also detect frequency of the homozygous IVS3-G>C mutation as high as 88.7% (49/55) among genetically diagnosed XP-A patients who visited the medical institution during 2010–2012, and

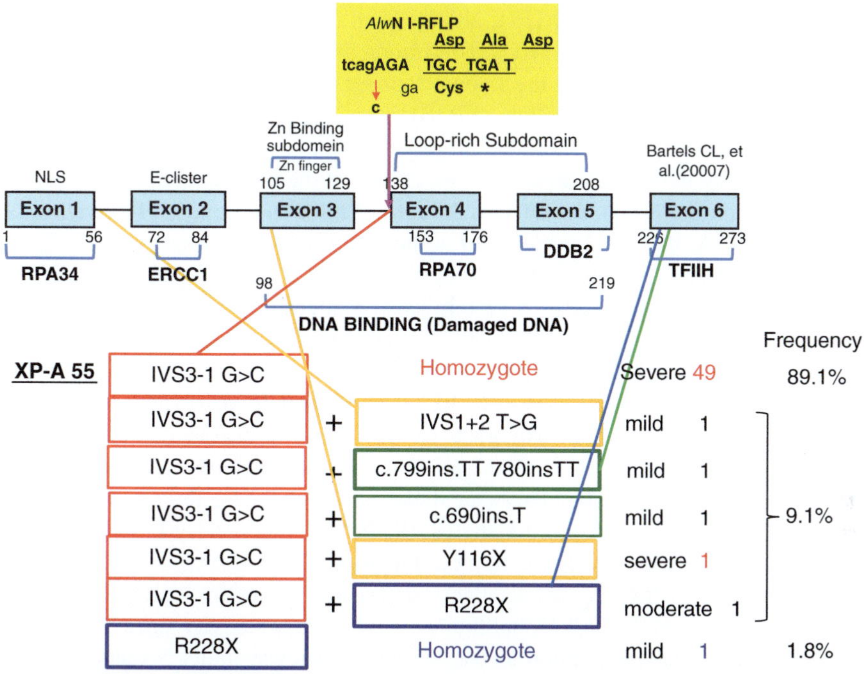

Fig. 5.5 Mutation sites detected in the 2012 survey were indicated with the scheme of XPA protein and its functional sites. Patients having homozygous IVS3-1G>C mutation are most frequently detected (89.1%), and patients having compound heterozygous mutation of IVS3-1G>C and other mutation account for 9.1%

9.1% of the patients (5/55) have compound heterozygote of the founder mutation and other mutations (Fig. 5.5). In this assay, the allele frequency of IVS3-1G>C among Japanese XP-A patients was 93.6% (49 × 2 + 5/55 × 2), almost the same as that of the previous study. This founder mutation at the 3′ splice acceptor site of intron 3 (IVS3-1G>C) induce alternative splicing, creating a stop codon at the second codon of exon 4, which results in no detectable protein production due to nonsense mediated decay and then markedly reduced DNA repair [15]. It goes along with the fact that typical Japanese patients with XP-A present with severe cutaneous symptoms and neurologic symptoms as described above. In the survey of 2012, we evaluated the severity scoring scale regarding neurologic and cutaneous symptoms for the 49 XP-A patients with IVS3-1G>C founder mutation. Among them we found that the patients' age and severity score were well correlated on "swallowing," "gait," "intellectual impairment," and "motivation," being the $R^2 = 0.70874$, 0.62437, 0.63781, and 0.75111, respectively [5]. The natural history of these XP-A patients with founder mutations coincides well with aforementioned typical clinical symptoms of XP-A. To date, several patients with XP-A have been reported in the literature who manifested milder symptoms among Japanese XP-A patients, including XP39OS [16]; two siblings XP3KR and XP4KR [9]; two siblings XP4KO and XP5KO [17]; XP2NI [18]; XP17HM, XP21HM, XP42HM, and XP43HM [19]; and XP113KO [20]. Among them all patients but XP39OS harbored IVS3-1G>C mutation, the known founder mutation, in one allele and the other mutations in the other allele. XP2NI, who manifested mild skin symptoms with slight sun sensitivity without skin cancer development or hearing loss at the age of 11 years, harbored G to C one base change at the last nucleotide of exon 5 in addition to the founder mutation, and the former mutation produced three types of aberrant mRNA, lacking 7 nucleotides at the end of exon 5, lacking entire exon 5, and lacking exons 3, 4, and 5. Western blot of XP2NI cells indicated that small but significant amount of a truncated protein was produced and the size of the protein coincides the protein lacking the seven nucleotides in exon 5 [18]. XP17HM, XP21HM, XP24HM, XP42HM, and XP43HM were newly diagnosed as XP-A at the age of 35 years old, 30 years old, 40 years old, and 45 years old, respectively. They presented with mild neurological symptoms and a history of moderate sun sensitivity. In addition to the IVS3-1G>C founder mutation in one allele, XP17HM had c.690insT in the exon 6 of the other allele and the rest of the three had 779insTT 780insTT in the exon6 of the other allele [19]. On the other hand, XP3KR, XP4KR, XP4KO, and XP5KO, having the founder mutation, IVS3-1G>C, in one allele and R228X in exon 6 in the other allele, manifested milder symptoms than typical IVS3-1G>C homozygotes [9, 17, 21], but they exhibited sun sensitivity since their childhood, and their fist consultation to the dermatologists was under 5 years old, and XP4KO developed BCC at the age of 13 years old, indicating these cases' severity was moderate. Note that the severity score of the XP5KO is far apart from the average score of the XP-A patients with founder mutation (Red circle in Fig. 5.6). XP39OS, who showed no neurological abnormalities at the age of 7 years, but areflexia of the patellar tendons at 11 years old, was diagnosed as XP-A by means of cell fusion technique before the discovery of *XPA* gene. In addition, fibroblasts form XP39OS revealed sensitivity to UVR but two times resistance as those of typical XP-A, XP3OS, homozygous IVS3-1G>C [16]. Genetic diagnosis indicated that XP39OS harbored homozygous R228X, which is the second frequent gene mutation identified among Japanese patients with XP-A and the only common mutation found in and outside Japan [22]. Previously, we reported that homozygous mutation of R228X is the most common type of mutation in Tunisian patients with XP-A, and they rather manifested milder cutaneous and neurologic symptoms despite their living environment was not so protective from UVR. Comparison of the genotype-phenotype correlation in patients with XP-A has suggested that those with mutations closer to the C-terminal coding region of the *XPA* have milder neurological and cutaneous symptoms. Figure 5.5 shows the putative function of its each domain of XPA protein and the mutation sites and their clinical severity of the XP-A patients enrolled in the survey of 2012. Cells from patients with milder symptoms in the literature, XP2NI, XP17HM, XP21HM, XP42HM, and XP43HM, showed a very small but traceable amount of XPA protein lacking a part of exon 5 or exon 6, which explains the residual repair functions observed in these patients. Furthermore recently, cases of XP-A having homozygous IVS4 + 8A>G manifesting unusually

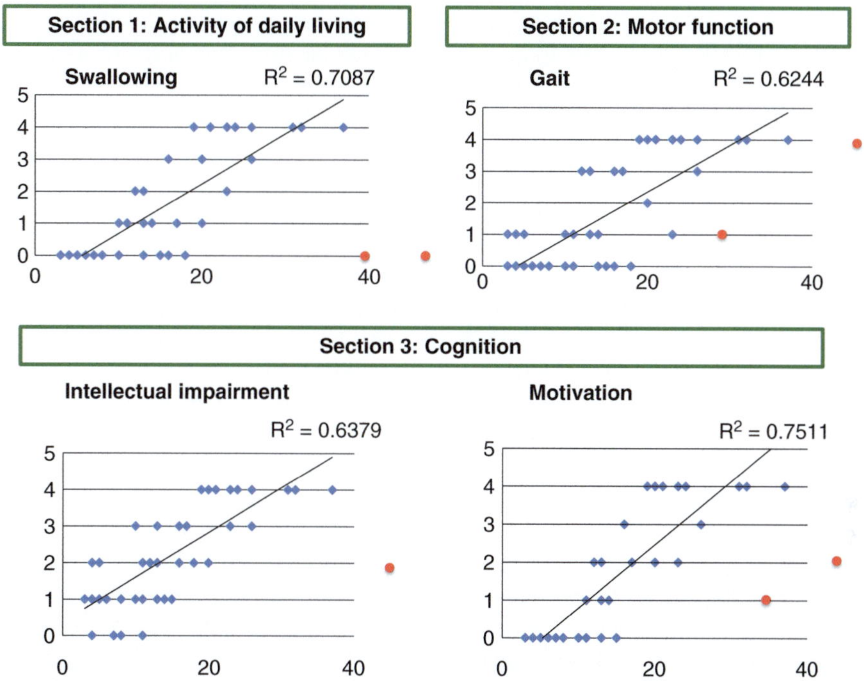

Fig. 5.6 Correlation of the severity index of patients with XP-A having homozygous Japanese XPA founder mutation, IVS3-1G>C, and patients' ages (Modified from the reference Masaki et al. [21]). The severity score of a milder case, a compound heterozygotes of the founder mutation, and R228X was plotted in red circle. Horizontal axis indicates years, and vertical axis indicates severity scores. Scores in swallowing stand for 0, normal; 1, rare choking; 2, occasional choking; 3, requires soft food; and 4, requires nasogastric or gastrostomy tube, respectively. Scores in walking stand for 0, normal; 1, mild difficulty; 2, moderate difficulty, but require little or no assistance; 3, severe disturbance of walking, requiring assistance; and 4, cannot walk at all, even with assistance. Scores in intellectual impairment stand for 0, normal; 1, mild (consistent impairment with partial recollection of events with no other difficulties); 2, moderate difficulty handling complex problems; 3, severe impairment with problems; and 4, unable to make judgments or solve problems. Scores in motivation stand for 0, normal; 1, lacking in energy, does not restrict activities; 2, lacking in energy, restricts hobbies, and interests; 3, lacking in energy, restricts routine activities; and 4, unable to carry out any task

mild symptoms have been reported [23]. Takahashi et al. [24] also reported an unusually mild XP-A patient having homozygous c.529G>A in exon 4, which creates a new cryptic donor site in exon 4, resulting in aberrant splicing. In both cases, the authors indicated that the patient's cells produce a very small amount of leaky normal XPA protein, although majority of the aberrant splicing product is nonfunctional truncated XPA protein. These findings indicate that even a very small amount of protein and even partially dysfunctional protein could ameliorate clinical symptoms. Analysis for those who manifested mild clinical symptoms gives us an important insight how we approach to treat XP-A patients.

Unlike the cases in Western countries, most Japanese patients with XP-D do not present with neurological symptom [4]. Before the identification of the responsible gene for XP-D, *ERCC2*, when genetic diagnosis for XP was not covered by health insurance, patients with XP-D without neurological abnormalities had failed to be diagnosed since we did not think of XP-D as a diagnosis for XP patients without neurologic abnormalities and often tentatively diagnosed as "possible XP-V" without detailed DNA repair test. However, after "genetic diagnosis for XP" became covered by health insurance, we were referred to many adult cases with possible XP, and among them several patients with XP-D were included. Thereafter, we gradually came to know that most Japanese patients with XP-D do not manifest neurologic abnormalities, which may increase the chance of diagnosis for XP-D greater than previously. To date, 11 Japanese patients with genetically diagnosed XP-D have been reported, and among them at least at present, few of them manifested neurologic abnormalities, and the onset age of skin cancer development was after 20 years [25]. Using molecular simulation, it has been hypothesized that the difference in the clinical symptoms between Japanese patients and patients from Western countries could be attributed to the difference in ATP binding ability of the mutated XPD proteins; R683W XPD cells, frequently observed in Western countries, do not bind with ATP, whereas R683Q, frequently observed in Japan and the Middle East, retain some, although reduced, binding ability with ATP [4]. Since XPD protein functions as ATP-dependent helicase, it may explain the heterogeneity within the same complementation group to a certain extent.

5.7 Cancer Frequency and Age Onset of Skin Cancer in Japanese Patients with XP: Results from the Survey for the XP Patients in 2012 vs 1988

Using the 1988 XP survey and the 2012 XP survey conducted in Japan with 25 years interval, we analyzed the cancer frequency and its onset age in Japanese patients with XP and compared the present results with those of 25 years ago. On the results of 1988 survey, we could exclude the effect of sun protection on the development of skin cancers, since most people at that time did not practice strict sun protection. And the survey 2012 may show the outcome of performance of strict sun protection. A survey for patients with XP was conducted by the Research Committee for Intractable Diseases supported by the Japanese Ministry of Health, Labour and Welfare. Finally, we could obtain clinical information of 170 patients from 57 institutions who visited the medical institutions, and they were analyzed and previously reported focusing on the neurologic symptoms [5]. Here using the same survey, we compared the cancer frequency and age onset of cancer development between the present results and those of 25 years ago [3]. In 1988, the frequency of skin cancer in XP patients was very high in all complementation groups and variant

Table 5.3 Frequency of skin cancer in xeroderma pigmentosum patients

		2012 [5]					1988 [3]				
		Total number of patients	Number of patients with cancer (%)	Numbers of patients with (%)			Total number of patients	Number of patients with cancer (%)	Numbers of patients with (%)		
				BCC	SCC	MM			BCC	SCC	MM
NER-deficit type	A	63	14 (22)	10 (16)	2 (3.2)	1 (1.8)	117	41 (34)	35 (85)	20 (49)	5 (12)
	C	2	0	0	0	0	5	7 (88)	5 (71)	4 (57)	3 (43)
	D	5	4 (80)	3 (60)	1 (20)	0	5	4 (67)	2 (50)	2 (50)	0
	F	1	1 (100)	1 (100)	0	0	12	3 (25)	2 (67)	1(33)	0
Variant type	V	22	19 (86)	10 (45)	7 (32)	8 (36)	87	40 (46)	29 (73)	14 (35)	9 (23)
Total		93	38 (41)	24 (26)	10 (11)	9 (10)					

(Table 5.3). In 1988, as high as 34% (41/117) of patients with XP-A developed skin cancers, and the onset was younger than 10 years old (Table 5.4), while the survey 2012 revealed that the frequency of skin cancer in XP-A strikingly decreased and age onset of developing skin cancer was later (Tables 5.3 and 5.4). This striking reduction of frequency of skin cancer in XP-A is largely attributed to the early detection and early diagnosis by dermatologists. Since the late 1980s, scientists and dermatologists have educated the patients and patients' families about sun protection, in addition to early diagnosis of the disease. Owing to these efforts, the frequency of skin cancer with XP-A has strikingly decreased.

Figure 5.4 represents the age distribution of patients with XP-A of the two surveys. It clearly shows that the cancer-developing age shifted toward the older age and the ratio of cancer-having patients reduced very much. Patients less than 5 years old in the 1988 survey are now patients in 20–29 in the present survey, and obviously the cancer frequency of these patients is lower than the patients in the 1988 survey, indicating the importance of educating sun protection. Also in patients with XP-D and variant, cancer-developing age is higher in the present results than that of 1988 results. In the survey 1988, the tendency can be observed that the higher frequency of melanoma in complementation group C and variant type. This tendency is more apparent in XP variant type in the present survey. The frequency of melanoma is strikingly high, reaching a level similar to or even more than squamous cell carcinoma (SCC), whereas in XP-A patients, the frequency of BCC is the highest, in both surveys, which frequency distribution is similar to that of normal control. Patients with XP-V do not manifest exaggerated sunburn, and in most cases, they

Table 5.4 Age of onset of skin cancer in xeroderma pigmentosum patients

		2010–2012					In 1988 [3]				
		Total number of patients	Number of patients with cancer (%)	Age of onset (year) and number of patients			Total number of patients	Number of patients with cancer (%)	Age of onset (year) and number of patients		
				BCC	SCC	MM			BCC	SCC	MM
NER-deficit type	A	63	14	16.8 (10)	17 (2)	22 (1)	117	41	9.3 (35)	8.2 (20)	7.5 (5)
	C	2	0	–	–	–	5	4	14.8 (5)	8.3 (4)	11.0 (3)
	D	5	4	40.3 (3)	55.0 (1)	–	5	3	31.0 (2)	42.5 (2)	–
	F	1	1	42.0 (1)	–	–	12	3	43.4 (2)	64 (1)	–
Variant type	V	22	19	46.9 (10)	56.2 (7)	54.2 (8)	89	41	40.8 (30)	42.0 (14)	46.8 (9)

are diagnosed as XP after adolescence, which may result in a high cumulative dose exposure to UV in the patients with XP-V. We should note that patients with XP-V have almost normal level of NER but are deficient in TLS.

5.8 Diagnosis, Treatment, and Patient Care

Early definite diagnosis by genetic analysis before skin cancers develop is crucial for the patient management, and performance of strict sun protection is essential. Brief summary of diagnostic procedure is shown in Fig. 5.7. Clinical symptoms are varied depending on the lifestyle that patients used to behave. Therefore, clinical inspection, history taking, and DNA repair test are important. Regarding DNA repair test, unscheduled DNA synthesis (UDS) has been used, and recently, UDS using non-RI such as bromodeoxyuridine (BrdU) or 5-ethynyl-2′-deoxyuridine (EdU) is also used. Recently flow cytometry-based quantification of removal of photolesions (6-4)PP was shown to correlate very well with UDS, and this method enables to measure cell cycle-specific NER, which is useful for the diagnosis of XP-V, since in XP-V cells, DNA repair is slower specifically in S-phase [26]. Phototesting is useful when the patient has possible "XP-D, or X′-F, and XP-G" to obtain objective sign of "exaggerated sunburn," since only history taking is not so reliable, but the highest dose for minumem erythema dose (MED) measurement for possible XP may be used below 100 mJ/cm^2. However, patients' clinical symptoms strongly suggest XP-A, the severest type, and we directly perform sequencing, without measuring MED. After the identification of the responsible gene for XP-A, and the presence of the founder mutation in Japanese patients with

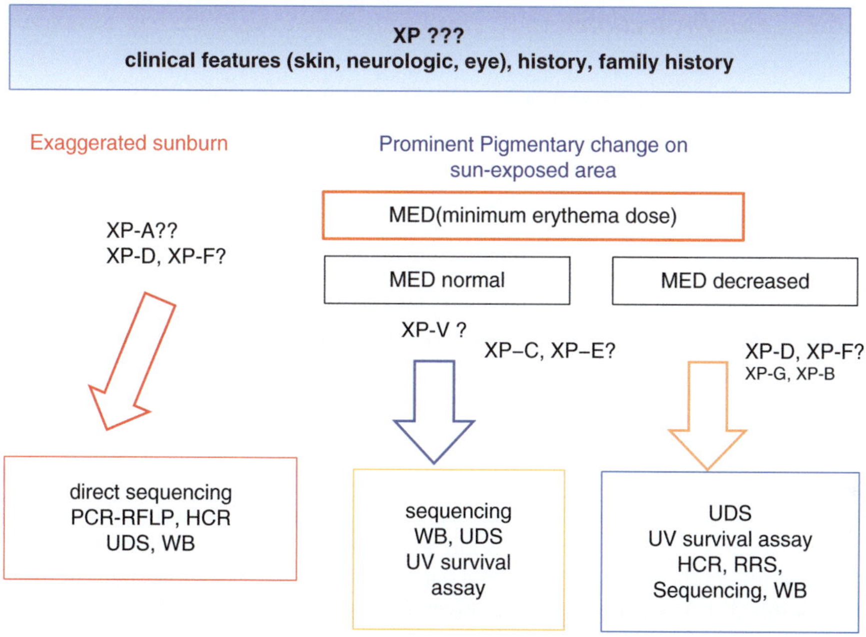

Fig. 5.7 Diagnostic procedures for each complementation groups of XP and variant type. When history of exaggerated sunburn exists, XP-A, XP-D, and XP-F are probable, and sunburn reaction is remarkable with blister formation, perform directly *XPA* sequencing, and otherwise perform measurement of MED and unscheduled DNA synthesis using fibroblasts derived from patients. If exaggerated sunburn is not obvious and freckle-like pigmentation is restricted on the sun-exposed area, then Western blot for POLH and POLH sequencing is required for the definite diagnosis

XP-A, the genetic diagnostic system for XP-A has been rapidly established. As for XP-V, patients do not manifest exaggerated sunburn but present only gradually increasing freckle-like pigmentation, and definite diagnosis before skin cancer develops is crucial. Diagnosis can be done by using immunoblot for POLH [27] and *POLH* direct sequencing because unscheduled DNA synthesis is within normal limit in XP-V cells. In Japan, genetic diagnosis for XP was approved as an advanced medical practice in 2008 based on its performance in our institution and eventually became covered by national health insurance starting from 2012. Accordingly, dermatologists became to refer to patients with possible XP for genetic diagnosis more easily, which increased the detection frequency of the disease for those that used to have overlooked 20 years ago. Previously, the frequency of the first-cousin marriage in parents of Japanese patients with XP was approximately 30% [31], while the frequency in the survey 2012 was only 11%, which reduced the frequency of XP slightly. However, the frequency of XP-A seems to remain at a certain level of frequency in Japan, because the frequency of the heterozygote carrier of XP-A founder mutation, IVS3-1G>C in *XPA*, is 1 out of 113 among the Japanese population [2].

5.8.1 *Sun Protection*

XP patients have to perform strict and complete protection from UV to prevent progression of dyschromatosis on the sun-exposed area and prevent the development of skin cancer. The eye and lips, especially lower lips, should be also protected from sun exposure. Specific protections are as follows:

1. Apply a sunscreen formulation with high sun protection factor and high protection grade of UV-A (PA) to the skin before going out; wear a cloth with long sleeves, trousers, protective clothing from UV, and glasses for UV protection.
2. Apply a film offering UV protection to windows and use a sunshade curtain to protect from light when the windows are opened.
3. When the patient is in school age, apply a film for UV protection to the windows in the school and be careful to avoid direct exposure to UV during outdoor activities and attending school.

Skin cancers, if developed, should be excised by like punch biopsy. Early case detection and early excision are principles. It has been also reported that imiquimod is useful for actinic keratosis and basal cell carcinoma and interferon are useful for melanoma [28]. Since there is also a report that delayed awakening from anesthesia occurs in XP patients, it is desirable to perform treatment early before general anesthesia is required. Probably because of the neurological dysfunction, patients in teen ages are mostly resistant to pain sensation, and local anesthesia using lidocaine can be used when excising small tumors.

For patients with XP, occasion in which chemotherapy is required seems to be infrequent, but sometimes patients need chemotherapy. Some investigators reported that cells derived from XP are sensitive to doxorubicin and other chemotherapeutic reagent [29]. Recently, side effect of cisplatin reportedly appeared to be severer in XP patients than non-XP patients [30].

5.8.2 *Care for Neurological Symptoms*

There is no useful evidence-based therapy since the pathogenesis for neurodegeneration of XP is still unknown. However, rehabilitation can be performed to deal with motor impairment and intellectual disability associated with neurodegeneration. Since patients' peak of neurologic development is achieved at about 5–6 years old, it is desirable to bring the peak of development to be higher by early rehabilitation. Since hearing loss often occurs around 4–7 years old, hearing test should be conducted at regular intervals to know an appropriate timing of wearing a hearing aid device.

Acknowledgment Most of the work in this chapter was supported by the Ministry of Health, Labour and Welfare, which is a research on intractable diseases. The authors appreciate patients' cooperation.

References

1. Takebe H, Nishigori C, Satoh Y. Genetics and skin cancer of xeroderma pigmentosum in Japan. Jpn J Cancer Res. 1987;78(11):1135–43.
2. Hirai Y, Kodama Y, Moriwaki S, et al. Heterozygous individuals bearing a non-functional allele at XPA gene exist in nearly 1% of Japanese populations. Mutat Res. 2006;601(1–2):171–8.
3. Sato Y, Nishigori C. Xeroderma pigmentosum: clinical aspects. Gann Monogr Cancer Res. 1988;35:113–26.
4. Nakano E, Ono R, Masaki T, et al. Differences in clinical phenotype among patients with XP complementation group D: 3D structure and ATP-docking of XPD *in silico*. J Investig Dermatol. 2014;134(6):1775–8.
5. Nakano E, Masaki T, Kanda F, et al. The present status of xeroderma pigmentosum in Japan and a tentative severity classification scale. Exp Dermatol. 2016;25(Suppl 3):28–33.
6. Ono R, Masaki T, Takeuchi S, et al. Three school-age cases of xeroderma pigmentosum variant type. Photodermatol Photoimmunol Photomed. 2013;29(3):132–9.
7. Hayashi M. Treatment, rehabilitation, and home care of xeroderma pigmentosum (XP); intractable disease and home care [in Japanese]. Home Health Care for the People with Intractable Diseases. 2008;14(9):58–61.
8. Hiroshima K, Inoue S. Symptoms of locomotorium in patients with xeroderma pigmentosum and the treatment; for maintenance of QOL [in Japanese]. J Clin Exp Med. 2009;228(2):147–53.
9. Nishigori C, Moriwaki S, Takebe H, et al. Gene alterations and clinical characteristics of xeroderma pigmentosum group A patients in Japan. Arch Dermatol. 1994;130(2):191–7.
10. Kanda T, Oda M, Yonezawa M, et al. Peripheral neuropathy in xeroderma pigmentosum. Brain. 1990;113(Pt4):1025–44.
11. Ueda T, Kanda F, Aoyama N, et al. Neuroimaging features of xeroderma pigmentosum group A. Brain Behav. 2012;2(1):1–5.
12. Moriwaki S, Nishigori C, Imamura S, et al. A case of xeroderma pigmentosum complementation group F with neurological abnormalities. Br J Dermatol. 1993;128(1):91–4.
13. Mimaki T, Tanaka K, Nagai A, et al. Neurological symptoms of group A of xeroderma pigmentosum and molecular genetic study [in Japanese]. Jpn J Clin Med. 1993;51(9):2488–93.
14. Imoto K, Nadem C, Moriwaki S, et al. Ancient origin of a Japanese xeroderma pigmentosum founder mutation. J Dermatol Sci. 2015;69(2):175–6.
15. Satokata I, Tanaka K, Miura N, et al. Characterization of a splicing mutation in group A xeroderma pigmentosum. Proc Natl Acad Sci U S A. 1990;87(24):9908–12.
16. Sato K, Watatani M, Ikenaga M, et al. Sensitivity to UV radiation of fibroblasts from a Japanese group A xeroderma pigmentosum patient with mild neurological abnormalities. Br J Dermatol. 1987;116(1):101–8.
17. Kondoh M, Ueda M, Nakagawa K, et al. Siblings with xeroderma pigmentosum complementation group A with different skin cancer development: importance of sun protection at an early age. J Am Acad Dermatol. 1994;31(6):993–6.
18. Sato M, Nishigori C, Yagi T, et al. Aberrant splicing and truncated-protein expression due to a newly identified XPA gene mutation. Mutat Res. 1996;362(2):199–208.
19. Takahashi Y, Endo Y, Sugiyama Y, et al. XPA gene mutations resulting in subtle truncation of protein in xeroderma pigmentosum group A patients with mild skin symptoms. J Investig Dermatol. 2010;130(10):2481–8.
20. Takeuchi S, Nakano E, Yamashita D, et al. A mild case of xeroderma pigmentosum type A. J Pediatr Dermatol. 2013;32(2):167–72.
21. Masaki T, Tsujimoto M, Kitazawa R, Funasaka Y, Ichihashi M, Kitazawa S, Kakita A, Kanda F, Nishigori C. Autopsy findings and clinical features of mild type xeroderma pigmentosum complementation group A. Siblings: 40 years follow up. J Am Acad Dermatol.
22. Nishigori C, Zghal M, Yagi T, et al. High prevalence of the point mutation in exon 6 of the xeroderma pigmentosum group A-complementing (XPAC) gene in xeroderma pigmentosum group A patients in Tunisia. Am J Hum Genet. 1993;53(5):1001–6.

23. Sidwell RU, Sandison A, Wing J, et al. A novel mutation in the *XPA* gene associated with unusually mild clinical features in a patient who developed a spindle cell melanoma. Br J Dermatol. 2006;155(1):81–8.
24. Takahashi Y, Endo Y, Kusaka-Kikushima A, et al. An *XPA* gene splicing mutation resulting in trace protein expression in an elderly patient with xeroderma pigmentosum group A without neurological abnormalities. Br J Dermatol. 2016;177(1):253–7. https://doi.org/10.1111/bjd.15051.
25. Ono R, Masaki T, Pozo FM, et al. TA 10-year follow-up of a child with mild cases of xeroderma pigmentosum complementation group D diagnosed by whole-genome sequencing. Photodermatol Photoimmunol Photomed. 2016;32(4):174–80.
26. Nakano E, Takeuchi S, Ono R, Tsujimoto M, Masaki T, Nishigori C. Xeroderma pigmentosum diagnosis using a flow cytometry-based nucleotide excise on repair assay. J Investig Dermatol. 2018;138(2):467–70.
27. Tanioka M, Masaki T, Ono R, et al. Molecular analysis of DNA polymerase eta gene in Japanese patients diagnosed as xeroderma pigmentosum variant type. J Investig Dermatol. 2007;127(7):1745–51.
28. Nagore E, Sevila A, Sanmartin O, et al. Excellent response of basal cell carcinoma and pigmentary changes in xeroderma pigmentosum to imiquimod 5% cream. Br J Dermatol. 2003;149:858–61.
29. Saffi J, Agnoletto MH, Guecheva TN, et al. Effect of the anti-neoplastic drug doxorubicin on XPD-mutated DNA repair-deficient human cells. DNA Repair (Amst). 2010;9(1):40–7.
30. Sumiyoshi M, Soda H, Sadanaga N, et al. Alert regarding Cisplatin-induced severe adverse events in cancer patients with xeroderma pigmentosum. Intern Med. 2017;56(8):979–82.

Chapter 6
Prenatal Diagnosis of Xeroderma Pigmentosum

Shinichi Moriwaki

Keywords Amniocentesis · Genetic counseling · Prenatal diagnosis · Xeroderma pigmentosum group A

6.1 Introduction

Xeroderma pigmentosum (XP) is an autosomal recessively transmitted, intractable photosensitive disorder with deficient post-ultraviolet (UV) DNA repair. Patients with XP are highly predisposed to skin cancers on sun-exposed areas. Approximately 30% of XP patients develop progressive neurological abnormalities, although the cause of these changes is unknown. XP is divided into eight genetically different clinical subtypes (A–G complementation groups of nucleotide excision repair deficient type and a variant type). The incidence of XP is higher in Japan (1/22,000) than in other countries, and XP group A (XP-A) accounts for 55% of Japanese XP cases. XP-A patients have the most serious cutaneous and neurological symptoms, resulting in a poor prognosis. Interestingly, of the XP-A patients in Japan, most have the homozygous genetic mutation IVS3-1 G>C in the *XP-A* gene, which is supposed to be a founder mutation. This mutation can be easily detected with PCR-restriction fragment length polymorphism (RFLP); therefore, most XP-A patients in Japan can receive a simple, rapid, definitive diagnosis through a genetic analysis, which can also be used for carrier and prenatal diagnoses [1, 2]. We have been employing XP genetic medicine for more than 20 years. In this article, we describe our experience with the XP-A prenatal diagnosis in Japan.

S. Moriwaki
Department of Dermatology, Osaka Medical College, Takatsuki, Osaka, Japan
e-mail: der002@osaka-med.ac.jp

C. Nishigori, K. Sugasawa (eds.), *DNA Repair Disorders*,
https://doi.org/10.1007/978-981-10-6722-8_6

6.2 What Is a Prenatal Diagnosis?

"Prenatal diagnosis" is a general term for tests conducted between 9 and 22 weeks' gestation to investigate the presence of fetal deformities, diseases, or chromosomal abnormalities. Routinely, ultrasound scans are performed, the presence of a fetal heartbeat is confirmed, and amniotic fluid volume is checked; however, when genetic diseases or congenital abnormalities are suspected, additional tests are performed for a prenatal evaluation of the health of the fetus, using the following methods [3]:

1. Maternal serum marker test
2. Noninvasive prenatal genetic testing (NIPT)
3. Amniocentesis
4. Chorionic villus sampling
5. Fetal blood and fetal skin sampling

Among these tests, 3, 4, and 5 are invasive and associated with a risk of miscarriage, infection, and bleeding in the uterus. The most commonly performed invasive tests are tests using amniotic fluid. However, in Japan, various societies associated with medical genetics, including the Japan Society of Obstetrics and Gynecology and the Japan Society of Human Genetics, have restricted the use of these tests to cases where there is a high risk of a serious single-gene disorder or chromosomal abnormality [4], according to the following criteria:

(a) One parent is a carrier of a chromosomal abnormality.
(b) History of pregnancy or childbirth involving a fetus affected by a chromosomal abnormality.
(c) Advanced maternal age.
(d) Mother is a heterozygote for a serious x-linked genetic disease that develops during the neonatal period or childhood.
(e) Both parents are heterozygotes for a serious autosomal recessive genetic disease that develops during the neonatal period or childhood.
(f) One or both parents are a heterozygote for a serious autosomal dominant genetic disease that develops during the neonatal period or childhood.
(g) Possibility that the fetus has a serious disease.

There are some merits to performing a prenatal diagnosis. If there is a high possibility that the child will be born with a serious disease and the mother chooses to deliver the child, the parents will be able mentally to prepare themselves before the child is born. In addition, medical treatment may be possible if the congenital abnormality is detected at an early stage. However, depending upon the country, there may also be socio-ethical concerns, such as an increase in the number of artificial terminations and the mental conflict afflicting both parents after electing to go through with a termination based on the test results. There has been an increase in the number of deliveries by women of advanced maternal age, and many parents are therefore concerned about the possibility of their fetus having a serious congenital abnormality. This also raises the ethical issue of whether or not it is acceptable for parents to choose to deliver a severely compromised fetus.

Therefore, a prenatal diagnosis should be considered on the premise that the client discusses this option within the family (particularly between the pregnant woman and her partner). A physician or a counselor familiar with genetic medicine in a facility approved by the ethics committee should perform the prenatal diagnosis only after genetic counseling.

6.3 XP Genetic Medicine

XP is caused by an abnormality in the mechanism that repairs DNA damage caused by ultraviolet (UV) exposure. There are eight genetically different XP groups, including seven groups (A–G) that show abnormal nucleotide excision repair (NER) and a variant (V) that has a normal NER function but an abnormality in the translesion synthesis of photodamaged DNA. Based on the previously reported cases, including our cases, XP-A, which is accompanied by serious cutaneous and neurological symptoms (neurological XP), accounts for 55% of cases in Japan, followed by XP-V, which manifests skin symptoms only (cutaneous XP; 25%), XP-D (8%), and XP-F (7%). XP-E and XP-G are extremely rare, and there have been no reports of XP-B cases in Japan [5]. The cases in Japan differ from those in the USA and Europe, and almost all Japanese XP-D and XP-G cases are cutaneous XP [6, 7]. Almost all XP-A cases also have progressive central and/or peripheral neuropathies, including psychomotor and developmental retardation, and the seriousness of this neurological abnormality is associated with the patient's prognosis. It also has been estimated that the type and location (genotype) of the genetic abnormality is related to the clinical characteristics and severity (phenotype) [8].

A definitive diagnosis of XP is mainly based on the results of DNA repair tests using cultured fibroblasts derived from the patient's skin [1], including:

1. Unscheduled DNA synthesis (UDS) measurement after exposure to UV rays
2. UV sensitivity test using colony-forming capability as an index
3. Genetic complementation test using plasmoid or viral vector host cell reactivation as an index
4. XP gene analysis

In the gene analysis for XP-A, which accounts for the majority of cases in Japan, 88% of cases have the same mutation in intron 3 of the *XP-A* gene, a homozygous mutation on the 3′ side of the splicing site (IVS3-1G>C). Another 9% of cases have a heterozygous mutation of IVS3-1G>C, and among these, 3% have a compound heterozygous mutation with a nonsense mutation in exon 6 (c.682 C>T [R228X]). These genetic abnormalities are known to be hotspots for *XP-A* mutations in Japanese XP-A patients. Each mutation can be detected by restriction fragment length polymorphism (RFLP) of PCR-amplified DNA using restriction enzymes:

AlwNI or *HphI* [1, 8], respectively. Based on this strong founder effect, most Japanese XP-A patients can be easily and rapidly identified with the PCR-RFLP method. This diagnostic procedure, despite its use being limited to Japanese XP-A patients, is also useful for carrier detection and the prenatal diagnosis of XP-A.

6.4 Prenatal Diagnosis of XP-A

Families with a proband who have serious neurological XP, such as XP-A (or XP-D), are eligible for an XP prenatal diagnosis [2]. In our facility, we perform XP prenatal diagnoses with ethics committee approval [9] under the following conditions:

1. The XP of the afflicted family member (proband) is serious (neurological XP, e.g., XP-A).
2. Both parents express a strong desire to perform the test after an agreement is reached following an in-family discussion.
3. Both parents fully understand the explanation of XP provided by the physician.
4. Genetic testing is technically possible.
   - The genetic mutation is detected in the afflicted family member (proband).
   - A reliable test can be performed easily, rapidly, and at low cost.
5. Both parents understand the risks of implementing the test as well as the sensitivity/specificity of the test (a false negative is possible) (Table 6.1).

Table 6.1 Procedures and characteristics of chorionic villus sampling and amniocentesis for the XPA prenatal diagnosis

	Chorionic villus sampling	Amniocentesis
Time for examination (weeks of pregnancy)	10–13 weeks	14–17 weeks
Approach	Transabdominal, transvaginal	Transabdominal
Technique	Difficult	Easy
Deleterious effects	Infection, bleeding	Infection, bleeding
	Abortion (1–3%)	Abortion (<0.3%)
	Teratogenesis	Premature rupture
Artificial abortion	Easy	Not easy
	Bearable	Mentally unbearable
DNA repair studies[a]	Impossible	Possible
DNA analysis[b]	Possible	Possible
	Trophoblasts, 1–3 days	Amniotic fluid, 1–3 days
		Cultured cells, 2–3 weeks
Contamination of maternal tissue	Considerable	Negligible

[a]UV survival, post-UV inhibition of DNA synthesis using cultured cells derived from amniotic fluid

[b]PCR-RFLP analysis

6. Both parents understand that there is a possibility of an outcome contrary to their expectations.
7. If any of family members need to cope with the outcome stipulated in 6, there are departments and staff members specialized in clinical psychology (or genetic counseling) available, and they will be able to receive care, including social and emotional support.

In our facility, if the mother of an XP-A proband with a mutation in the *XP-A* gene (e.g., IVS3-1G>C) wishes to have a prenatal test performed for XP-A, we proceed in the following manner:

1. Confirm the proband's *XP-A* mutation.
2. Confirm that both parents of the proband understand the prenatal diagnosis, and confirm the parents' intentions through genetic counseling.
3. Determine whether or not both parents are *XP-A* carriers (Fig. 6.1a).
4. Obtain written consent from both parents in accordance with the ethics committee guidelines in our facility.
5. Collect amniotic fluid (9 mL) from the pregnant woman (proband's mother) at 14–16 weeks' gestation, conduct genetic testing using floating cells in the amni-

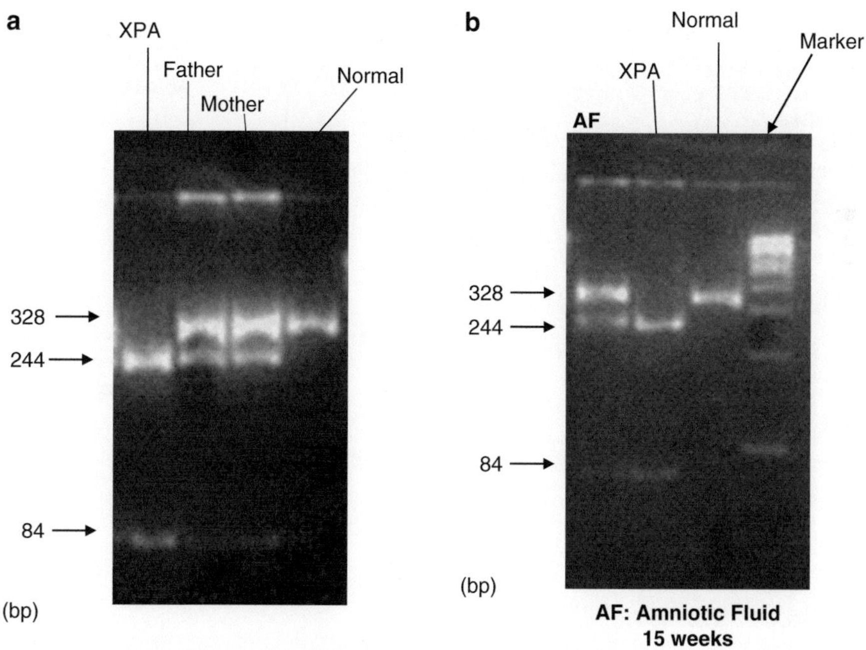

Fig. 6.1 (**a**) PCR-amplified, *AlwNI*-digested DNA fragments from normal cells (328 bp), XPA cells (244 bp + 84 bp), and blood cells from the parents of case 8 (328 bp + 244 bp + 84 bp). This result shows that both parents are XPA carriers. (**b**) PCR-amplified, *AlwNI*-digested DNA fragments from normal cells (328 bp), XPA cells (244 bp + 84 bp), and amniotic floating cells (328 bp + 244 bp + 84 bp). This result shows that the fetus is an XPA carrier

otic fluid (4.5 mL; Fig. 6.1b), and simultaneously establish a primary culture of the amniotic fluid cells from the remaining 4.5 mL of amniotic fluid. Then, conduct UV sensitivity tests (Fig. 6.2a, b) and another round of *XP-A* genetic analyses (Fig. 6.2c) using the cultured amniotic fluid cells 2–3 weeks after beginning the primary culture.

6. A comprehensive judgment is made to determine whether or not the fetus has XP based on the genotype and phenotype (UV sensitivity) analyses, and the parents are informed of the results and given genetic counseling if necessary.
7. After the fetus is born, blood is collected, and an *XP-A* genetic analysis is performed to determine whether or not the prenatal diagnosis was correct.

We conduct fetal XP-A diagnoses extremely carefully by performing an *XP-A* genetic analysis of amniotic fluid floating cells, testing again for the *XP-A* genetic mutation using cultured amniotic fluid cells, and evaluating the phenotype by testing the UV sensitivity of the amniotic fluid cells. The results of the prenatal diagnoses performed for 9 XP-A families (13 fetal cases) in our department to date are summarized in Table 6.2. For the first case, both chorionic villus sampling and amniocentesis were performed;

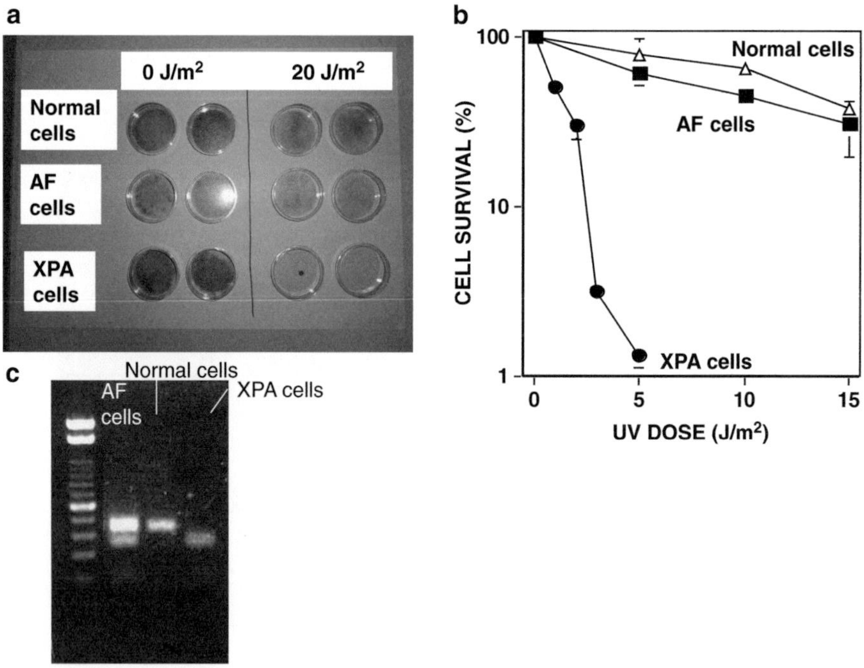

Fig. 6.2 (**a**, **b**) Qualitative and quantitative cell survival analyses of cultured amniotic fluid cells, respectively. These results show that the amniotic cells have a normal response to UV irradiation, which is compatible with XPA carrier cells. (**c**) PCR-amplified, *AlwNI*-digested DNA fragments from normal cells (328 bp), XPA cells (244 bp + 84 bp), and cultured amniotic fluid cells (328 bp + 244 bp + 84 bp). This result, along with the cell survival data, confirms that the fetus is an XPA carrier

Table 6.2 List of the families that underwent prenatal testing for XPA by chorionic villi sampling (case 1) and amniocentesis (all cases) in our laboratory (Modified from Ref. [9])

Case no.	XPA family no.	Age of father (y)	Age of mother (y)	Age of proband at visit (y)	XPA mutation in proband	Sample taken for analysis (weeks of pregnancy)	Result of prenatal analysis (final diagnosis and course)
1	1	32	30	XPA (first child, 3 y)	IVS3-1G>C, homozygous	CV (10 wk)	IVS3-1G>C, heterozygous
						AF (15 wk)	IVS3-1G>C (XPA carrier, intrauterine death)
2	2	41	37	Three XPA children (11,7,2 y)	IVS3-1G>C, homozygous	AF (16 wk)	IVS3-1G>C, homozygous (XPA, born)
3	3	36	33	XPA (second child, 2 y)	IVS3-1G>C, homozygous	AF (15 wk)	IVS3-1G>C, heterozygous (XPA carrier, born)
4	4	36	33	XPA (first child, 2 y)	IVS3-1G>C, homozygous	AF (14 wk)	No mutation (wild type, born)
5	5	35	32	XPA (first child, 3 y)	IVS3-1G>C, homozygous	AF (14 wk)	IVS3-1G>C, heterozygous (XPA carrier, born)
6	6	34	31	XPA (first child, 3 y)	IVS3-1G>C, homozygous	AF (14 wk)	No mutation (wild type, born)
7	7[a]	33	32	XPA (first child, 2 y)	IVS3-1G>C, homozygous	AF (14 wk)	IVS3-1G>C, heterozygous (XPA carrier, born)
8	8	33	30	XPA (first child, 1 y)	IVS3-1G>C, homozygous	AF (15 wk)	IVS3-1G>C, heterozygous (XPA carrier, born)
9	9[b]	34	29	XPA (first child, 2 y)	IVS3-1G>C, homozygous	AF (15 wk)	IVS3-1G>C, heterozygous (XPA carrier, born)
10	7[a]	36	35	XPA (first child, 5 y)	IVS3-1G>C, homozygous	AF (15 wk)	IVS3-1G>C, homozygous (XPA, abortion)
11	7[a]	37	36	XPA (first child, 6 y)	IVS3-1G>C, homozygous	AF (15 wk)	No mutation (wild type, born)
12	10	33	32	XPA (first child, 2 y)	IVS3-1G>C, heterozygous	AF (15 wk)	No IVS3-1G>C mutation
					c.682C>T heterozygous		c.682C>T heterozygous (XPA carrier, born)
13	9[b]	38	33	XPA (first child, 6 y)	IVS3-1G>C, homozygous	AF (15 wk)	No mutation (wild type, born)

y years old, *wk* weeks of pregnancy, *CV* chorionic villi, *AF* amniocentesis fluid
[a]Same family; the mother in this family experienced XPA prenatal diagnosis three times
[b]Same family; the mother in this family experienced XPA prenatal diagnosis twice

however, the fetus suffered an intrauterine death of unknown cause after amniocentesis. Beginning with the second case, only amniocentesis was performed. There were two families for which prenatal testing was performed two or more times.

6.5 Current Status of XP Prenatal Diagnosis

This article described the current state of the XP-A prenatal diagnosis in Japan. Of course, XP prenatal diagnoses are also conducted in other countries around the world, including the USA [10, 11], the UK [12], France [13, 14], the Netherlands [15, 16], China [17, 18], and Tunisia [19]. However, the legal restrictions related to prenatal diagnoses vary among these countries. In Japan, there are no regulations restricting a prenatal diagnosis. Therefore, the only regulations related to the implementation of prenatal testing are the guidelines of each society associated with genetic medicine [4]. In addition, no special clauses regarding fetuses have yet been incorporated into the Maternal Health Act, so when an abnormality is found in a fetus, the pregnancy may be terminated at less than 22 weeks' gestation based on the reasons indicated in Article 14 of the current law, which are based on the gestational age at which the unborn child cannot survive outside the mother's body (up to 21 weeks). Even in the USA, there are no substantial regulations on fetal medicine at the national level. In China, an ordinance that prohibits the birth of children with serious genetic diseases has been established, and there are no legal restrictions on artificial abortion. Termination of pregnancy is also legally recognized in Tunisia. However, in the UK and France, prenatal diagnoses are performed based on independent criteria outlined in each country's legal restrictions, but artificial abortion is permitted when at least two physicians approve it in both countries.

References

1. Moriwaki S, Kraemer KH. Xeroderma pigmentosum—bridging a gap between clinic and laboratory. Photodermatol Photoimmunol Photomed. 2001;17(2):47–54.
2. Matsumoto N, Saito N, Harada K, Tanaka K, Niikawa N. DNA-based prenatal carrier detection for group A xeroderma pigmentosum in a chorionic villus sample. Prenat Diagn. 1995;15(7):675–7.
3. Wieacker P, Steinhard J. The prenatal diagnosis of genetic diseases. Dtsch Arztebl Int. 2010;107(48):857–62.
4. Guidelines for genetic tests and diagnosis in medical practice. The Japanese Association of Medical Sciences. 2011.
5. Nakano E, Masaki T, Kanda F, Ono R, Takeuchi S, Moriwaki S, Nishigori C. The present status of xeroderma pigmentosum in Japan and a tentative severity classification scale. Exp Dermatol. 2016;25(Suppl 3):28–33.
6. Nakano E, Ono R, Masaki T, Takeuchi S, Takaoka Y, Maeda E, Nishigori C. Differences in clinical phenotype among patients with XP complementation group D: 3D structure and ATP-docking of XPD in silico. J Invest Dermatol. 2014;134(6):1775–8.

7. Moriwaki S, Takigawa M, Igarashi N, Nagai Y, Amano H, Ishikawa O, Khan SG, Kraemer KH. Xeroderma pigmentosum complementation group G patient with a novel homozygous mutation and no neurological abnormalities. Exp Dermatol. 2012;21(4):304–7.
8. Nishigori C, Moriwaki S, Takebe H, Tanaka T, Imamura S. Gene alterations and clinical characteristics of xeroderma pigmentosum group A patients in Japan. Arch Dermatol. 1994;130(2):191–7.
9. Moriwaki S, Yamashita Y, Nakamura S, Fujita D, Kohyama J, Takigawa M, Ohmichi M. Prenatal diagnosis of xeroderma pigmentosum group A in Japan. J Dermatol. 2012;39(6):516–9.
10. Cleaver JE, Volpe JP, Charles WC, Thomas GH. Prenatal diagnosis of xeroderma pigmentosum and Cockayne syndrome. Prenat Diagn. 1994;14(10):921–8.
11. Kogan SC, Doherty M, Gitschier J. An improved method for prenatal diagnosis of genetic diseases by analysis of amplified DNA sequences. Application to hemophilia A. N Engl J Med. 1987;317(16):985–90.
12. Drury S, Boustred C, Tekman M, Stanescu H, Kleta R, Lench N, Chitty LS, Scott RH. A novel homozygous ERCC5 truncating mutation in a family with prenatal arthrogryposis—further evidence of genotype-phenotype correlation. Am J Med Genet A. 2014;164A(7):1777–83.
13. Alapetite C, Benoit A, Moustacchi E, Sarasin A. The comet assay as a repair test for prenatal diagnosis of Xeroderma pigmentosum and trichothiodystrophy. J Invest Dermatol. 1997;108(2):154–9.
14. Sarasin A, Blanchet-Bardon C, Renault G, Lehmann A, Arlett C, Dumez Y. Prenatal diagnosis in a subset of trichothiodystrophy patients defective in DNA repair. Br J Dermatol. 1992;127(5):485–91.
15. Kleijer WJ, van der Sterre ML, Garritsen VH, Raams A, Jaspers NG. Prenatal diagnosis of xeroderma pigmentosum and trichothiodystrophy in 76 pregnancies at risk. Prenat Diagn. 2007;27(12):1133–7.
16. Graham JM Jr, Anyane-Yeboa K, Raams A, Appeldoorn E, Kleijer WJ, Garritsen VH, Busch D, Edersheim TG, Jaspers NG. Cerebro-oculo-facio-skeletal syndrome with a nucleotide excision-repair defect and a mutated XPD gene, with prenatal diagnosis in a triplet pregnancy. Am J Hum Genet. 2001;69(2):291–300.
17. Zhou EY, Wang H, Lin Z, Xu G, Ma Z, Zhao J, Feng C, Duo L, Yin J, Yang Y. Clinical and molecular epidemiological study of xeroderma pigmentosum in China: a case series of 19 patients. J Dermatol. 2017;44:71. https://doi.org/10.1111/1346-8138.13576.
18. Yang Y, Ding B, Wang K, Bu D, Tu P, Zhu X. DNA-based prenatal diagnosis in a Chinese family with xeroderma pigmentosum group A. Br J Dermatol. 2004;150(6):1190–3.
19. Messaoud O, Ben Rekaya M, Jerbi M, Ouertani I, Kefi R, Laroussi N, Bouyacoub Y, Benfadhel S, Yacoub-Youssef H, Boubaker S, Zghal M, Mrad R, Amouri A, Abdelhak S. The experience of a Tunisian referral centre in prenatal diagnosis of Xeroderma pigmentosum. Public Health Genomics. 2013;16(5):251–4.

Chapter 7
Neurological Disorders and Challenging Intervention in Xeroderma Pigmentosum and Cockayne Syndrome

Masaharu Hayashi

Abstract Xeroderma pigmentosum (XP) is a genetic disorder in DNA nucleotide excision repair and is characterized by skin disorders and progressive neurological impairment, which is complicated in some complementation groups, especially in XP group A (XP-A). Cockayne syndrome (CS) is caused by abnormalities in genes of transcription-coupled repair, and patients with CS develop growth failure, characteristic facial features, skin symptoms, and neurological disorders. Model animals have not reproduced neurological disorders in either XP-A or CS. We have performed immunohistochemistry for oxidative stress markers, antioxidant enzymes, neurotransmitters, and markers of glial cells in autopsy brains. We have also performed enzyme-linked immunosorbent assay for oxidative stress in the urines and cerebrospinal fluid, isolated from patients with XP-A and CS. It has been demonstrated that oxidative DNA damage, disturbed metabolism of monoamines and melatonin, vascular changes in the brain, and/or pathology of oligodendrocytes and microglial cells may be involved in neurodegeneration, suggesting the possibility of treatments with free radical scavengers, monoaminergic agents, and/or melatonin. We reported that the therapy with low dose of levodopa ameliorated laryngeal dystonia and involuntary movements in the arm in some patients with XP-A. In addition, it is speculated that melatonin may be a therapeutic option in patients with XP-A and CS.

Keywords Xeroderma pigmentosum · Cockayne syndrome · Immunohistochemistry · Enzyme-linked immunosorbent assay · Oxidative stress · Monoamine · Melatonin · Glial cells

---

M. Hayashi
Department of Brain Development and Neural Regeneration, Tokyo Metropolitan Institute of Medical Science, Tokyo, Japan
e-mail: hayashi-ms@igakuken.or.jp

C. Nishigori, K. Sugasawa (eds.), *DNA Repair Disorders*,
https://doi.org/10.1007/978-981-10-6722-8_7

7.1 Introduction

Xeroderma pigmentosum (XP) is a genetic disorder of nucleotide excision repair (NER), in which global genome repair (GGR) and/or transcription-coupled repair (TCR) is impaired. XP is divided into eight complementation groups (XP-A to XP-G and XP-V) that are associated with mutations in eight genes. XP patients bear molecular defects either in NER or in translesion synthesis, in the latter of which *XPV* encodes DNA polymerase η that is responsible for bypassing unrepaired lesions during DNA replication [1]. Clinically, XP is characterized by skin hypersensitivity to sunlight, predilection of skin cancers, and progressive neurological impairment in several groups. XP-A is common in Japan, showing severe neurological disorders such as mental deterioration, cerebellar ataxia, extrapyramidal abnormalities, and neuronal deafness [2]. Cockayne syndrome (CS) is caused by abnormalities in two CS genes (*CSA* and *CSB*), which are involved in transcription-coupled repair, another step of NER. Patients with CS develop severe growth failure with reduced subcutaneous fat, characteristic facial features (sunken eyes and sharp noses), mild photosensitive skin symptoms, and neurological disorders such as demyelinating neuropathy, ataxia, spasticity, deafness, and cognitive deterioration [3]. Protection from ultraviolet light is effective for amelioration of cutaneous disorders, whereas it will not prevent development of neurodegeneration. In addition, $Xpa^{-/-}$ mice fail to exhibit obvious neurological symptoms and neuropathological changes observed in human XP-A [4]. Likewise, $Csa^{-/-}$ and $Csb^{-/-}$ mouse models for CS, except for photoreceptor loss, do not show overt neurological abnormalities [4]. In order to clarify the pathomechanism of neurodegeneration in XP-A and CS, we have examined brains using immunohistochemistry (IM) for oxidative stress markers, antioxidant enzymes, neurotransmitters, and markers of glial cells in autopsy cases. We have also performed enzyme-linked immunosorbent assay (ELISA) for oxidative stress in the urines and cerebrospinal fluid (CSF), isolated from patients with XP-A and CS. Ethical committee of Tokyo Metropolitan Institute of Medical Science approved the study, and the family provided informed consent for all studies.

7.2 The Involvement of Oxidative Stress in Neurodegeneration in XP-A and CS

Decreased DNA repair and persistent DNA damage may result in augmented oxidative DNA damage in XP and CS. Incidentally, oxidative DNA damage and/or disturbed antioxidant system, such as superoxide dismutase (SOD), have been shown in isolated skin and blood cells or their cell lines [5–7]. We have also performed IM on the deposition of oxidative stress markers and SOD, one of the main antioxidant enzymes in the brain, in autopsy cases of XP-A and CS [5, 8]. Cases of XP-A and, to a lesser extent, those of CS demonstrated nuclear deposition of markers for oxidative DNA damage, 8-hydroxy-2'deoxyguanosine (8-OHdG) and thymidine glycol, in neurons and glial cells in the globus pallidus and cerebellar cortex [8]. XP-A cases exhibited reduced cytoplasmic immunoreactivity for Cu/ZnSOD in the neurons of the basal ganglia and cerebellar cortex, whereas CS cases demonstrated comparatively preserved immunoreactivity for SODs, suggesting that oxidative DNA damage with disturbed SOD expression may be involved in the degeneration of the basal ganglia and cerebellum predominantly in XP-A.

Simultaneously, we tried ELISA analysis on 8-OHdG and hexanoyl lysine adduct (HEL), one of the markers for lipid peroxidation, in urine samples from 7 XP-A patients, 1 XP-D patient, 5 CS patients, and 17 healthy controls aged 3–81 years [8]. XP-A patients aged over 20 years with long disease duration, suffering from respiratory insufficiency, showed a robust increase in both urinary levels of 8-OHdG and HEL. Twin CS patients aged over 20 years with long disease duration demonstrated an increase in urinary levels of HEL but not of 8-OHdG. We also performed a preliminary analysis of CSF levels of 8-OHdG and HEL in three and one patients of XP-A and XP-D, respectively. One XPA patient showed a significant increase in CSF level of 8-OHdG, whereas the CSF levels of HEL were not elevated in either of four patients. Taking these data together, it is likely that oxidative stress may be implicated in neurodegeneration in XP-A and/or CS, indicating the possibility of treatment with edaravone, which is one of the free radical scavengers, modifies lipid peroxidation, and has been applied in patients with various disorders in Japan, in addition to cerebral infarction at acute disease stage [9].

7.3 Disturbed Metabolism of Monoamine in Xeroderma Pigmentosum Group A and Therapeutic Challenge

Although XP-A cases show widespread neuronal loss throughout the CNS [10], we found a selective impairment of catecholaminergic neurons (CANs), including dopamine neurons, in the basal ganglia and brainstem in autopsy cases, being related to the occurrence of extrapyramidal symptoms and brainstem dysfunction, respectively [11]. In this point, we have performed the treatment with low-dose levodopa (TLL) in some patients with XP-A. In the original trial in three patients, two showed amelioration of laryngeal dystonia and tremor-like movements in the arms during the TLL [12]. On the other hand, the TLL was ineffective for mental disorders, such as attention deficit and/or hyperkinesis, in infants with XP-A. In all patients, the TLL caused mild muscle hypotonia, which lasted for a few months after the start of treatment and vanished gradually. The TLL has been reported to reduce the frequency of tics in some patients with Tourette's syndrome [13]. Neurophysiological data suggested that the TLL may alleviate the receptor supersensitivity in dopamine neurons in the basal ganglia, leading to the decrease of tics in Tourette's syndrome. It is likely that the TLL may be effective for the selective lesion of dopamine neurons in the basal ganglia [11].

Patients with XP-A suffer from disturbed mental abilities and a worsening of cerebral atrophy, according to MRI studies [14]. Additionally, Japanese XP-A patients showed a higher incidence of epileptic seizures than age-matched children, not suffering from XP-A [15]. We performed IM analysis in autopsy cases of XP-A, regarding acetylcholinergic neurons in the nucleus basalis of Meynert and the pedunculopontine tegmental nucleus, which are involved in mental development and learning abilities, in addition to GABAergic interneurons in the cerebral cortex [16]. The density and percentages of GABAergic interneurons were reduced selectively in the cerebral cortex in cases of XP-A, although the pyramidal neurons were spared. XP-A cases also showed a significant reduction of acetylcholinergic neurons in both the nucleus basalis of Meynert and the pedunculopontine tegmental nucleus. It is suggested that the selective lesions in the GABAergic interneurons in the cerebral cortex and acetylcholinergic neurons may be involved in the mental disturbances, the higher occurrence of epilepsy, and/or the abnormalities in rapid eye movement sleep in patients with XP-A [11]. Donepezil, a representative acetylcholinesterase inhibitor, has been used globally in patients with Alzheimer's disease [17]. Clinical practice with donepezil has also been done in children with Down syndrome [18], because the acetylcholinergic neurons in the nucleus basalis of Meynert and/or the pedunculopontine tegmental nucleus are impaired in the brain in both Alzheimer's disease and Down syndrome. Our findings in the autopsy brains strongly indicate the possibility of treatment with donepezil in patients with XP-A.

7.4 Disturbed Metabolism of Melatonin in Xeroderma Pigmentosum Group A and Cockayne Syndrome

Melatonin is a functionally pleiotropic and neuroendocrine molecule and is produced by the pineal gland under the control of the suprachiasmatic nucleus (SCN) [19]. Melatonin is primarily synthesized in the pineal and secreted at night, and the circadian rhythm of melatonin production is determined by the prevailing light-dark cycle. Consequently, nocturnal circulating melatonin levels are higher at night than during the day [20]. Melatonin regulates circadian rhythm and plays a role in the transduction of the chronobiological actions of multiple hormones. It is speculated that melatonin may reset feedback signal to the SCN and influence cells in other areas of the central nervous system and in the periphery as well as an SCN-dependent output signal [19]. In addition, melatonin has the ability to scavenge toxic free radicals directly and may reduce the upregulation of pro-inflammatory cytokines, preventing the translocation of the nuclear factor-kappa B (NF-kB) to the nucleus [21]. Accordingly, melatonin has been discussed as one of therapeutic tools for neurological disorders, such as multiple sclerosis and Huntington's disease. At first, we performed a questionnaire survey in the families of patients with CS; enzyme-linked immunosorbent analyses of the melatonin metabolite, 6-sulphatoxymelatonin

(6-SM), in the patients' urine; and IM analysis in the hypothalamus, the basal nucleus of Meynert, and the pedunculopontine tegmental nucleus in autopsy cases [22]. Patients with CS demonstrated disturbed circadian rhythms of sleep-wakefulness and body temperature regulation. The hypothalamic nuclei were comparatively preserved in the autopsy cases, the acetylcholine neurons were severely decreased in the nucleus basalis of Meynert and the pedunculopontine tegmental nucleus, just like in cases of XP-A [16]. The urinary levels of melatonin metabolite in patients with XP-A and patients with CS were measured, and the level was predominantly reduced in those with CS. It is suggested that the combination of lesions of acetylcholine neurons and disturbed melatonin metabolism may be involved in the disturbed circadian rhythms of sleep wakefulness.

Recently, redox regulation and/or oxidative stress has been reported to have diurnal variation, and the circadian rhythms of oxidative stress markers and antioxidant enzymes were examined in healthy subjects and patients with neurological disorders. The production of antioxidants and protective enzymes has been reported to be regulated in rhythmic fashions, and oxidative stress seems to have a connection with the circadian rhythm of sleep wakefulness [23]. Although melatonin has antioxidant activity as aforementioned, the relationships in diurnal variation between melatonin and oxidative stress markers still remain to be investigated. Herein, we analyzed the circadian rhythms of oxidative stress markers and melatonin metabolites in urine of patients with XP-A. We confirmed the diurnal variation of melatonin metabolites, oxidative stress markers, and antioxidant power (TAO) in urine of XP-A patients and age-matched healthy controls, using enzyme-linked immunosorbent assay (ELISA) [24]. The peak of 6-SM, a metabolite of melatonin, was seen at 6:00 in both the XP-A patients and controls, though the peak value is lower, specifically in the younger patients with XP-A under 15 years of age. The older patients with XP-A, over 15 years of age, demonstrated an increase in the urinary levels of 8-OHdG and HEL, having a robust peak at 6:00 and 18:00, respectively. In addition, the urinary level of TAO was decreased in the older patients. Accordingly, it is speculated that the administration of melatonin in childhood may prevent the development of oxidative stress in adulthood in patients with XP-A. In preliminary analysis on the relationships between oxidative stress and melatonin metabolism in patients with CS (Table 7.1), six younger patients under 15 years of age showed a significant reduction of 6-SM in the urine, when compared with 12 age-matched controls. On the other hand, six older patients over 15 years of age demonstrated a significant reduction of TAO but not of 6-SM in the urine, compared with 11 age-matched controls. Clinical practice with melatonin has been done in children with developmental disorders globally, and some treatment trials resulted in partial improvement of sleep disturbances in the absence of severe adverse effects in patients with autism spectrum disorder, Rett syndrome, and/or Angelman syndrome [25]. In Japan, clinical practice with melatonin in children with developmental disorders is under progress, and we are planning preclinical studies, using iPS cells isolated from patients with XP-A, in order to explore the antioxidant effect of melatonin.

Table 7.1 The relationship between oxidative stress and melatonin metabolism in patients with Cockayne syndrome

Subjects	Hexanoyl lysine (nmol/mg)	8-OHdG (ng/mg)	Total antioxidant power (μM uric acid equiv.)	6-sulphatoxymelatonin (ng/mgCre)
Under 15 years of age				
Control ($n = 12$)	111.5 (25.6)	14.2 (5.2)	3133.1 (1061.6)	149 (139)
Cockayne syndrome ($n = 6$)	111.8 (43.7)	17.2 (4.5)	2471.2 (939.1)	23.8 (7.6)
Mann–Whitney (p)				<0.01
Over 15 years of age				
Control ($n = 11$)	107.9 (76)	9 (2.42)	3651.1 (681)	28 (19.3)
Cockayne syndrome ($n = 6$)	221.9 (207.8)	12.72 (8.8)	2615.5 (887.1)	10 (3.1)
Mann–Whitney (p)			<0.01	

Data are shown as mean (standard deviation, SD)
8-OHdG 8-hydroxy-2′deoxyguanosine

7.5 Vascular Changes in Autopsy Brains in Xeroderma Pigmentosum Group A and Cockayne Syndrome

Arteriosclerosis in the brain and subdural hemorrhage has been reported in a few cases of CS [3]. We performed elastica van Gieson staining and IM analysis for collagen type IV and CD34 to evaluate the brain vessels in autopsy cases of CS and XP-A [26]. Small arteries without arteriosclerosis in the subarachnoid space increased in cases of CS but not of XP-A, and such increase may be related to the predilection for subdural hemorrhage in patients with CS. The string vessels (twisted capillaries), which were pointed out in the previous paper [3], were identified in the cerebral white matter, and the density of CD34-immunoreactive vessels was increased in cases of CS but not of XP-A (Fig. 7.1). It is noteworthy that such increase of density of CD34-immunoreactive vessels was found only in cases of CS, and it did not seem to be caused by brain atrophy itself. It is possible that such facilitated increase of vascular density by age may be one of progeroid manifestation in CS.

7.6 Specific Pathology of Oligodendroglia and Microglia in Autopsy Brains in Xeroderma Pigmentosum Group A and Cockayne Syndrome

It is well known that patients with CS show severe loss of myelinated fibers in both the central and peripheral nervous systems [2], although the exact pathogenesis of selective vulnerability of oligodendroglia and Schwann cells still remains to be investigated [27]. In addition, biallelic *ERCC6* mutations include static hypomyelination, microcephaly, mild growth failure, and psychomotor developmental delay

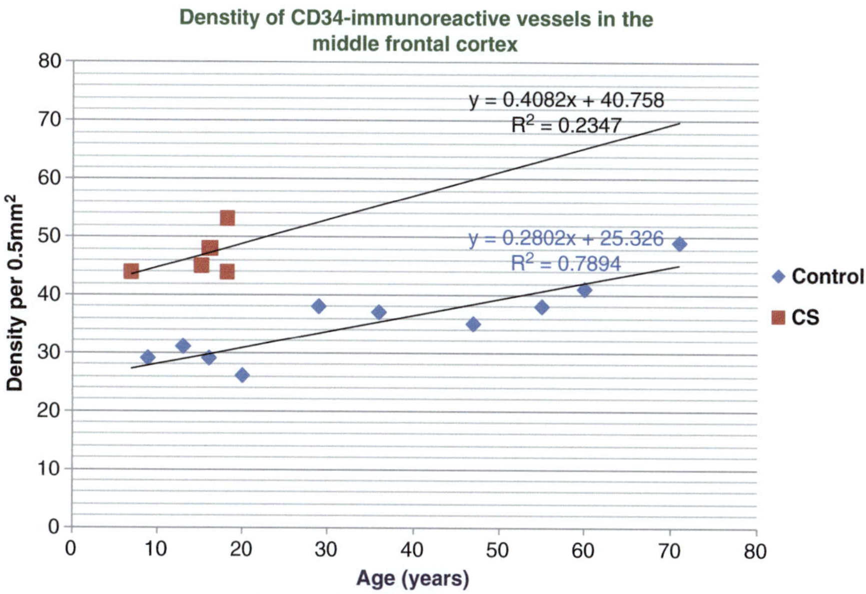

Fig. 7.1 The density of CD34-immunoreactive vessels in the controls and the cases of Cockayne syndrome (CS). In the middle frontal cortex, the numbers of vessels immunoreactive for CD34 were determined in five non-overlapping microscopic subfields at 200-fold magnification by using a counting box (0.5 mm^2). Data were analyzed by nonparametric Mann–Whitney *U* test to compare the results in different subjects. The linear approximation by age was drawn in the controls and the cases of CS

in the absence of photosensitivity, progeria, or leukodystrophy classically associated with CS [28]. On the other hand, patients with XP-A were reported to reveal fibrillary gliosis in the cerebral white matter [10]. Recently, MRI and MR spectroscopy demonstrated age-dependent decline in multimodalities in patients with XP-A, such as decreased fractional anisotropy value in diffusion tensor imaging (DTI) and reduced NAA/Cre ratio [14]. DTI also demonstrated reduced directionality of the white matters in the thalamus, the corticospinal tracts, and the dorsal corpus callosum in patients with XP-A, XP-C, XP-D, XP-F, and XP-V [29]. Accordingly, we started pathological analysis in autopsy brains, focusing on glial cells, such as oligodendrocytes and microglial cells. We tried IM analysis on oligodendrocytes in six cases each of XP-A and CS, in addition to six controls, lacking neurological disorders (Table 7.2). Serial sections in the cerebrum and cerebellum were treated with antibodies against myelin basic protein (MBP), proteolipid protein (PLP), and GFAP. Two aged cases of XP-A showed reduction of myelinated fibers immunoreactive for MBP and PLP throughout the deep white matters in both the cerebrum and cerebellum. On the other hand, all six cases of CS demonstrated patchy reduction of myelinated fibers immunoreactive for MBP and PLP in the deep white matters in the cerebrum, irrespective of calcification. Interestingly, three cases of CS additionally had small foci of complete loss of immunoreactivity for both MBP and PLP in the cerebral cortex (Fig. 7.2).

Table 7.2 The decrease of myelinated fibers immunoreactive for myelin basic protein (MBP) and proteolipid protein (PLP) in the white matters

Age (years)/sex	Cerebral white matter	Subcortical perforating fiber	Cerebellar white matter
Xeroderma pigmentosum group A			
19/male	(−)	(−)	(−)
19/male	(−)	(−)	(−)
21/male	(−)	(−)	(−)
23/female	1+	(−)	1+
24/female	2+	(−)	2+
26/female	(−)	(−)	(−)
Cockayne syndrome			
7/female	1+	(−)	(−)
15/male	1+	Patchy	(−)
16/female	1+	Patchy	(−)
18/male	2+	(−)	(−)
18/male	1+	(−)	(−)
35/female	1+	Patchy	1+

The degree of decrease or pathology in the myelinated fibers is denoted in the following: (−), absent; 1+, mild decrease; 2+, severe decrease; patchy, occurrence of small foci lacking immunoreactivity for MBP and PLP

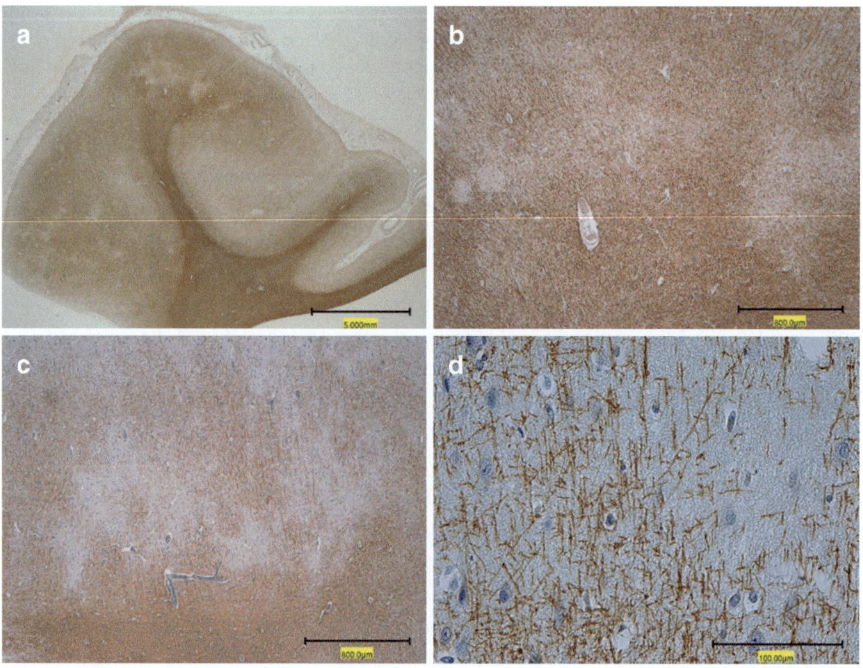

Fig. 7.2 Small foci lacking immunoreactivity for proteolipid protein (PLP) (**a** and **b**), and myelin basic protein (**c** and **d**) in the cerebral cortex in the case of Cockayne syndrome. Bars = 5 mm (**a**), 800 μm (**b** and **c**), and 100 μm (**d**)

Microglia are tissue-resident macrophages in the CNS, and they have been known to be critical effectors and regulators of changes in CNS homeostasis during development and in health and disease [30]. The involvement of microglia in synaptic pruning and remodeling has been found at noninflammatory states during both development and adulthood, via complement receptor 3, CX3CR1, and DAP12. Clinically, PET, using [^{11}C] (R)-PK11195 binding potential as a measure of microglial activation, showed that the binding values were higher in multiple brain regions in young adults with autism spectrum disorder, compared with those of controls [31]. Both Nasu–Hakola disease and diffuse leukoencephalopathy with spheroids (HDLS) are characterized by the white matter degeneration. They are caused by the mutation of genes, being related to microglial functions (*DAP12* and *TREM2* in the former and *CSFR1* in the latter). Accordingly, "microgliopathy" has been proposed [30]. Recently, we performed IM analysis on microglia markers in XP-A and CS autopsy brains. Subjects included six cases each of XP-A and CS, in addition to six controls, lacking neurological disorders (Table 7.3). Serial sections in the cerebrum

Table 7.3 The density of microglia immunoreactive for CD68, HLA-DR, and CD163 in the cortex and white matter in the temporal lobe and cerebellum

	Temporal lobe		Cerebellum	
	Cortex	White matter	Cortex	White matter
CD68				
Control	19.2 (11.3)	32.8 (19.1)	16.6 (8.8)	32.8 (21.2)
Xeroderma pigmentosum- group A	30.2 (25.8)	67.6 (35.6)	21.4 (9.9)	65 (26.4)
Cockayne syndrome	19.1 (15.2)	51.1 (20.5)	15.9 (8.5)	80.1 (27.6)
Kruskal–Wallis	(NS)	(NS)	(NS)	$p < 0.05$
Mann–Whitney ($p < 0.05$)				Ct vs. CS
HLA-DRα				
Control	13.1 (14.3)	21.4 (7)	15.1 (18.2)	21.1 (6.1)
Xeroderma pigmentosum group A	21.4 (28)	51.6 (35.7)	12.2 (15.5)	48.6 (31.4)
Cockayne syndrome	9 (4)	41.9 (23.2)	17.9 (8.3)	55 (23.8)
Kruskal–Wallis	(NS)	(NS)	(NS)	$p < 0.05$
Mann–Whitney ($p < 0.05$)		Ct vs. XP-A		Ct vs. XP-A Ct vs. CS
CD163				
Control	0.8 (2)	2.7 (4.3)	1.3 (1.3)	3.2 (2.2)
Xeroderma pigmentosum group A	16.4 (23.8)	24.3 (31.4)	13.3 (19.9)	28.4 (35.2)
Cockayne syndrome	7.4 (9.7)	14 (12.4)	6.7 (9.1)	10 (13.1)
Kruskal–Wallis	(NS)	(NS)	(NS)	(NS)

The number of microglia immunoreactive for each marker was determined in five nonoverlapping microscopic subfields at 200-fold magnification by using a counting box (0.5 mm^2). Data are shown as mean (SD). Bartlett's test judged that samples lacked homogeneity of variances. Independent samples were examined by Kruskal–Wallis test to compare nonparametric variables among controls (Ct), xeroderma pigmentosum group A (XP-A), and Cockayne syndrome (CS). Mann–Whitney's U test was used to compare nonparametric variables between two groups each. The level of significance was set at $p < 0.05$

NS not significant, *vs.* versus

and cerebellum were treated with antibodies against CD68, HLA-DR alpha, and CD163. Density of amoeboid microglia immunoreactive for CD68 and HLA-DR alpha was increased in the cerebellar white matter in XP-A and/or CS cases. A few CD163-immunoreactive microglia were scattered in controls, and they seemed to be increased in both disorders, but not significantly. Increase of amoeboid microglia immunoreactive for CD68 and HLA-DR alpha in the cerebellar white matter suggested the possible microglial activation in the XP-A and CS brains. The future ELISA on microglial markers in the cerebrospinal fluid and IM analysis for DAP12, TREM12, and CSF1R may give us a hint for clarifying microglial pathology in XP-A and CS.

7.7 Conclusion

Through the clinical studies using the autopsy brains and patient samples in XP-A and CS, we have demonstrated the involvement of oxidative stress, disturbed metabolism of monoamine and melatonin, vascular changes, and glial cell pathology in the neurological disorders, although how various factors may interrelate and interact still remain to be investigated. Our findings have provided us with clues for neurological intervention, and the TLL has been shown to relieve the extrapyramidal abnormalities in some patients with XP-A. We believe that donepezil and melatonin may also be therapeutic options in patients with XP-A and CS.

References

1. Fu L, Xu X, Ren R, Wu J, Zhang W, Yang J, Ren X, Wang S, Zhao Y, Sun L, Yu Y, Wang Z, Yang Z, Yuan Y, Qiao J, Izpisua Belmonte JC, Qu J, Liu GH. Modeling xeroderma pigmentosum associated neurological pathologies with patients-derived iPSCs. Protein Cell. 2016;7(3):210–21. https://doi.org/10.1007/s13238-016-0244-y.
2. Rapin I. Disorders of nucleotide excision repair. Handb Clin Neurol. 2013;113:1637–50. https://doi.org/10.1016/B978-0-444-59565-2.00032-0.
3. Rapin I, Weidenheim K, Lindenbaum Y, Rosenbaum P, Merchant SN, Krishna S, Dickson DW. Cockayne syndrome in adults: review with clinical and pathologic study of a new case. J Child Neurol. 2006;21(11):991–1006. https://doi.org/10.1177/08830738060210110101.
4. Jaarsma D, van der Pluijm I, de Waard MC, Haasdijk ED, Brandt R, Vermeij M, Rijksen Y, Maas A, van Steeg H, Hoeijmakers JH, van der Horst GT. Age-related neuronal degeneration: complementary roles of nucleotide excision repair and transcription-coupled repair in preventing neuropathology. PLoS Genet. 2011;7(12):e1002405. https://doi.org/10.1371/journal.pgen.1002405.
5. Hayashi M. Role of oxidative stress in xeroderma pigmentosum. Adv Exp Med Biol. 2008;637:120–7.
6. Nishigori C, Miyachi Y, Imamura S, Takebe H. Reduced superoxide dismutase activity in xeroderma pigmentosum fibroblasts. J Invest Dermatol. 1989;93:506–10. https://doi.org/10.1111/1523-1747.ep12284060.

7. Parlanti E, Pietraforte D, Iorio E, Visentin S, De Nuccio C, Zijno A, D'Errico M, Simonelli V, Sanchez M, Fattibene P, Falchi M, Dogliotti E. An altered redox balance and increased genetic instability characterize primary fibroblasts derived from xeroderma pigmentosum group A patients. Mutat Res. 2015;782:34–43. https://doi.org/10.1016/j.mrfmmm.2015.10.002.
8. Hayashi M, Tanuma N, Miyata R. Oxidative stress in developmental brain disorders. Adv Exp Med Biol. 2012;724:278–90.
9. Kikuchi K, Takeshige N, Miura N, Morimoto Y, Ito T, Tancharoen S, Miyata K, Kikuchi C, Iida N, Uchikado H, Miyagi N, Shiomi N, Kuramoto T, Maruyama I, Morioka M, Kawahara KI. Beyond free radical scavenging: beneficial effects of edaravone (Radicut) in various diseases (review). Exp Ther Med. 2012;3(1):3–8. https://doi.org/10.3892/etm.2011.352.
10. Itoh M, Hayashi M, Shioda K, Minagawa M, Isa F, Tamagawa K, Morimatsu Y, Oda M. Neurodegeneration in hereditary nucleotide repair disorders. Brain Dev. 1999;21(5):326–33.
11. Hayashi M, Araki S, Kohyama J, Shioda K, Fukatsu R, Tamagawa K. Brainstem and basal ganglia lesions in xeroderma pigmentosum group A. J Neuropathol Exp Neurol. 2004;63(10):1048–57. https://doi.org/10.1093/jnen/63.10.1048.
12. Miyata R, Sasaki T, Hayashi M, Araki S, Shimohira M, Kohyama J. Low dose of levodopa is effective for laryngeal dystonia in xeroderma pigmentosum group A. Brain Dev. 2010;32(8):685–7. https://doi.org/10.1016/j.braindev.2009.09.008.
13. Nomura Y, Segawa M. Neurology of Tourette's syndrome (TS) TS as a developmental dopamine disorder: a hypothesis. Brain Dev. 2003;25(Suppl 1):S37–42.
14. Ueda T, Kanda F, Aoyama N, Fujii M, Nishigori C, Toda T. Neuroimaging features of xeroderma pigmentosum group A. Brain Behav. 2012;2(1):1–5. https://doi.org/10.1002/brb3.22.
15. Kohyama J, Furushima W, Sugawara Y, Shimohira M, Hasegawa T, Hayashi M, Moriwaki S, Iwakawa Y. Convulsive episodes in patients with group A xeroderma pigmentosum. Acta Neurol Scand. 2005;112(4):265–9. https://doi.org/10.1111/j.1600-0404.2005.00478.x.
16. Hayashi M, Ohto T, Shioda K, Fukatsu R. Lesions of cortical GABAergic interneurons and acetylcholine neurons in xeroderma pigmentosum group A. Brain Dev. 2012;34(4):287–92. https://doi.org/10.1016/j.braindev.2011.06.015.
17. Galimberti D, Scarpini E. Old and new acetylcholinesterase inhibitors for Alzheimer's disease. Expert Opin Investig Drugs. 2016;25:1181. https://doi.org/10.1080/13543784.2016.1216972.
18. Kondoh T, Kanno A, Itoh H, Nakashima M, Honda R, Kojima M, Noguchi M, Nakane H, Nozaki H, Sasaki H, Nagai T, Kosaki R, Kakee N, Okuyama T, Fukuda M, Ikeda M, Shibata Y, Moriuchi H. Donepezil significantly improves abilities in daily lives of female down syndrome patients with severe cognitive impairment: a 24-week randomized, double-blind, placebo-controlled trial. Int J Psychiatry Med. 2011;41:71–89. https://doi.org/10.2190/PM.41.1.g.
19. Hardeland R, Cardinali DP, Srinivasan V, Spence DW, Brown GM, Pandi-Perumal SR. Melatonin—a pleiotropic, orchestrating regulator molecule. Prog Neurobiol. 2011;93(3):350–84. https://doi.org/10.1016/j.pneurobio.2010.12.004.
20. Reiter RJ, Tan DX, Fuentes-Broto L. Melatonin: a multitasking molecule. Prog Brain Res. 2010;181:127–51. https://doi.org/10.1016/S0079-6123(08)81008-4.
21. Escribano BM, Colín-González AL, Santamaría A, Túnez I. The role of melatonin in multiple sclerosis, Huntington's disease and cerebral ischemia. CNS Neurol Disord Drug Targets. 2014;13(6):1096–119.
22. Okoshi Y, Tanuma N, Miyata R, Hayashi M. Melatonin alterations and brain acetylcholine lesions in sleep disorders in Cockayne syndrome. Brain Dev. 2014;36(10):907–13. https://doi.org/10.1016/j.braindev.2014.01.004.
23. Wilking M, Ndiaye M, Mukhtar H, Ahmad N. Circadian rhythm connections to oxidative stress: implications for human health. Antioxid Redox Signal. 2013;19(2):192–208. https://doi.org/10.1089/ars.2012.4889.
24. Miyata R, Tanuma N, Sakuma H, Hayashi M. Circadian rhythms of oxidative stress markers and melatonin metabolite in patients with xeroderma pigmentosum group A. Oxidative Med Cell Longev. 2016;2016:5741517. https://doi.org/10.1155/2016/5741517.

25. Schwichtenberg AJ, Malow BA. Melatonin treatment in children with developmental disabilities. Sleep Med Clin. 2015;10(2):181–7. https://doi.org/10.1016/j.jsmc.2015.02.008.
26. Hayashi M, Saito-Miwa N, Tanuma N, Kubota M. Brain vascular changes in Cockayne syndrome. Neuropathology. 2012;32(2):113–7. https://doi.org/10.1111/j.1440-1789.2011.01241.x.
27. Gitiaux C, Blin-Rochemaure N, Hully M, Echaniz-Laguna A, Calmels N, Bahi-Buisson N, Desguerre I, Dabaj I, Wehbi S, Quijano-Roy S, Laugel V. Progressive demyelinating neuropathy correlates with clinical severity in Cockayne syndrome. Clin Neurophysiol. 2015;126(7):1435–9. https://doi.org/10.1016/j.clinph.2014.10.014.
28. Shehata L, Simeonov DR, Raams A, Wolfe L, Vanderver A, Li X, Huang Y, Garner S, Boerkoel CF, Thurm A, Herman GE, Tifft CJ, He M, Jaspers NG, Gahl WA. ERCC6 dysfunction presenting as progressive neurological decline with brain hypomyelination. Am J Med Genet A. 2014;164A(11):2892–900. https://doi.org/10.1002/ajmg.a.36709.
29. Kassubek J, Sperfeld AD, Pinkhardt EH, Unrath A, Müller HP, Scharffetter-Kochanek K, Ludolph AC, Berneburg M. The cerebro-morphological fingerprint of a progeroid syndrome: white matter changes correlate with neurological symptoms in xeroderma pigmentosum. PLoS One. 2012;7(2):e30926. https://doi.org/10.1371/journal.pone.0030926.
30. Prinz M, Priller J. Microglia and brain macrophages in the molecular age: from origin to neuropsychiatric disease. Nat Rev Neurosci. 2014;15(5):300–12. https://doi.org/10.1038/nrn3722.
31. Suzuki K, Sugihara G, Ouchi Y, Nakamura K, Futatsubashi M, Takebayashi K, Yoshihara Y, Omata K, Matsumoto K, Tsuchiya KJ, Iwata Y, Tsujii M, Sugiyama T, Mori N. Microglial activation in young adults with autism spectrum disorder. JAMA Psychiat. 2013;70(1):49–58. https://doi.org/10.1001/jamapsychiatry.2013.272.

Chapter 8
Xeroderma Pigmentosum in the UK

Hiva Fassihi, Isabel Garrood, Natalie Chandler, Shehla Mohammed, Alan R. Lehmann, and Robert Sarkany

8.1 Xeroderma Pigmentosum: Multidisciplinary Model of Care in the UK

The xeroderma pigmentosum (XP) population in the UK is around 100 patients. Since 2010, their medical care has been provided by a single national multidisciplinary clinical service, which cares for patients of all ages.

8.1.1 Background

In the United Kingdom (UK), medical care operates within the government-funded, government-managed 'National Health Service'. Prior to 2010, XP patients were looked after by their local dermatologist, ophthalmologist or neurologist depending on the clinical presentation of the disease. The lack of a centralised service meant that most patients were treated by physicians with no expertise in this disease.

In the UK, recognition of the special challenges of managing patients with severe rare multisystem diseases has led the National Health Service to fund national centralised 'highly specialist' multidisciplinary services for some of these diseases.

H. Fassihi (✉) · I. Garrood · R. Sarkany
National Xeroderma Pigmentosum Service, St John's Institute of Dermatology, Guy's and St Thomas' NHS Foundation Trust, London, UK
e-mail: hiva.fassihi@gstt.nhs.uk

N. Chandler · S. Mohammed
Genetics Department, 5th Floor Tower Wing, Guy's Hospital, London, UK

A. R. Lehmann
Genome Damage and Stability Centre, University of Sussex, Brighton, UK

C. Nishigori, K. Sugasawa (eds.), *DNA Repair Disorders*,
https://doi.org/10.1007/978-981-10-6722-8_8

By the year 2000, non-clinical elements were already in place for a National XP Service. ARL at the University of Sussex had provided a national diagnostic DNA repair assay service since the 1970s, and Mrs. Sandra Webb, the mother of an XP patient, had established a National Patient Support Group ('XP Support Group') in 1999. RS looked after 15 XP patients in the specialist Photosensitivity Dermatology Clinic at St. John's Institute of Dermatology in London and became aware of the poor clinical services for XP patients and of the need for an expert multidisciplinary service. A few small 'pilot' multidisciplinary XP clinics were set up, and quickly their success became clear with high levels of physician and patient satisfaction. In 2009, a team, led by RS, applied to the Government Department of Health to fund a multidisciplinary National XP Service to provide medical care for patients of all ages, from anywhere in the UK.

8.1.2 Applying for Government Funding for a National XP Service

The application was successful because the need for a national service and the 'cost-effectiveness' of that service were demonstrated.

There was a clear need for the service, and the low standard of medical care for many XP patients was demonstrated: details of the histories of the medical care of individual patients were submitted, which catalogued poorly organised, inefficient, non-expert and clinically poor care.

In order to show that the proposed service would be 'cost-effective', it had to meet four criteria:

1. Improve key clinical outcomes
2. 'Geographical equity': Provide an equally good service to all patients regardless of where they live in the UK
3. Patients' needs and wishes taken into account
4. An overall cost saving for the National Health Service

Improving Clinical Outcomes Five measurable clinical outcomes were chosen which could be improved by medical intervention, leading to better overall health outcomes:

- Improved skin protection against ultraviolet radiation (UVR) (which prevents skin cancers from developing)
- Earlier detection of skin cancers (which reduces mortality from advanced cancers)
- Improved eye UVR protection and early detection of ocular surface scarring (which reduce blindness)
- Early detection of hearing loss (which enables earlier fitting of hearing aids)
- Detection of early cognitive impairment (which enables adjustment of schooling)

'Geographical Equity': Providing an Equal Service to All XP Patients Regardless of Where They Live It was made easier and cheaper for patients to travel and attend the multidisciplinary clinic in London by funding and booking travel and accommodation for patients. An XP nurse-led national 'outreach network' (see below) was also set up to enable as much care as possible to be delivered to patients near to their homes.

Taking Patients' Needs and Wishes into Account Attendance of a support group representative at the multidisciplinary clinic and involvement of a support group representative at meetings and in decision-making were incorporated into the service. The design of the multidisciplinary clinic was altered because of the advice from the XP Support Group to include a clinical psychologist. Patient support group involvement has enabled the service to flexibly adapt to the needs and wishes of patients and their families.

The Proposed Service Had to Create an Overall Cost Saving for the National Health Service The National Health Service allocates each UK citizen a unique identification number, which was used to track the total current NHS annual treatment costs of 15 XP patients. The result was that the existing costs were estimated to be £200,000 higher per year than the cost of the proposed service. The cost saving was probably because treatment by non-expert doctors involved expensive and unnecessary skin and eye surgery, scans and other tests, which would be avoided by a more expert clinical team.

The Government Department of Health agreed to fully fund a new National XP Service for an unlimited period beginning in 2010 with RS as the service lead. Every year, a full 'annual report' about the service is produced, containing clinical outcome data, and government inspectors meet RS and the team to assess the performance of the service.

8.1.3 Design of the Service

The service was designed to achieve the objectives required by the government: improved clinical outcomes, geographical equity, patient involvement and cost efficiency. Two further objectives were added: integration between the team's specialties and translational research.

The service is a partnership between:

- The XP multidisciplinary clinic
- The nurse-led outreach service
- The laboratory diagnostic service
- The patient support group

8.1.3.1 Multidisciplinary XP Clinic

The multidisciplinary clinic is based in London at St. Thomas' Hospital. The clinic is held twice a month and lasts a full day for the five patients and six doctors. The dermatologist is the overall clinic coordinator.

At every clinic, patients are seen by a dermatologist, a dermatological surgeon, an ophthalmologist, a neurologist and a neuropsychologist. In the morning, each patient has a 45 min consultation with each doctor. A clinical geneticist (SM) attends some clinics for specialised genetic counselling. The head of the DNA repair assay diagnostic laboratory (ARL) attends the dermatology consultations, and the patient group representative (Mrs Sandra Webb) stays in the clinic area to talk to patients and staff.

The lunchtime team meeting, chaired by the dermatologist (HF), aims to integrate the specialties. Each physician, and the laboratory scientist, presents their findings, and a clinical plan is created for each patient. Following this, the patient representative joins the meeting for discussion of service changes and plans.

In the afternoon, patients have special investigations (nerve conduction studies, audiometry, MRI), and excision of skin tumours identified in the morning. Finally, the team meets the patients and their families to discuss the clinical plan.

A multidisciplinary clinic report is created by the coordinating dermatologist containing a section written by each doctor and the clinical agreed plan at the lunchtime meeting. This is sent to relevant doctors and copied to the patient. Urgent requests for local action are communicated by the XP nurses.

The National XP Service helps patients travel to the clinic in London. The service, the support group and the company 'Virgin Trains' provide financial assistance, and the support group provides organisational help for travel plans.

A multidisciplinary clinic, where patients spend a whole day and where physicians from many specialties work and meet together, has specific and unusual accommodation requirements, which it shares with those of other multidisciplinary services for patients with rare genetic diseases. Since 2018, the National XP Service has been rehoused in a new Rare Disease Centre at St. Thomas' Hospital, the purpose built to fit the needs of multidisciplinary clinics, caring for patients with XP and other rare genetic diseases.

8.1.3.2 Nurse-Led 'Outreach' Service

The two XP specialist nurses (one paediatric and one adult) ensure implementation of the recommendations of the specialist clinic. They coordinate patients' care by communicating with local nurses and physicians (specialists and family doctors) and with patients in their homes. They visit patients' homes, schools and work places, measure UV levels and give advice and training to help patients, families and staff to create a UVR-free environment. Improved UVR protection is a clinical outcome measure on which we provide data in the 'annual report' for the government inspectors.

8.1.3.3 Laboratory Diagnostic Service

The laboratory diagnostic service consists of:

- The pre-existing DNA repair assay service (ARL, University of Sussex)
- A new DNA repair gene mutation analysis service (SM and NC, Guy's Hospital)

8.1.3.4 Translational Research

Translational research is an integral part of the UK National XP Service. All patients are asked to consent for the use of their clinical and laboratory data for research. Each specialty has a one-page clinical data collection form, which is completed by the doctor at every consultation, and the data is entered into a database to enable 'deep phenotyping'. All the nucleotide excision repair genes are sequenced for every patient to determine complementation group and pathogenic mutation(s). Research projects and data generated from the clinic are discussed each year at the whole-day annual team workshop at the University of Sussex.

The integration into the clinic of the non-clinical diagnostic laboratory head and the patient representative (within the limits of patient confidentiality) has had a profound impact. It has enabled the clinicians to develop a holistic understanding of patient experience and disease pathogenesis and the non-clinicians to develop a deeper understanding of clinical issues. This has created a culture of cross-fertilisation of ideas and breaking down boundaries between specialties and professions to create an atmosphere conducive to innovative research. The National XP Service has been awarded over £1.5 million in research grants over the past 3 years. Issues around authorship of published papers are avoided by a clear arrangement which emphasises fairness and acknowledges how much each individual contributes to each project.

8.2 DNA Repair Testing

The DNA repair diagnostic laboratory of ARL at the University of Sussex has been providing cellular diagnoses for XP, Cockayne syndrome and trichothiodystrophy (TTD) for many years. The testing for XP and TTD is, since 2010, an integral part of the National XP Service.

Either a 4 mm skin biopsy or fibroblast culture is sent to the laboratory. Skin biopsies are obtained with consent according to the regulations of the UK human tissue act. Fibroblast cultures are established by methods also described in other chapters and frozen in liquid nitrogen for permanent storage. The first test is to measure unscheduled DNA synthesis (UDS) following UVC irradiation of fibroblasts. Having established the procedures many years ago, it remains more convenient to continue to use incorporation of radioactive thymidine for these studies. Cells are serum-starved for 4 days to bring them into a non-proliferating state. They are then pretreated with the inhibitor hydroxyurea for 30 min in order to abolish any residual replicative DNA synthesis, prior to UVC irradiation with doses of 0, 2.5, 5 and 10 Jm^{-2}. Following UVR irradiation, cells are incubated with ^{3}H-thymidine for 2 h (maintaining the serum starvation and hydroxyurea) and then processed for measurement of incorporated thymidine using liquid scintillation counting. Procedures for using the analog 5-ethynyl-2′-deoxyuridine (EdU), now in use in many laboratories (e.g. see Chap. 2 by T. Ogi in this volume), are being developed in combination with FACS analysis to assess UDS without using radioactivity or expensive microscopes.

If the cells show a normal UDS response, a test for XP variants is carried out based on their hypersensitivity to UVC irradiation in the presence of caffeine. Dividing cells are UVC irradiated with doses of 1, 3 or 5 Jm^{-2} followed by incubation in the presence of caffeine for 2 days. Cells killed by this treatment will no longer be able to replicate their DNA and incorporate ^{3}H-thymidine into their DNA, whereas normal cells are barely affected by this treatment and incorporation of thymidine is affected much less. A typical result is shown in Fig. 8.1. This test is not perfectly specific, and occasionally cell lines give a response intermediate between that of normal cells and bona fide XP variants. These individuals do not have mutations in the *POLH* gene, which is defective in XP variants. Therefore, a test has been recently developed that not only dispenses with radioactive precursors but also appears to be more specific. Briefly, cells are treated with UVC (0, 3 and 5 Jm^{-2}) and incubated with caffeine as described above. Cells are then incubated for 3 h with EdU, harvested and treated with Click-iT reagent and subjected to FACS analysis. Most DNA-damaging agents kill cells or arrest the cell cycle either in the G1 or G2 phase of the cell cycle. Uniquely, however, XP variant cells, treated with UV and caffeine, arrest the cell cycle in S phase. These arrested cells are identified by FACS analysis as cells not incorporating EdU but with a DNA content between that of G1 and G2 populations (Fig. 8.2). In the normal cells, only 3% fall into this gate after irradiation with 5 Jm^{-2} UVC. In contrast, 25% of the XP-V population is arrested in S phase. This S-phase arrested population is seen specifically in XP variant cells and this test is now being used diagnostically in the UK.

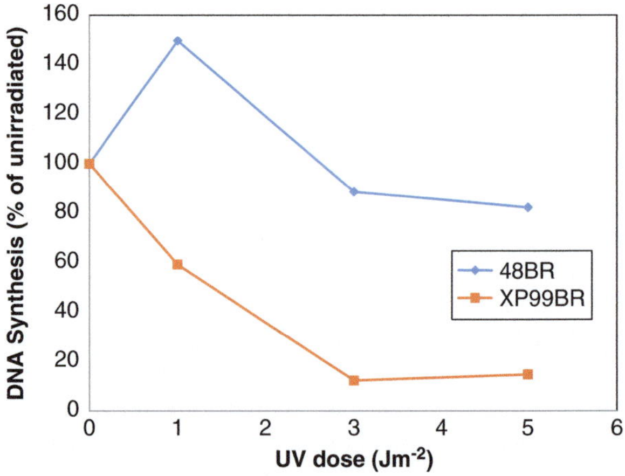

Fig. 8.1 XP-V cells are killed by UV + caffeine treatment. Dividing cells from normal donor 48BR or XP-V donor XP99BR were UVC-irradiated and incubated with caffeine and their ability to incorporate ^{3}H- thymidine measured 2 days later

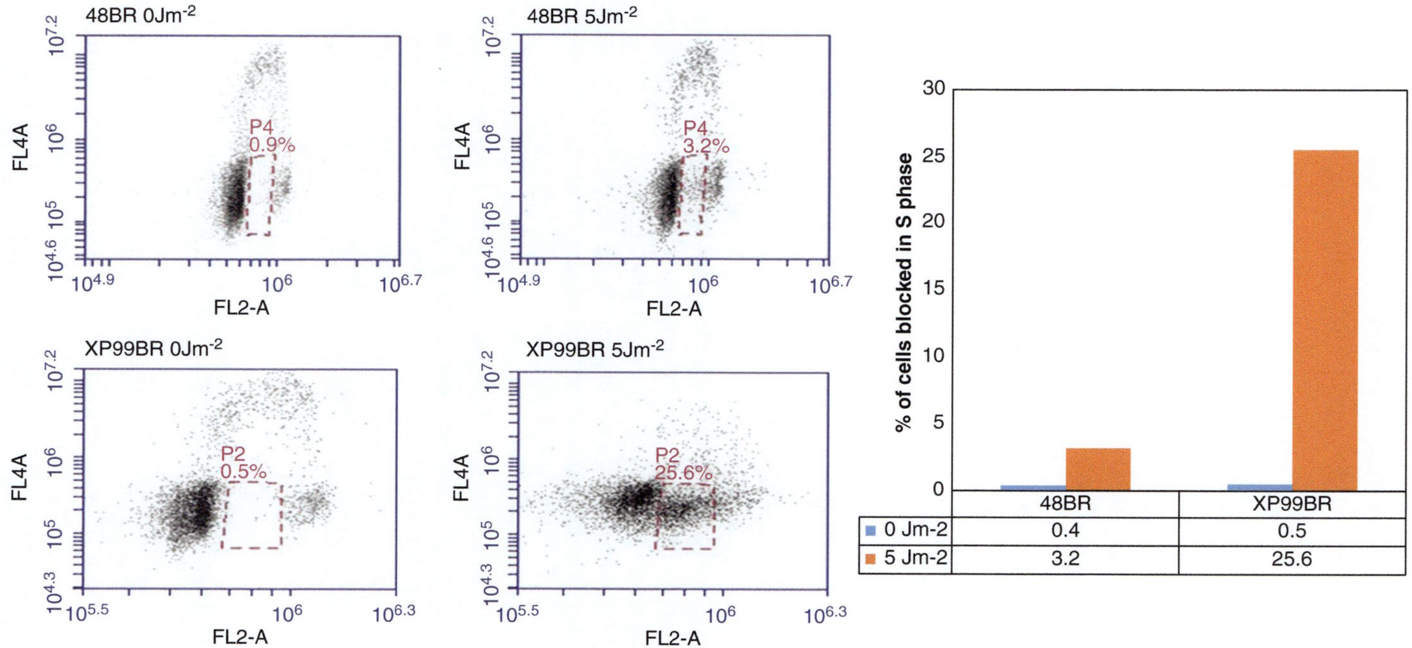

Fig. 8.2 XP-V cells are blocked in S phase following UV + caffeine treatment. Dividing cells treated as above were analysed by flow cytometry. FACS profiles of normal 48BR cells (upper panels) or XP99BR XP-V cells (lower panels) either unirradiated (left) or irradiated with 5 Jm^{-2} UVC are shown. The population blocked in S phase is indicated by the boxes, and the fraction of the population within the boxes is plotted in the bar chart at the right of the figure

If either of these tests is consistent with a diagnosis of XP (or TTD), the results are sent both to the referring clinician and to HF, clinical lead of the National XP Service. The patient is then invited to attend the clinic.

8.3 DNA Repair Gene Mutation Analysis

In 2012, the Viapath Genetics Laboratory, Guy's Hospital, was asked to introduce a sequencing service for XP and related nucleotide excision repair (NER) disorders to support the XP multidisciplinary clinic. The service was introduced to provide detailed genotyping information for each patient and then to be able offer familial carrier testing and prenatal diagnosis when required.

After appropriate consent, blood samples are sent to the laboratory and DNA is extracted. The DNA sequences of interest are then captured using Agilent SureSelect XT technology. Briefly, the patient's DNA is broken into smaller fragments and adaptor sequences are added to the ends allowing for amplification. The DNA fragments that originated from 14 XP- and NER-related disorders (see Table 8.1) are then captured using complementary RNA baits which bind to magnetic beads. The beads are added to bind these RNA-bound DNA fragments, and then washes are performed on a magnet to remove the uncaptured fragments from the rest of the genome. The captured fragments are amplified to create a library using a PCR primer with a unique sequence for each sample so that multiple patient samples can be sequenced together. The patient samples are then pooled together and sequenced using an Illumina massively parallel sequencing platform.

The resulting sequences are aligned against the human genome, and any differences in the coding regions and surrounding intronic sequences of the genes of interest are recorded. Each of these differences is assessed to determine whether it is likely to be a benign sequence variant or a disease-causing mutation. Additionally, large deletions and duplications in these genes are looked for by comparing the number of sequencing reads for each coding region against known normal patient samples. Any variant that is classified as a likely disease-causing mutation is then confirmed either using PCR and standard Sanger sequencing or quantitative fluorescent PCR. This confirms that these mutations are real and that they are present in the correct patient sample.

8.4 The Multidisciplinary Team (Table 8.2)

There are over 90 patients with XP living in the UK, and the multidisciplinary team works closely together to deliver the highest standard of care to them and their families. Although the main focus is on clinical care, close involvement of ARL facilitates ongoing translational research, encouraged and supported by patients and the XP Support Group.

Table 8.1 Genes included in the XP and related NER disorder NGS panel

Gene symbol	Protein name	Transcript	Exons	Coding sequence	Chromosomal location	OMIM number: gene
XPA	XPA	NM_000380.3	6	821	9q22.33	611153
ERCC3	XPB	NM_000122.1	15	2348	2q14.3	133510
XPC	XPC	NM_004628.4	16	2822	3p25.1	613208
ERCC2	XPD	NM_000400.3	23	2282	19q13.32	126340
DDB1	DDB1	NM_001923.4	27	3423	11q12.2	600045
DDB2	XPE	NM_000107.2	10	1283	11p11.2	600811
ERCC4	XPF	NM_005236.2	11	2750	16q13.12	133520
ERCC5	XPG	NM_000123.3	15	3560	13q33.1	133530
POLH	Pol η (XPV)	NM_006502.2	11	2141	6p21.1	603968
ERCC8	CSA	NM_000082.3	12	1190	5q12.1	609412
ERCC6	CSB	NM_000124.2	21	4481	10q11.23	609413
GTF2H5	TTDA	NM_207118.2	3	215	6q25.3	608780
ERCC1	ERCC1	NM_001983.3	10	893	19q13.2	126380
MPLKIP	MPLKIP	NM_138701.3	2	537	7p14.1	609188

8.4.1 Dermatologist and Dermatological Surgeon

There is a significant emphasis on photoprotection measures to reduce UVR exposure and thereby the risk of skin cancers. Outreach visits, by the clinical nurse specialist in the community to homes, schools and work places, ensure that photoprotective measures (such as UVR-protective window films and full-face visors) are in place and allow education of family members, school staff and work place colleagues.

The main role of the dermatologist and dermatological surgeon is to diagnose and treat skin cancers early. Patients are closely monitored using clinical photographs and dermoscopic images to identify new or changing lesions. Confocal microscopy is used to further assess suspicious lesions, especially in children. Topical immune modulators, such as imiquimod cream, and appropriate surgical procedures, including Mohs micrographic surgery, are used to treat skin cancers, most commonly observed on the face and neck. Although there is significant morbidity from repeated surgical excisions, there has been no mortality associated with metastatic skin cancer in the UK XP population over the last 5 years.

8.4.2 Ophthalmologist

Ophthalmological review in patients with XP focuses on three aspects. Firstly, UVR-induced DNA damage of the eyelids and periocular skin plays a role in the development of cicatricial skin changes as well as skin cancers. Secondly, the ocular

Table 8.2 The XP multidisciplinary team

Dermatologist
Dermatological surgeon
Ophthalmologist
Neurologist
Paediatrician
Clinical psychologist
Clinical geneticist
XP imaging and medical photography
Clinical nurse specialists (paediatric and adult)
DNA repair scientist
Research nurse
Patient support group representative

surface (conjunctiva and cornea) can develop UVR-related damage including dry eye, conjunctival injection and inflammation (without infection), as well as development of premature pinguecula and pterygia. Prolonged corneal exposure can result in corneal scarring and visual impairment. Ocular surface cancers, mainly squamous cell carcinomas, have also been reported in patients with significant UVR exposure and poor ocular photoprotection. Even patients with few ophthalmic signs commonly describe photophobia, which is the earliest presenting ophthalmic symptom of XP. Finally, patients with XP-related neurodegeneration may also develop neuro-ophthalmological features, including sluggish pupils, nystagmus and strabismus. Ocular CT may play a role in early detection of neurodegeneration.

Ocular photoprotection is emphasised to reduce the morbidity associated with ocular surface disease and cancer. Full-face UVR-protective visors, peaked hats, UVR-protective sunglasses and clear non-prescription glasses with a UVR-protective coat are recommended and encouraged.

8.4.3 Neurodevelopmental Paediatrician and Neurologist

The neurodevelopmental paediatrician and neurologist play an important role in assessment and detection of neurological disease in XP. Patients undergo regular clinical examination as well as investigations such as audiometry, nerve conduction studies and MRI brain and spine, as clinically indicated. Evidence of sensorineural hearing loss in children allows the early introduction of hearing aids, which can have a significant impact on the child's communication, interactions and schooling. Early detection of neurocognitive impairment allows supportive measures from special school and community mental health teams to be put in place for the XP patients. Towards the end of their life, XP patients with severe neurodegeneration will require maximal assistance with all activities of daily living, and careful management planning by all members of a multidisciplinary team is required.

8.4.4 Clinical Neuropsychologist

The role of the clinical neuropsychologist (IG) in the XP clinic is twofold. It involves the assessment and monitoring of patient's psychological well-being and of their cognitive status. The nature of the input depends on the age of a patient, their cognitive status and the severity of other XP-related symptoms. However, it can be broken down into several components.

The neuropsychologist takes the patient's psychiatry history and assesses their current psychological status. Where a patient may meet criteria for formal psychiatric diagnosis or where subthreshold symptoms are present, the severity of symptoms may be measured using questionnaires, e.g. the Hospital Anxiety and Depression Scale (HADS) [1] or Beck Depression Inventory (BDI-II) [2]. This is important as symptoms may affect engagement with health protection measures, or in the case of procedural phobias, they may affect investigations.

An assessment of factors may affect quality of life, e.g. occupational status, sleep quality and an investigation of the patient's understanding of XP. This is assessed via interview, but quality-of-life measures (e.g. PedsQL) or questionnaires such as the Illness Perception Questionnaire can also be useful [3, 4]. For children play-based measures may be used. Again these factors may have an impact on patient well-being and engagement with treatment.

Where patients are in a complementation group which can be associated with cognitive deficits (A, B, D, F, G) or have symptoms that may be suggestive of cognitive difficulties, a screening assessment may be carried out during clinic to decide the appropriateness of a formal neuropsychological assessment. Tools used include the Mini-Mental State Examination (MMSE) for very impaired patients [5] or the Repeatable Battery for the Assessment of Neuropsychological Status (RBANS) for less impaired patients [6]. These measures can also be used to monitor cognitive status when there is already evidence of significant impairment.

The subsequent role of the clinical neuropsychologist can then include carrying out a formal neuropsychological assessment, referring patients onto the appropriate local services (e.g. cognitive behavioural therapy for depression) and providing psychoeducation about XP in conjunction with the clinical nurse specialists or providing some with in-session counselling support. For children there may also be liaison with school and provision of emotional support for parents.

8.4.5 Clinical Geneticist

Genetic counselling by the clinical geneticist is an essential part of the management of patients and families with XP. The recessive inheritance of this disease and the exact XP complementation group and pathogenic mutation(s) in the relevant XP gene are discussed. Reproductive risk is considered and families can consider prenatal and pre-implantation genetic testing if appropriate. Testing of cord blood immediately after birth in siblings of affected individuals allows early diagnosis and introduction of photoprotective measures.

8.5 Summary of Findings from 89 XP Patients: 2010–2016

The findings on 89 XP patients examined regularly by the UK XP team have been recently published [7]. The key findings are summarised here.

XP-C was the most common complementation group (n = 28, 31%) followed by XP-A (n = 18, 20%) and XP-D (n = 14, 16%). There were only two patients with XP-B. Fifty percent were female and mean age was 36 years (mean range 17–56 years). 83% of XP-A and 75% of XP-C were Asian, with a higher proportion of Caucasian patients in XP-D (85%), XP-E (75%), XP-F (100%), XP-G (62%) and XP-V (83%) groups.

Table 8.3 shows a comparison between complementation groups of the mean values for different parameters that have been measured either in the clinic or in the laboratory. Several scoring systems have been designed in an attempt to provide a semi-quantitative assessment of clinical severity.

Sunlight sensitivity (Sunburn Severity Score) is assessed on a scale of 1–3 based on responses to three questions: (1) Has the patient had sunburn requiring medical consultation/treatment? (2) Does the patient burn outside of the months of March–September in the UK? (3) Does the patient have an abnormal sunburn reaction time course? Each patient was given a numerical Sunburn Severity Score—0, normal sunburn; 1, mildly exaggerated sunburn; 2, moderately exaggerated sunburn; and 3, severely exaggerated sunburn [9].

Ophthalmology Severity Score was measured on a scale of 1–12 based on the following parameters: photophobia (1 point), interpalpebral conjunctival melanosis (2 points awarded if a patient has fair skin), conjunctival injection (1 point), conjunctival corkscrew vessels (1 point), pterygia (1 point), pinguecula occurring under age 50 years (1 point), lagophthalmos/ectropion (1 point), keratopathy (1 point), corneal scarring/neovascularisation (1 point), and ocular surface cancer (2 points).

Neurological disease (Neurology Severity Score) is assessed on a scale of 1–8 based on the neurological systems affected: peripheral (motor or sensory) neuropathy (1 point), cerebellar signs (1 point), sensorineural hearing loss (1 point), impaired cognition (1 point), dysphagia/percutaneous endoscopic gastrostomy feeding (1 point), abnormal pupillary response/abnormal eye movements (1 point), and assisted gait (1 point) (or wheelchair-bound 2 points).

The A and D groups have been sub-divided into two sub-groups, with A' and D' being milder as a consequence of their specific mutation, as discussed below. The UDS values measured (Column 4, Table 8.3) are in accordance with similar data published from other labs over many years, as are the levels of neurological abnormalities (column 11), these being found in groups A, B, D, F and G, but note that

Table 8.3 Means of different phenotype scores in XP complementation groups

Complementation group[a]	No. of patients	Mean age	Mean UDS	Mean SSS[b]	Age first cancer (no. of patients with cancers)[c]	Mean no. of skin cancers	Mean age-corrected no. of skin cancers[d]	Mean ocular score	Mean age-corrected ocular score[d]	Mean neuro score	Mean age-corrected neuro score[d]
A	8	18	4	2.5		0.25	0.03	3.25	0.2	5.25	0.36
A'	10	33	11.5	1.3		5.2	0.025[e]	2.1	0.08	0	0
B	2	45	17.5	2.5		4	0.07	3	0.07	2.5	0.056
C	28	17	14	0	15 (15)	2.25	0.12	4.3	**0.3**	0	0
D	12	29	25	2.6	18.5 (9)	13	0.31	2.2	0.12	4.6	0.2
D'	2	27.5	45	3		0	0	2	0.07	0	0
E	4	51	65	0.75	7.7 (3)	152	**2.02**	3	**0.05**	0	0
F	3	30	19	2.7		0	0	1.3	0.035	2	0.04
G	8	26	4	2.25		0.125	0.003	2.75	0.18	3.9	0.23
V	12	50	109	0	32 (9)	26	**0.43**	2.2	**0.05**	0	0

[a] A', XP-A patients with mutation in intron 4 [8], D', XP-D patients with Arg683Gln mutation
[b] Sunburn severity score (see text and [9])
[c] Mean age of first skin cancer for those patients who have developed one or more skin cancers and in parenthesis, number of patients in the group who have developed skin cancers. Blank spaces indicate either no skin cancers in this group or only one or two patients with skin cancers
[d] For each patient, the number of skin cancers, ocular score or neuro score, respectively, was divided by the age of the patient—data show means of these calculations for each group
[e] 81-year-old patient excluded from calculation

the sub-groups A' (10 patients) and D' (2 patients) are neurologically unaffected. However a new finding, reported in detail elsewhere, is that sunburn severity is found only in groups A, B, D, D', F and G with scores of between 2.25 and 3 out of a maximum of 3 (column 5). In contrast groups C, E and V show a normal sunburn reaction and that of the A' group is only slightly elevated (see [9] for more details).

As the patient groups have different age ranges and since skin cancers, ocular and neurological abnormalities all increase with age, we have attempted to improve comparisons between complementation groups by taking age into account. For each patient we have divided the number of cancers and ocular and neurological scores by their ages and calculated mean 'age-related' parameters. Columns 7, 9 and 11 show the mean scores for these parameters without age correction, and columns 8, 10 and 12 show them with age correction.

Examination of columns 7 and 8 shows that XP-E and XP-V patients have the highest numbers of skin cancers, even correcting for age (indicated in bold in column 8). Patients in these groups, classically considered the mildest, show few symptoms until the age of 20, resulting in a late diagnosis, by which in time many precancerous mutations will have been generated. Typically, from the age of 20 to 30, these develop into frank cancers at high frequency. An additional complicating factor is that since these patients are usually not diagnosed until adulthood, it is more difficult to persuade them to take appropriate protective measures. Groups D and C are the next most cancer-prone groups in the UK cohort.

There are 10 patients in our A' sub-group. They all have the same mutation in *XP-A*, namely, a mutation in intron 4 which generates a new splice donor site and results in mis-splicing of the *XP-A* mRNA. However 5% of the mRNA is correctly spliced resulting in a small amount of functional XP-A protein. This seems sufficient to substantially alleviate the skin hypersensitivity and completely protects from the neurological problems usually associated with XP-A patients. These mildly affected patients are discussed in detail in [8]. The oldest patient, who is in the XP-A' sub-group and is now aged 82, has been excluded from the cancer column in Table 8.3. This is because his 45 cancers would distort the numbers in the A' group, and he had no cancers before the age of 46, despite having been exposed to high levels of sunlight in India before he came to the UK. This suggests that the A' group, though displaying very mild features, may, without appropriate protection, develop skin cancers in later life.

All but one of the XP-D patients are mutated in Arg683 in at least one allele. In the majority the mutation is to Trp, and these patients are very sunburn-sensitive, and nearly all of them developed neurological problems. In two patients however, Arg683 is mutated to Gln, and these patients, assigned to the D' sub-group, maintain sunburn hypersensitivity, but neither has shown any sign of neurological abnormalities. The likely cause of this difference is that the hydrophilic glutamine is able to maintain some of the intramolecular interactions associated with the correct Arg residue, whereas tryptophan is hydrophobic and much bigger and so is likely to produce a severe local distortion of the XP-D structure.

Interestingly, no skin cancers were found in the three XP-F patients and only one in the eight XP-G patients, despite their all having a severe sunburn reaction. This suggests that individuals in these two groups may be somehow protected from the high incidence of skin tumours associated with other groups. Scanning of the literature reveals that low incidences of cancer are associated with these two groups in other studies [10–12].

Columns 9 and 10 show that ocular abnormalities are most prevalent in XP-C patients (see [7] for more details). It is particularly instructive to compare the age-corrected skin cancers and ocular scores of the XP-C, XP-E and XP-V groups, none of whom have an abnormal sunburn reaction. Although the latter 2 have 16- and 3.5-fold more skin cancers than the former, the reverse is true for ocular abnormalities, for which XP-C have an age-related score 6 times higher than XP-E and XP-V (indicated in bold in column 10). Thus although the exposure to sunlight in early years prior to diagnosis in the XP-E and XP-V groups results in a high incidence of skin cancers as discussed above, it does not have a similar effect on ocular abnormalities. This susceptibility to ocular abnormalities appears to be confined to the XP-C group, even though they are in general diagnosed much earlier. This intriguing observation remains unexplained.

The mild XP-A cohort has been discussed in which the pathological mutation generates a new splice site, but a small amount of normally spliced product dramatically ameliorates the phenotype. Patients in the XP-C, XP-D and XP-G groups with mutations resulting in abnormally spliced products and milder than expected phenotypes have also been identified. Though we have less direct evidence, we have suggested that in these cases also, a small amount of normally spliced product or in-frame splicing could account for the milder phenotypes.

An important aspect of these studies is that in several cases, we are now able to provide improved prognoses for patients, based on molecular analyses. For example, we are now able to predict with confidence that the splicing mutation in intron 4 of *XP-A* and Arg683Gln in *XP-D* will not result in neurological abnormalities, or at the worst they will be late onset. In the UK, where we are in general able to provide early diagnoses and advice for protection from UVR, skin cancers can be minimised, and neurological problems become the over-riding concern for many families.

In the case of UVR-induced skin cancers, all patients are encouraged to protect well, but for the mild XP-As, XP-Fs and XP-Gs, the level of protection may perhaps not need to be as rigorous for the other groups.

References

1. Zigmond AS, Snaith RP. The hospital anxiety and depression scale. Acta Psychiatr Scand. 1983;67:361–70.
2. Beck AT, Steer RA, Brown GK. Manual for the Beck depression inventory-II. San Antonio, TX: The Psychological Corporation; 1996.
3. Broadbent E, et al. The brief illness perception questionnaire. J Psychosom Res. 2006;60(6):631–7.
4. WHOQOL Group. Development of the World Health Organization WHOQOL-BREF quality of life assessment. Psychol Med. 1998;28(3):551–8.

5. Folstein MF, Folstein SE, McHugh PR. "Mini-mental state." a practical method for grading the cognitive state of patients for the clinician. J Psychiatr Res. 1975;12:189–98.
6. Randolph C. Repeatable battery for the assessment of neuropsychological status manual. San Antonio, TX: The Psychological Corporation; 1998.
7. Fassihi H, et al. Deep phenotyping of 89 xeroderma pigmentosum patients reveals unexpected heterogeneity dependent on the precise molecular defect. Proc Natl Acad Sci U S A. 2016;113(9):E1236–45.
8. Sethi M, et al. A distinct genotype of XP complementation group A: surprisingly mild phenotype highly prevalent in northern India/Pakistan/Afghanistan. J Invest Dermatol. 2016;136(4):869–72.
9. Sethi M, et al. Patients with xeroderma pigmentosum complementation groups C, E and V do not have abnormal sunburn reactions. Br J Dermatol. 2013;169(6):1279–87.
10. Kondo S, et al. Late onset of skin cancers in 2 xeroderma pigmentosum group F siblings and a review of 30 Japanese xeroderma pigmentosum patients in groups D, E and F. Photo-Dermatology. 1989;6(2):89–95.
11. Nishigori C, Fujisawa H, Uyeno K, Kawaguchi T, Takebe H. Xeroderma pigmentosum patients belonging to complementation group F and efficient liquid-holding recovery of ultraviolet damage. Photodermatol Photoimmunol Photomed. 1991;8(4):146–50.
12. Schafer A, et al. Characterization of three XPG-defective patients identifies three missense mutations that impair repair and transcription. J Invest Dermatol. 2013;133(7):1841–9.

Chapter 9
Cockayne Syndrome: Clinical Aspects

Masaya Kubota

9.1 Introduction and History

Cockayne syndrome (CS) is a rare autosomal recessive degenerative disorder caused by deficient nucleotide excision repair. It is characterized by a distinctive facial appearance (sunken eyes and cheeks), postnatal poor growth, and progressive multisystem dysfunction, including those of the central and peripheral nervous systems. Cockayne first described the condition in two siblings that exhibited small heads, small faces with sunken eyes, and intellectual disabilities in a paper entitled "Dwarfism with Retinal Atrophy and Deafness" in 1936 [1]. Ten years later, he reported follow-up data for the same patients, which showed that the siblings had suffered from progressive hearing loss, visual dysfunction, and joint contracture [2]. In the first paper, he had also described skin symptoms, which worsened with sun exposure according to the patients' mother, although he could not reach the idea of specific symptom associated with CS pathophysiology.

In 1977, Sugarman et al. [3] described the clinical and pathological findings of CS patients and reported eight major characteristics (dwarfism, intellectual disability, microcephaly, ataxia, retinal pigmentation, neural deafness, progeroid features, and intracranial calcification) and seven minor characteristics (photosensitivity; kyphosis; ankylosis; optic atrophy; carious teeth; large, cold hands and feet; and hypogonadism). After that, an analysis of 140 CS patients by Nance et al. [4], the first comprehensive review of the condition, demonstrated that CS could be divided into three types based on age at onset, clinical severity, and the disease progression rate; i.e., into type 1 (the classic form), type 2 (the severe form), and type 3 (the late-onset milder form). Some rare cases involve a combination of the symptoms of CS and xeroderma pigmentosum (CS/XP). Nance et al. [4] also described the age-related complications of each type of CS in detail. In 2011, Natale [5] proposed a

M. Kubota
Division of Neurology, National Center for Child Health and Development, Tokyo, Japan
e-mail: kubota-ms@ncchd.go.jp

C. Nishigori, K. Sugasawa (eds.), *DNA Repair Disorders*,
https://doi.org/10.1007/978-981-10-6722-8_9

revised CS classification including a fourth group (ultraviolet [UV] sensitivity alone/adult onset cases), which was based on disease severity. The new classification included extensive descriptions of disease severity as a continuous spectrum (very severe (cerebrooculofacioskeletal (COFS) syndrome), severe, moderate, mild, very mild, and UV sensitivity syndrome/adult-onset) instead of type 3 grouping. These descriptions of the links between disease severity and signs and symptoms are very useful for clinicians. However, the conventional 3-category grouping is also useful for obtaining a rough initial diagnosis in spite of the broad phenotypic variation and mixture of features seen in different types of CS. As far as the core patients in each group are concerned, the 3-category system is still useful for helping clinicians to recognize such rare conditions. In 2015, Kubota et al. [6] reported the incidence, prevalence, and clinical findings of CS patients in a nationwide survey of Japan, in which a comparative analysis between the surviving and deceased patients demonstrated that the management of renal failure and nutrition is very important for ensuring good quality of life throughout the long-term course of CS. Recently, Wilson et al. [7] reported detailed information about the prevalence of CS and the onset of clinical symptoms, the achievement of neurodevelopmental milestones, and patient management based on comprehensive clinical data obtained directly from patents' families and their clinicians. Laugel [8] reported modified clinical diagnostic criteria comprising developmental delay, progressive growth failure, and progressive microcephaly as major signs and cutaneous photosensitivity, retinopathy and/or cataracts, progressive sensorineural hearing loss, enamel hypoplasia, and enophthalmia as minor signs (three out of five). Thus, numerous clinical assessments are required to support a diagnosis of CS, and comprehensive investigations of the needs of patients and their families are also necessary.

The pathogenesis of CS has been considered to be related to premature or accelerated physiological aging for a long time; however, the main mechanisms responsible for the various disease courses seen at the individual level remain unclear. Oxidative stress, microangiopathy, hypoxic-ischemic changes, and mitochondrial dysfunction associated with altered autophagy are also speculated as causative factors [9–11].

9.2 Epidemiology

In the first nationwide survey of clinically confirmed CS cases (a few patients had genetically confirmed CS) in Japan, 45 institutions reported that they had experience of diagnosing or treating patients with CS ($n = 76$) [6]. It is estimated that in Japan the minimum incidence of CS is 2.77 per million births (95% confidence interval: 2.19–3.11) and the prevalence of CS is about 1 in 2,500,000. There were no significant regional differences in the distribution of CS throughout Japan. Kleijer et al. [12] reported that the minimum incidence of genetically confirmed CS (including XP/CS complex) in Western Europe (France, Italy, the UK, and the Netherlands) is 2.7 per million births. These similar results suggest that any racial differences in the incidence of CS are not very large. They also estimated the minimum incidence of CS to be 1.8 per million in the autochthonic Western-European population since immigrant populations were disproportionately represented in their patient population. Although the minimum incidence of CS has not been estimated in other countries, it might be broadly similar around the world, except in regions and groups with high incidence rates (e.g., some indigenous Canadians and Saudi Arabians).

9.3 Common Symptoms, Clinical Profiles, and Developmental Aspects

The most common symptoms of CS type 1 are listed in Table 9.1. Growth failure, i.e., height, weight, and head circumference are significantly lower than the −2 standard deviation (SD) value (a cardinal symptom), becomes evident during the first 2 years of life in CS type 1 (moderate form). Thus, if a patient's body weight and height are normal at the age of 2 years, CS can be excluded. In the abovementioned nationwide survey conducted in Japan, the mean body weight of 21 CS type 1 survivors (mean age: 16.1 years old) [6] was 15.3 kg, while that of 14 deceased patients (mean age at death: 18.6 years old) was 12.6 kg. Likewise, the mean height of the 20 CS type 1 survivors was 102.5 cm, while that of the 10 deceased patients was 96.2 cm, and the difference was significant. The short stature of CS patients is not due to growth hormone deficiency.

Microcephaly is another prominent finding of diagnostic value. In the abovementioned nationwide survey, 13 CS type 1 patients exhibited a mean head circumference of 44.9 ± 1.8 cm. Usually, CS patients display a normal head circumference at birth (Table 9.1), but patients with severe type CS or COFS syndrome exhibit congenital microcephaly [4]. Although there are rare exceptions, if no microcephaly is present at the age of 2, other differential diagnoses should be considered [4]. The degree of microcephaly correlates with disease severity.

In the abovementioned nationwide survey, the following 7 signs and symptoms were observed in >90% of the 41 CS type 1 (moderate form) patients: a profound failure to thrive, photosensitivity, deafness, a distinct facial appearance (such as sunken eyes due to a lack of subcutaneous orbital fat), foot joint contracture, intellectual disability, and the detection of basal ganglia calcification on computed tomography (CT) [6] (Table 9.1). Of these seven symptoms, failure to thrive, photosensitivity, and intellectual disability (language delays) were seen before the age of 2 or 3 years. In addition, abnormal auditory brainstem responses (ABR), retinitis pigmentosum, and decayed teeth were found in >70% of patients [6].

Photosensitivity, a cardinal symptom of CS, becomes apparent around the cheeks, especially in summer. It is not usually very severe, but appropriate environmental settings and the use of sunscreen are recommended. The absence of clinical photosensitivity does not preclude a problem at the biochemical level [5]. Skin cancer, which can occur in xeroderma pigmentosum, does not develop in CS.

Table 9.1 Common symptoms of the 41 CS type 1 patients

	Surviving patients ($n = 21$)	Deceased patients ($n = 20$)	P	Total ($n = 41$)
Age (years)/sex	16.1 ± 8.6/14 males Range: 4–38	18.9 ± 4.0/10 males Range: 14–27	NS	
Head circumference at birth (cm)	32.8 (9)	32.4 (4)		
Body weight (kg) at survey	15.3 (21) Range: 9–27	12.6 (14) Range: 7–20	0.04	
Height (cm) at survey	103.5 (20) Range: 85–125	96.2 (10) Range: 87–103	0.02	
Failure to thrive	20 (20)	16 (16)	NS	36/36 (100%)
Able to walk Mean age started to walk (months)	11 (20) 25	10 (18) 19	NS	21/38 (55.3%)
Able to speak	18 (20)	12 (19)	0.046	30/39 (76.9%)
Intellectual disability	21 (21)	20 (20)	NS	41/41 (100%)
Oral nutrition	19 (20)	6 (17)	0.0001	25/37 (67.6%)
Photosensitivity	17 (19)	17 (18)	NS	34/37 (91.9%)
Sunken eyes	19 (20)	19 (19)	NS	38/39 (97.4%)
Thinning hair	6 (20)	9 (17)	NS	15/37 (40.5%)
Deafness	16 (19)	12 (12)	NS	28/31 (90.3%)
Cataracts	11 (17)	9 (14)	NS	20/31 (64.5%)
Retinopigmental changes	14 (17)	11 (11)	NS	25/28 (89.3%)
Optic atrophy	6 (14)	6 (10)	NS	12/24 (50%)
Scoliosis	9 (19)	3 (12)	NS	12/31 (38.7%)
Foot joint contracture	18 (19)	12 (13)	NS	30/32 (93.4%)
Decayed teeth	10 (16)	8 (9)	NS	18/25 (72%)
Hypertension	3 (10)	6 (10)	NS	9/20 (45%)
Renal failure	1 (21)	9 (13)	<0.001	10/34 (29.4%)
Anemia	2 (21)	8 (15)	0.004	10/36 (27.8%)
Lacrimal hyposecretion	2 (12)	6 (9)	0.02	8/21 (38.1%)
Hypohydrosis	10 (15)	8 (10)	NS	18/25 (72%)
Finger tremors	8 (18)	9 (12)	NS	17/30 (56.7%)
Nystagmus	3 (19)	4 (13)	NS	7/32 (21.9%)
Dystonia	0 (19)	4 (10)	0.009	4/29 (13.8%)
Myoclonus	0 (19)	1 (9)	NS	1/28 (3.6%)
Brisk DTR	12 (19)	6 (9)	NS	18/28 (64.3%)
Seizures	4 (20)	2 (13)	NS	6/33 (18.2%)
Sleep disorder	2 (17)	5 (11)	0.04	7/28 (25%)
BT dysregulation	8 (18)	4 (7)	NS	12/25 (48%)
Diarrhea	6 (19)	5 (9)	NS	11/28 (39.3%)

The numbers in parentheses indicate the number of patients for which data were available
DTR deep tendon reflex, *BT* body temperature, *NS* not significant

High-tone hearing loss appears earlier than visual dysfunction. Appropriate follow-up by an otolaryngologist is important because many patients initially respond to the use of hearing aids. A previous report described the insertion of cochlear implants in two CS patients [13]. One patient benefited from the surgery, but the other did not. Such outcomes depend upon the patient's neurodevelopmental state because the neural degeneration seen in CS is essentially progressive.

Some CS patients exhibit myotic pupils and do not react markedly to mydriatic agents. Visual dysfunction due to retinitis pigmentosa, optic nerve atrophy, and/or cataracts progresses with age.

Concerning gross motor development, especially locomotion ability, clear differences in the course of the disease exist among the three types of CS. In the above-mentioned nationwide survey of CS, the CS type 1 (the moderate form) patients were initially able to walk alone [6] (21/38 cases (55.3%), Table 9.1) or with assistance, but thereafter their conditions gradually deteriorated. Conversely, the CS type 2 (the severe form) patients could only move by crawling and could not stand up alone. As for the CS type 3 (the late-onset mild form) patients, they exhibited almost normal motor development and were usually able to walk by themselves, although they demonstrated ataxia, until their late 20s, but their conditions subsequently slowly worsened [6]. As the condition progresses, even patients with mild CS experience the same problems as other CS patients. Although the severity of CS forms a continuous spectrum, locomotion ability is reflective of the overall degree of development because locomotion development requires the broad neuronal circuits responsible for postural adjustment and central pattern generation. Even when CS type 1 (the moderate form) patients acquire the ability to walk, they exhibit an ataxic and unsteady gait combined with lordosis and foot contracture from an early stage. Detailed developmental milestones taken from Wilson et al.'s report are shown in Fig. 9.1 [7].

Clinically, the deep tendon reflexes of the extremities are often brisk in the early stage, but they become weakened due to the development of peripheral neuropathy, joint contracture, and muscle wasting.

Patients with CS often experience hand or finger tremors, which can occur during action or (in rare cases) at rest. Nystagmus is observed less frequently than finger tremors (Table 9.1).

Somewhat hoarse and husky voices are common in CS patients, regardless of gender, which are probably due to the presence of a small mandible, limited mouth opening, a markedly restricted space within the oral cavity, and the possible maldevelopment of the larynx [14].

COFS syndrome is considered to be a distinct genetic entity but represents an allelic (the clinically most severe) form of CS [15].

In UV-sensitive syndrome (UVSS), as with CS, the repair of UV-induced DNA damage via transcription-coupled nucleotide excision repair is completely lost, and the same phenotype as is seen in CS is observed at the cellular level, i.e., UV hypersensitivity and reduced recovery from RNA synthesis errors [16]. There are three complementation groups in UVSS. There is no doubt that UVSS is a mild condition, but the nosological position of UVSS is yet to be established.

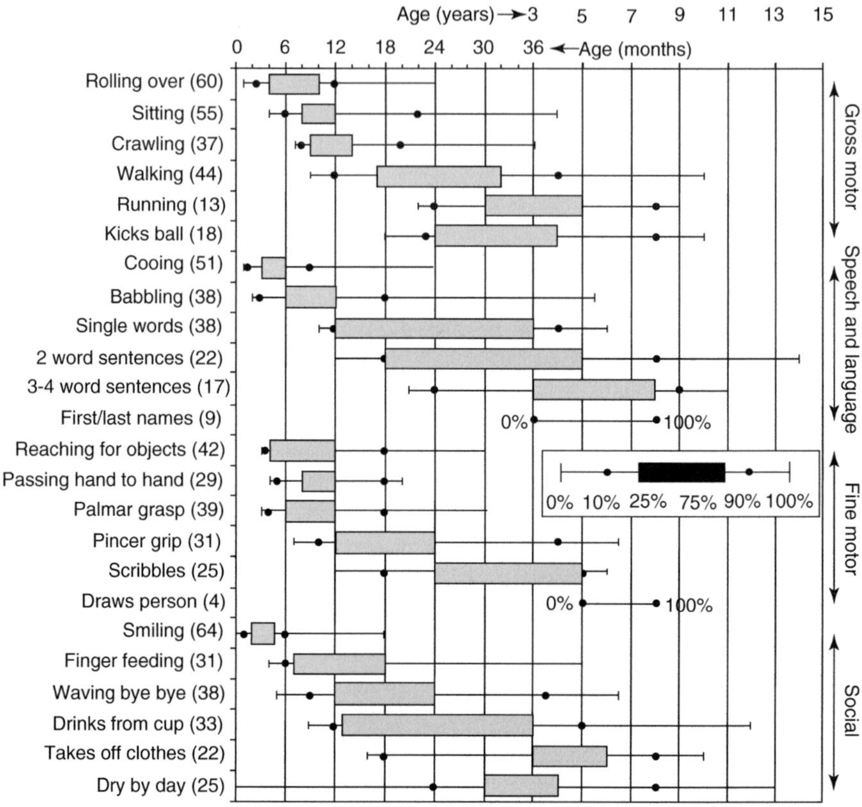

Fig. 9.1 Developmental milestones in Cockayne syndrome (Wilson et al. [7])

9.4 Neurophysiology

Regarding auditory evoked potentials (ABR) (Table 9.2), prolonged latency, and wave amplitude reduction are seen in the early stages of CS, and such abnormalities might progress from waves V to waves I, i.e., from the upper brainstem to the cochlear nerve [17], as shown in Fig. 9.2. The lesions that cause hearing loss in CS are located in the brainstem and at peripheral sites [17].

Reductions in nerve conduction velocity and the amplitude of compound motor action potentials (CMAP) and/or sensory nerve action potentials (SNAP) are caused by demyelination and axonal lesions of the peripheral nerves [9, 18] (Table 9.2). Needle electromyography can be used to detect the characteristic features of denervation (Fig. 9.3). In addition, the reduction or absence of N20 deflection and the prolongation of its latency in short latency somatosensory

Table 9.2 Laboratory data of the 41 CS type 1 patients

	Surviving patients ($n = 21$)	Deceased patients ($n = 20$)	P	Total ($n = 41$)
Age (years)/sex at survey	16.1 ± 8.6/14 males Range: 4–38	18.9 ± 4.0/10 males Range: 14–27	NS	
Peripheral nerve conduction abnormalities	6 (13)	3 (6)	NS	9/19 (47.4%)
Abnormal ABR	14 (17)	4 (4)		18/21 (85.7%)
ALT (IU/dL)	63.9 (20)	62.7 (19)	NS	
AST (IU/dL)	114.8 (20)	119.8 (19)	NS	Abnormal value 31/36 (79.5%)
BUN (mg/dL)	18.2 (20)	40.5 (17)	0.002	
Cr (mg/dL)	0.49 (20)	1.67 (16)	0.03	
Uric acid (mg/dL)	6.31 (16)	7.11 (10)	NS	
Total cholesterol (mg/dL)	173.7 (16)	212.1 (13)	NS	
Triglyceride (mg/dL)	127.8 ($n = 12$)	205.9 ($n = 8$)	NS	
Proteinuria	2 (16)	9 (11)	0.0003	11/28 (39.3%)

The numbers in parentheses indicate the number of patients for which data were available
BUN blood urea nitrogen, *Cr* creatinine, *AST* aspartate aminotransferase, *ALT* alanine aminotransferase (normal <35 IU/dL), *ABR* auditory brainstem responses, *NS* not significant

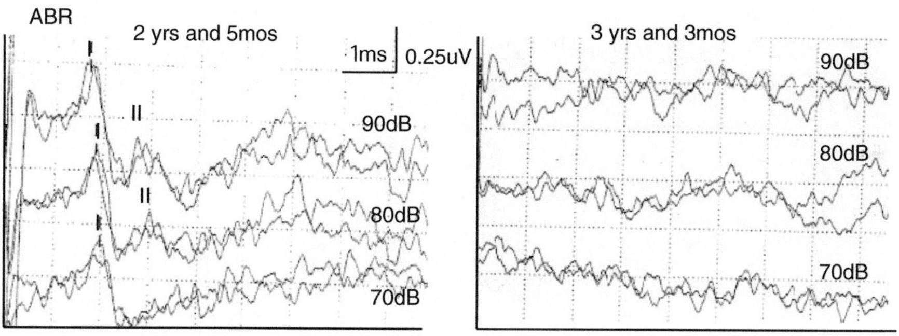

Fig. 9.2 Auditory brainstem responses in a patient with CS type 2 (severe form)

evoked potentials (SSEP) (Fig. 9.4), and the prolongation of P100 latency in visual evoked potentials (VEP) are indicative of hypomyelination and further demyelination of the peripheral and central nervous systems. Thus, studies of combined multimodal evoked potentials and nerve conduction are diagnostically highly sensitive, even in the early stages of CS [18], although CS patients often require sedation during such tests.

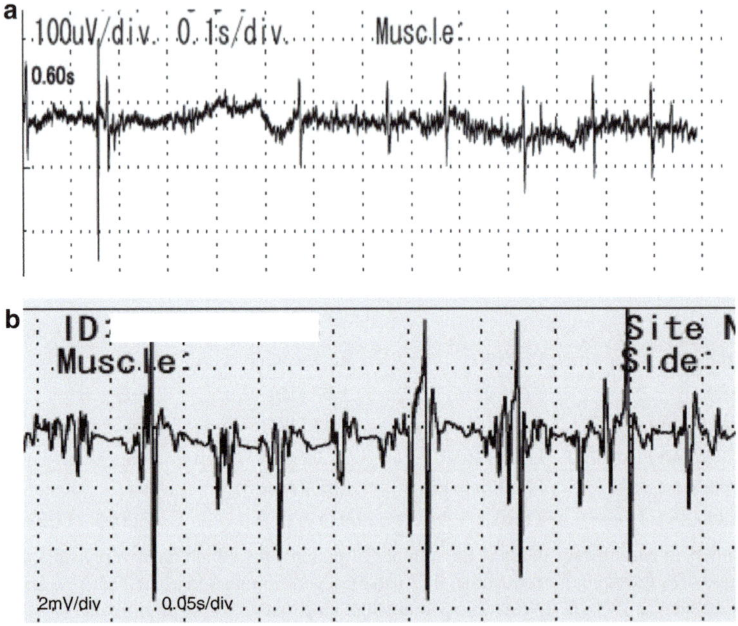

Fig. 9.3 Needle electromyography of the anterior tibial muscle in a patient with CS type 3 (late-onset mild form) showing a denervation pattern ((**a**) fibrillation potentials and (**b**) high-amplitude, polyphasic motor unit potentials)

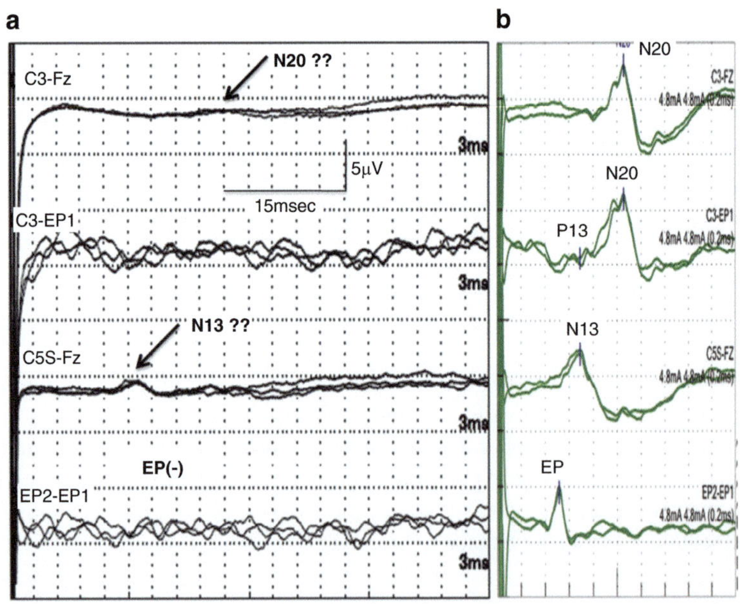

Fig. 9.4 Short-latency somatosensory evoked potentials produced by right median nerve stimulation in a patient with CS type 1 (moderate form) at 5 years (**a**) and a control (**b**) *EP* Erb's point potential, *N13* cervical potentials, *N20* the first cortical potentials from the primary somatosensory area

9.5 Neuroimaging

Hypomyelination, calcifications, and brain atrophy are the main imaging features of CS [19].

On brain CT scans, calcification is observed in the basal ganglia (most frequently in the putamen), cortex, subcortical white matter, and the dentate nucleus of the cerebellum (Figs. 9.5, 9.6, and 9.7). The cortical calcifications are larger at the depths of the sulci (Fig. 9.5) and are typically bilateral and symmetrical, ranging from punctate to severe in extent [19]. High signal intensity on T1-weighted images (WI) and low signal intensity on T2WI are indicative of calcification on CT scans of the basal ganglia (Figs. 9.6 and 9.7). A few patients with genetically proven CS who lacked calcification have been reported, but calcification is still a very useful diagnostic factor. Progressive volume reduction of the bilateral hemispheres (predominantly the white matter) and cerebellum are also evident (Figs. 9.5 and 9.6). Basal ganglia calcification is recognizable at 3 years of age, at the latest, in nearly all cases except for milder late onset cases (type 3). White matter hypoattenuation is also noted in the semioval center and/or in the periventricular white matter, mainly in the frontal regions, from 2 years of age [19]. There are no strict correlations between the

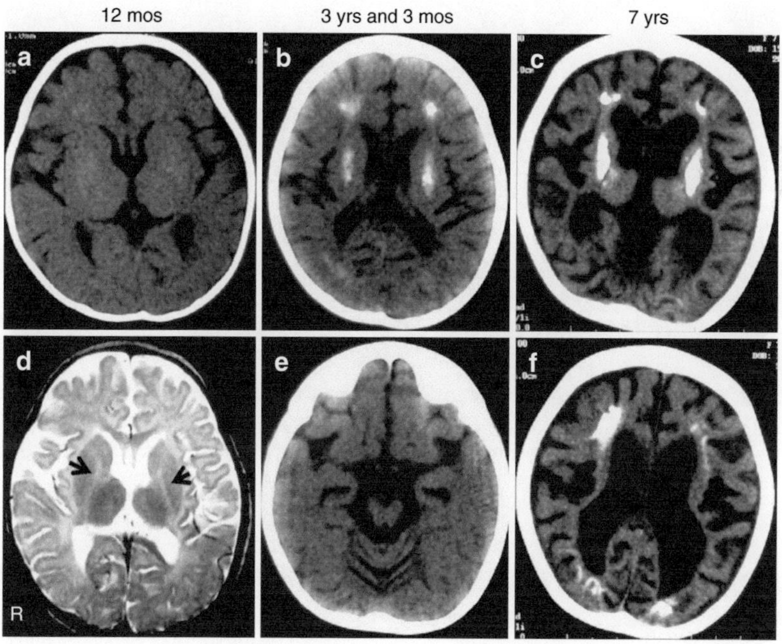

Fig. 9.5 Brain CT (**a–c**, **e**, and **f**) and MRI (**d**) scans of a patient with CS type 2 (severe form). Putaminal and subcortical calcifications appeared at 3 years of age. Hypomyelination was evident on T2WI ((**d**), high intensity in the posterior limbs, arrows). Brain atrophy progressed with age

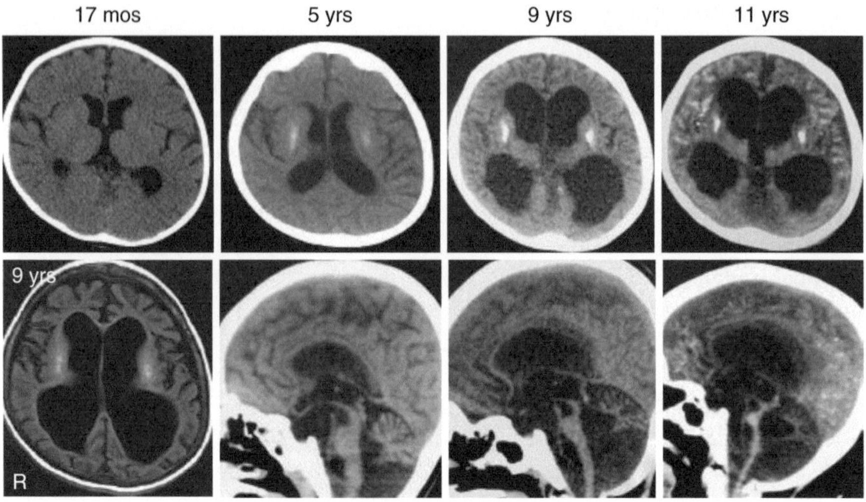

Fig. 9.6 Brain CT and MRI (the lower left panel) scans obtained in a patient with CS type 1 (moderate form). Putaminal calcification was evident at 5 years of age, and brain atrophy (predominantly in the white matter) progressed with age. Progressive brainstem and cerebellar atrophy were also clearly demonstrated

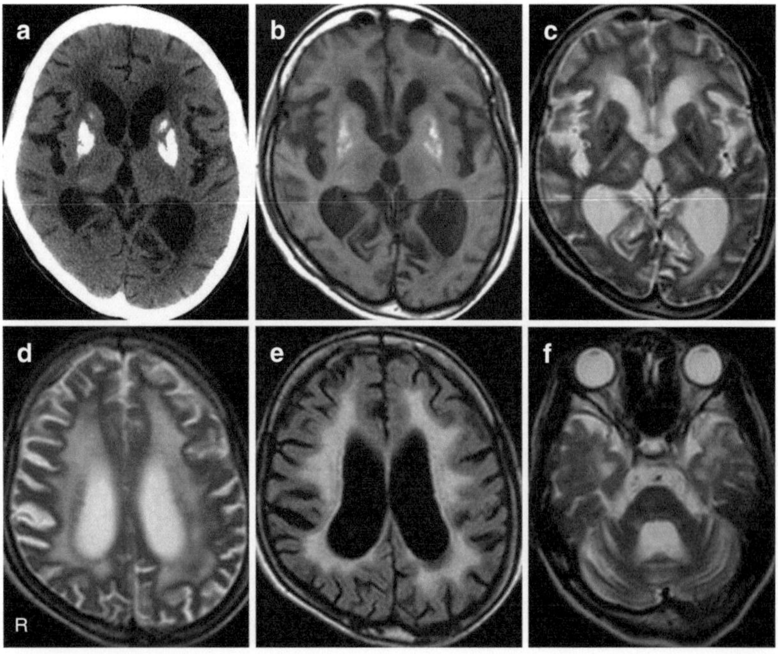

Fig. 9.7 Brain CT (**a**) and MRI (**b–f**) scans obtained at the age of 25 in a patient with CS type 3 (late-onset mild form). The putaminal calcifications detected on CT (**a**) corresponded to the areas of high intensity seen on T1WI (**b**) and low intensity on T2WI (**c**). Cortical atrophy and relatively well-preserved U-fibers were seen (**c**, **d**, and **e**)

extent or intensity of such calcifications and age, the severity of neurological symptoms, and the degree of cerebral atrophy [19]. Once calcifications develop, they never become smaller. Cerebral supratentorial atrophy is related to progressive and extensive white matter loss and subsequent ventricular dilation (Figs. 9.5 and 9.6). Such brain atrophy might be largely due to leukoencephalopathy [9]. A gross correlation was detected between age and the degree of atrophy within each subtype. Cerebral and cerebellar atrophy clearly progress with age (Fig. 9.6). The relative preservation of the cortex seen until the late stages of the disease might be associated with the maintenance of social skills (in comparison with the observed motor impairment), which has been noted since the first reports by Cockayne himself [1, 19].

On 1H-magnetic resonance spectroscopy, double lactate peaks are found in almost all cases of CS, and the *N*-acetylaspartate to creatine ratio is decreased in the cortex and white matter, which is indicative of a lower number of viable cells, and the choline/creatine ratio of the white matter is also reduced, which is indicative of hypomyelination. On T2WI, a mixture of low (normal myelination) and high (progressive demyelination) signal intensity is seen, which is indicative of tigroid leukodystrophy.

9.6 Management and Therapy

9.6.1 Nutrition

In CS, poor weight gain due to poor feeding is observed from an early stage. Feeding dysfunction appears during the progression of neurological symptoms, primarily due to the presence of pseudobulbar palsy. As the malnutrition worsens, the patient's general condition deteriorates (exacerbating their cachexia), and switching to the use of a nasogastric tube or gastrostomy or the combined use of such techniques and oral feeding is important. More than 50% of patients are being fed through a nasogastric tube or gastrostomy before they reach the age of 5 years [7]. According to Wilson et al. [7], >60% of patients who started being fed via a nasogastric tube transitioned to percutaneous gastrostomy feeding after a mean interval of 7 months. In the abovementioned nationwide survey conducted in Japan, the body weight of the surviving CS type 1 patients was greater than that of the deceased patients [6] (Table 9.1). It is worth noting that most of the surviving patients could be fed orally. It is also important to note that children with moderate to severe CS are prone to weight loss during various infections.

There are no special diets that have been shown to affect the prognosis of CS patients, but in cases involving renal failure, protein, salt, or potassium restriction will be necessary. Video fluorography (VF)-based evaluations of the feeding function of CS patients can be used to determine the most appropriate form of food for each patient. VF is also useful for ensuring safe oral feeding and assessing the necessity of introducing tube feeding. The use of antioxidants (vitamins C and E) against oxidative stress has been attempted, but there is no definitive evidence that it has a beneficial effect.

9.6.2 *Renal Failure*

Although renal failure is not included in the diagnostic criteria for CS, it is considered to be an important symptom that can affect survival. The impact of renal failure on the general condition of CS patients might have been overlooked [6, 8]. In the abovementioned nationwide survey of Japan, it was found that the cause of death of CS patients is often associated with renal failure as a primary cause or as a significant complication (9/13 patients, 69.2%) [6] (Table 9.1). In contrast, Nance et al. [4] reported that renal complications developed in approximately 10% of CS patients and that most of these patients did not require any medical interventions. The reason for the difference in the incidence of renal failure between our patients and Nance et al. [4] is uncertain, but we consider that the variations in the mean age at death (our CS1 patients [$n = 20$], 18.9 years vs. Nance et al. [4] CS1–3 patients [$n = 37$], 12.3 years) is one possible cause. In addition, the genetic backgrounds of CS patients from Japan might differ from those of CS patients from other countries. For example, *ERCC6* (CSB) mutations (~65%) are more prevalent than *ERCC8* (CSA) mutations (~35%) in other countries [8, 9], while the situation is reversed in Japan; i.e., *ERCC8* (CSA) mutations predominate, especially exon 4 deletion of *ERCC8* (CSA), which might represent a founder effect (personal communication from Dr. Ogi T., Nagoya University, 2016). Genetic heterogeneity in the renal pathology of CS might also exist. In our series, serum markers of renal failure, i.e., the serum levels of blood urea nitrogen (BUN) and creatinine (Cr), were significantly higher in the deceased patients ($n = 20$) than the surviving patients ($n = 21$) (Table 9.2). The kidney function of CS patients can be divided into two subgroups. Although some cases of renal failure in CS have been described in the literature, there have not been any detailed reports about when markers of renal failure become apparent.

In CS type 1 (the moderate form) patients in Japan, hyperuricemia appeared earlier and was followed by elevated serum Cr levels and high blood pressure (before the age of 10) [20]. Nakajima et al. [20] emphasized that the early evaluation of renal function is essential in cases of CS involving patients of preschool age. Motojima et al. [21] demonstrated that in CS type 1 patients, the serum Cr level started to gradually increase around the age of 6–8 years, and then renal failure progressed from around the age of 10 years. Serum Cr levels corrected for height (height (m) × 0.3) (mg/dL) are very useful as a reference value for evaluating renal function [21]. Since the serum Cr levels of CS patients tend to be underestimated because of muscle wasting, cystatin C (CysC) measurement is recommended during renal function checkups. CysC levels are less dependent on age, sex, and muscle bulk than Cr. CysC is also considered to be a better marker of the glomerular filtration rate than the serum Cr level [22]. Proteinuria is an important sign of the initial stages of renal dysfunction (Table 9.2), but periodic urinalysis alone has limitations for detecting mild renal dysfunction.

Figure 9.8 shows the kidney pathology of a girl with CS type 1 at 8 years of age, who suffered from severe renal failure (serum BUN, 40.8 mg/dL; serum Cr, 1.68 mg/dL; estimated glomerular filtration rate, 18.4 mL/min), stage 5 chronic kidney disease, anorexia, body weight loss (8.0 kg; SD, −3.7), and hypertension (systolic blood pressure, 140 mmHg) [6]. Global glomerulosclerosis, a tortuous and thickened basement membrane, the disappearance of podocytes, and renal tubular narrowing combined with interstitial fibrosis were detected. These histological features were compatible with previously reported renal lesions in CS patients. After her vital condition had been stabilized, she underwent gastrostomy followed by catheter surgery to enable peritoneal dialysis. Before her renal failure became apparent, her oral food intake reduced significantly. These changes should be a warning sign. In such cases, the introduction of tube feeding should be considered to prevent dehydration and to maintain nutritional status, and clinicians should coordinate with various specialists to ensure appropriate blood pressure management; that the patient retains sufficient iron, calcium, and phosphorus levels, and that peritoneal dialysis is introduced as appropriate. This approach will ensure that pediatric CS patients and their families will be able to spend meaningful time together while the patient retains a good quality of life [21]. In the abovementioned case, the patient continued to exhibit a stable postoperative systemic condition for 3 years, but she

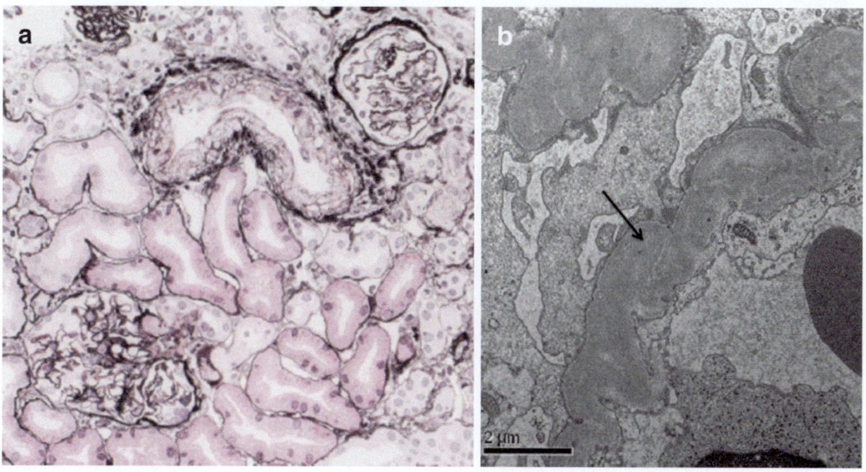

Fig. 9.8 Kidney biopsy findings of CS type 1. (**a**) Sclerotic changes were seen, and narrowed tubules were also observed (periodic acid methenamine silver stain). (**b**) Almost all of the basement membrane (arrow) exhibited thickening, torturous changes, and the disappearance of podocytes

died of heart failure at the age of 11 years. Genetic testing revealed a homozygous exon 4 deletion of the *ERCC8* (CSA) gene. At present, there are no treatments that can be employed to prevent the progression of CS-related renal disorders. In Natale's report [5], the leading cause of death among 42 patients with severe CS and 21 patients with moderate CS was pneumonia/respiratory ailments followed by renal failure. Analyses of the genetic backgrounds and follow-up data of CS patients that do and do not suffer from renal failure should be conducted in the future.

9.6.3 Sleep Disorders and Thermoregulation Problems

Sleep disorders and thermoregulation problems are often observed in CS patients, possibly due to hypothalamic and/or peripheral autonomic dysfunction (Table 9.1). According to a questionnaire survey of the families of CS patients in Japan [23], thermoregulation problems (abnormal changes in body temperature, hypothermia, or high fever in summer) and sleep disorders (abnormal shifts in sleep wakefulness rhythms, frequent arousal and excitement during night sleep, and daytime drowsiness) are frequently observed. These circadian rhythm abnormalities should be taken into consideration as one of the major health problems seen in patients with CS. Concerning thermoregulation problems, the appropriate setting of the environmental temperature and the selection of suitable clothing are important.

Okoshi et al. [23] reported that CS patients exhibit reduced urinary 6-sulphatoxymelatonin (6-SM) (melatonin metabolite) levels. They also detected marked reductions in the numbers of acetylcholine neurons (AchNs) in the basal nucleus of Meynert (NbM) and the pedunculopontine tegmental nucleus (PPN) in four autopsy cases of CS. AchN modulates both arousal and rapid eye movement sleep, and selective lesions of the AchN in the PPN and/or NbM in combination with disturbed melatonin metabolism might be involved in the sleep disorders seen in CS. As CS patients display reduced urinary secretion of 6-SM irrespective of the presence/absence of visual impairments, melatonin or a melatonin receptor agonist (such as ramelteon) might ameliorate their sleep problems.

According to a questionnaire survey of the families of CS patients in Japan, pediatric CS patients like coffee or cafe au lait. The reason for this is unknown, but they might achieve raised daytime alertness by drinking it.

9.6.4 Movement Disorders

Various movement disorders and postural abnormalities are seen in CS patients (Table 9.1). Gait difficulties are ascribed to a mixture of spasticity, ataxia, and peripheral neuropathy due to white matter (pyramidal tract), cerebellar, and peripheral nervous system lesions, respectively, although the degree of such problems varies widely among patients. Foot joint contracture (equinovarus foot) is a feature of CS that is commonly seen in the early stages of the condition. The early assessment of such deformities followed by physiotherapy and the provision of an orthosis (a special shoe or auxiliary equipment) is necessary in such cases, but the long-term efficacy of this approach is limited. Surgical management, including Achilles tendon lengthening or muscle-release surgery, is also occasionally indicated along with ankle foot orthoses. Patients' innate abilities will initially allow them to overcome such problems, but disease progression will result in gait disturbance even in the moderate group (type 1 CS). Environmental adjustments, such as reducing the presence of steps and obstacles, might be necessary to avoid falls and injuries at home and school.

Finger or hand tremors are seen at rest or intentionally induced in more than half of patients (Table 9.1). TRH (thyrotropin-releasing hormone) derivative was somewhat effective against tremors in our mild CS patients (type 3 CS), but the evidence for its use is still anecdotal. Neilan et al. [24] reported the cases of three adolescents with mild CS, all of whom had tremors and motor difficulties that responded to therapy with carbidopa-levodopa, which had a significant beneficial effect on their ability to perform daily activities. They speculated that dopaminergic pathways might be damaged in CS. Although the basal ganglia (predominantly the putamen) are degenerated by the development of calcifications, the administration of carbidopa-levodopa might restore dopaminergic influence on motor pathways.

Hebb et al. [25] reported that deep brain stimulation (DBS) of the bilateral ventral intermediate nuclei (VIM) of the thalamus brought about a marked response in a 17-year-old CS patient (type 1, moderate form) with intractable hyperkinetic movements, such as severe chorea, ballismus, and myoclonus; a bilateral intention tremor; and dystonic features. DBS improved the patient's hyperkinetic movements as well as his quality of life and reduced the burden on his caregivers. Before the DBS, the ballistic movement necessitated strapping the patient's arms and trunk to his wheelchair to avoid injury. Interestingly, the improvement in the hyperkinetic movement persisted after the discovery of a non-functioning pulse generator at 4 postoperative years. The authors speculated that some neuroplastic changes might have contributed to this process.

9.6.5 *Tooth Decay*

As tooth decay develops in >70% of CS patients, it must be followed up from an early stage (within the first few years of life) [4]. The teeth of CS patients are of normal size and, hence, are too large for their jaws. Nance et al. [4] demonstrated the diagnostic significance of early tooth decay. The tooth decay seen in CS is considered to be due to saliva hyposecretion, poor mandibular movement, and autonomic dysfunction associated with tooth blood flow. It is necessary to check the oral hygiene and feeding function of CS patients.

Hypohydrosis, saliva hyposecretion, and mild hypoalgesia are observed in CS patients, and neurogenic bladder develops in rare cases. These symptoms are

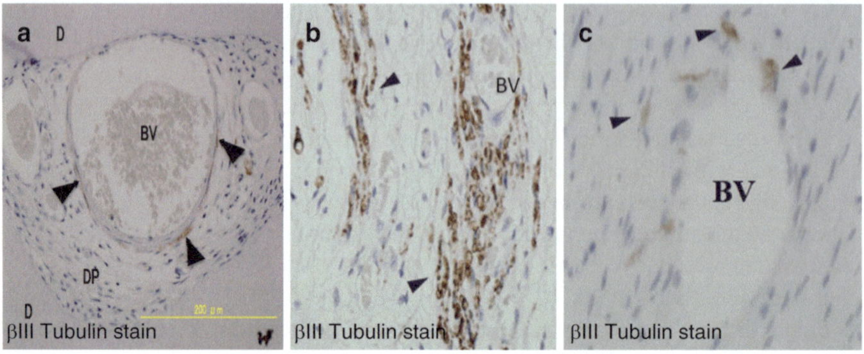

Fig. 9.9 Pathology of the dental pulp CS type 3 at the age of 25, (**b**) a normal control, and (**c**) a patient with congenital insensitivity to pain with anhydrosis (CIPA) Compared with the normal control (**b**), autonomic nerves (arrowheads) were distributed sparsely around the blood vessels in the CS (**a**) and CIPA (**c**) patients. (Prepared by Professor Sato T, Tsurumi University) *BV* blood vessel, *DB* dental pulp, *D* dentine

explained by autonomic nervous system dysfunction, mainly due to reductions in the numbers of C and Aδ fibers. Figure 9.9 shows the neuropathology of dental pulp tissue obtained from a CS patient during a tooth extraction procedure. Compared with the control, few autonomic nerves were noted in the CS tooth. Interestingly, similar findings are seen in patients with congenital insensitivity to pain with anhydrosis (CIPA, hereditary sensory, and autonomic neuropathy type IV). CIPA is characterized by recurrent episodic fevers, anhidrosis (an inability to sweat), the absence of a reaction to noxious (or painful) stimuli, self-mutilating behavior, and an intellectual disorder caused by mutations in the *TRKA* gene that results in a lack of C and Aδ fibers. Abnormalities of the dental pulp nerve might reflect the systemic autonomic condition of CS patients.

9.6.6 Non-autistic Tendency

Cockayne described in his first report [1] that CS patients are friendly and playful, invariably good tempered, and laugh with obvious enjoyment at the slightest provocation. Generally, even late in life, CS patients exhibit good sociability. The reason why CS patients maintain a stable rapport with others is uncertain. The relative lack of cortical lesions in the frontocentral cortices might account for this tendency. The absence of definite cortical pathological lesions, such as the cortical tubers seen in tuberous sclerosis, which is frequently associated with autism, might be one of the reasons why CS does not cause autism.

9.6.7 Epilepsy

Nance et al. [4] reported that the incidence of epilepsy ranged from 5 to 10% in 140 CS patients. Our nationwide survey disclosed that 18.2% (8/21) of CS patients experienced seizures (Table 9.1). Compared with the general population, CS patients suffer from epileptic disorders more often, but they are not intractable and are well controlled.

9.6.8 Guidelines for CS

Currently, only symptomatic and palliative treatments are administered for the various symptoms of CS. We produced care guidelines for the daily management of CS based on how individual problems affect patients' daily lives, and these are available on our homepage (Cockayne Research and Care; http://www.cockayneresearchcare.jp/). These guidelines include information regarding how to deal with skin symptoms, decayed teeth, ophthalmological problems, movement disorders, renal failure, hypertension, epilepsy, nutrition, and sleep disorders. At present, the guidelines are only available in Japanese, but an English version is being prepared.

References

1. Cockayne EA. Dwarfism with retinal atrophy and deafness. Arch Dis Child. 1936;11:1–8.
2. Cockayne EA. Dwarfism with retinal atrophy and deafness. Arch Dis Child. 1946;21:52–4.
3. Sugarman GI, Landing BH, Reed WB. Cockayne syndrome: clinical study of two patients and neuropathologic findings in one. Clin Pediatr. 1977;16:225–32.
4. Nance MA, Berry SA. Cockayne syndrome: review of 140 cases. Am J Med Genet. 1992;42:68–84.
5. Natale V. A comprehensive description of the severity groups in Cockayne syndrome. Am J Med Genet A. 2011;155A:1081–95.
6. Kubota M, Ohta S, Ando A, Koyama A, Terashima H, Kashii H, Hoshino H, Sugita K, Hayashi M. Nationwide survey of Cockayne syndrome in Japan: incidence, clinical course and prognosis. Pediatr Int. 2015;57:339–47.
7. Wilson BT, Stark Z, Sutton RE, Danda S, Ekbote AV, Elsayed SM, Gibson L, Goodship JA, Jackson AP, Keng WT, King MD, McCann E, Motojima T, Murray JE, Omata T, Pilz D, Pope K, Sugita K, White SM, Wilson IJ. The Cockayne syndrome natural history (CoSyNH) study: clinical findings in 102 individuals and recommendations for care. Genet Med. 2016;18:483–93.
8. Laugel V. Cockayne syndrome: the expanding clinical and mutational spectrum. Mech Ageing Dev. 2013;134:161–70.

9. Rapin I, Weidenheim K, Lindenbaum Y, Rosenbaum P, Merchant SN, Krishna S, Dickson DW. Cockayne syndrome in adults: review with clinical and pathologic study of a new case. J Child Neurol. 2006;21:991–1006.
10. Hayashi M, Miwa-Saito N, Tanuma N, Kubota M. Brain vascular changes in Cockayne syndrome. Neuropathology. 2012;32:113–7.
11. Scheibye-Knudsen M, Ramamoorthy M, Sykora P, Maynard S, Lin PC, Minor RK, Wilson DM 3rd, Cooper M, Spencer R, de Cabo R, Croteau DL, Bohr VA. Cockayne syndrome group B protein prevents the accumulation of damaged mitochondria by promoting mitochondrial autophagy. J Exp Med. 2012;209:855–69.
12. Kleijer WJ, et al. Incidence of DNA repair deficiency disorders in western Europe: Xeroderma pigmentosum, Cockayne syndrome and trichothiodystrophy. DNA Repair (Amst). 2008;7:744–50.
13. Morris DP, Alian W, Maessen H, Creaser C, Demmons-O'Brien S, Van Wijhe R, Bance M. Cochlear implantation in Cockayne syndrome: our experience of two cases with different outcomes. Laryngoscope. 2007;117:939–43.
14. Cook S. Cockayne's syndrome. Another cause of difficult intubation. Anaesthesia. 1982;37:1104–7.
15. Meira LB, Graham JM Jr, Greenberg CR, Busch DB, Doughty AT, Ziffer DW, Coleman DM, Savre-Train I, Friedberg EC. Manitoba aboriginal kindred with original cerebro-oculo- facio-skeletal syndrome has a mutation in the Cockayne syndrome group B (CSB) gene. Am J Hum Genet. 2000;66:1221–8.
16. Nakazawa Y, Sasaki K, Mitsutake N, Matsuse M, Shimada M, Nardo T, Takahashi Y, Ohyama K, Ito K, Mishima H, Nomura M, Kinoshita A, Ono S, Takenaka K, Masuyama R, Kudo T, Slor H, Utani A, Tateishi S, Yamashita S, Stefanini M, Lehmann AR, Yoshiura K, Ogi T. Mutations in UVSSA cause UV-sensitive syndrome and impair RNA polymerase IIo processing in transcription-coupled nucleotide-excision repair. Nat Genet. 2012;44:586–92.
17. Iwasaki S, Kaga K. Chronological changes of auditory brainstem responses in Cockayne's syndrome. Int J Pediatr Otorhinolaryngol. 1994;30:211–21.
18. Scaioli V, D'Arrigo S, Pantaleoni C. Unusual neurophysiological features in Cockayne's syndrome: a report of two cases as a contribution to diagnosis and classification. Brain and Development. 2004;26(4):273–80.
19. Koob M, Laugel V, Durand M, Fothergill H, Dalloz C, Sauvanaud F, Dollfus H, Namer IJ, Dietemann JL. Neuroimaging in Cockayne syndrome. AJNR Am J Neuroradiol. 2010;31:1623–30.
20. Nakajima E, Hattori A, Ito T, Kobayashi S, Nakajima Y, Ando N, Sugita K, Saitoh S. Onset of complications and management of Cockayne syndrome type 1. J Jpn Pediatr Soc. 2013;117:760–5. (in Japanese).
21. Motojima T, Sugita K, Omata T, Fujii K. Clinical examination of renal function in Cockayne syndrome. No To Hattatsu. 2014;46:311–4. (in Japanese).
22. Zappitelli M, Parvex P, Joseph L, Paradis G, Grey V, Lau S, Bell L. Derivation and validation of cystatin C-based prediction equations for GFR in children. Am J Kidney Dis. 2006;48(2):221–30.
23. Okoshi Y, Tanuma N, Miyata R, Hayashi M. Melatonin alterations and brain acetylcholine lesions in sleep disorders in Cockayne syndrome. Brain Dev. 2014;36:907–13.
24. Neilan EG, et al. Response of motor complications in Cockayne syndrome to carbidopa-levodopa. Arch Neurol. 2008;65:1117–21.
25. Hebb MO, Gaudet P, Mendez I. Deep brain stimulation to treat hyperkinetic symptoms of Cockayne syndrome. Mov Disord. 2006;21:112–5.

Chapter 10
Trichothiodystrophy

Donata Orioli and Miria Stefanini

Abstract Trichothiodystrophy (TTD) is a rare autosomal recessive multisystem disorder characterized by hair abnormalities and a wide spectrum of clinical manifestations including physical and mental retardation, ichthyosis, proneness to infections, signs of premature ageing and, in about half of the reported cases, cutaneous photosensitivity. Both the photosensitive and the non-photosensitive form of TTD show similar clinical outcome and include patients who differ in type and severity of symptoms. Here we discuss the cellular and genetic defects associated with TTD, the functions of the disease genes identified so far and the genotype-phenotype relationships. The three genes responsible for the photosensitive form of TTD encode distinct subunits of the general transcription factor TFIIH, which plays a key role also in DNA repair. Subtle defects in transcription can easily explain the spectrum of TTD clinical symptoms except for clinical and cellular photosensitivity that, however, in TTD does not result in increased carcinogenesis. The recent finding that alterations in the β-subunit of the basal transcription factor TFIIE result in the non-photosensitive form of TTD highlights the relevance of transcriptional alterations for the TTD pathological phenotype. Besides reporting recent research advances, we discuss how alterations in distinct pathways may result in specific TTD clinical manifestations, namely, cutaneous photosensitivity, lack of skin cancer, ageing signs and neurological alterations.

10.1 Introduction

Trichothiodystrophy (TTD) was first introduced as a distinct clinical entity in 1980 [1] to describe a group of autosomal recessive neuroectodermal disorders whose defining feature is sulphur-deficient brittle hair. The designation derives from Greek (*tricho*, hair; *thio*, sulphur; *dys*, faulty; *trophe*, nourishment) and accounts for the reduced cysteine content of the hair that in TTD is less than half of the normal

D. Orioli · M. Stefanini (✉)
Istituto di Genetica Molecolare CNR, Pavia, Italy
e-mail: orioli@igm.cnr.it; stefanini@igm.cnr.it

C. Nishigori, K. Sugasawa (eds.), *DNA Repair Disorders*,
https://doi.org/10.1007/978-981-10-6722-8_10

amount. Besides the brittle and fragile hair, TTD patients experience a wide range of symptoms with variable degrees of severity, including physical and mental retardation, small stature, ichthyotic skin, nail dysplasia, decreased fertility, unusual facial features, cataracts, proneness to infections, signs of premature ageing and dental caries. Skin photosensitivity is reported in about 50% of patients and is associated with an altered cellular response to ultraviolet (UV) light caused by a defect in nucleotide excision repair (NER), the DNA repair pathway that removes a wide spectrum of DNA lesions, including UV-induced damage. Even though the accumulation of unrepaired DNA lesions might account for the ageing features in TTD, developmental delay and mental retardation cannot be ascribed to defective NER. The finding that the photosensitive form of TTD is caused by mutations in *ERCC2/XPD*, *ERCC3/XPB* or *GTF2H5/TTDA* genes, which encode distinct subunits of the general transcription factor IIH (TFIIH), provides a rationale to explain the pathological phenotype. TFIIH is a multi-protein complex that plays crucial roles in both NER and transcription. Thus, subtle transcriptional alterations also contribute to TTD clinical features. Evidence to this claim have been recently provided by research studies on the non-photosensitive form of TTD, which is also genetically heterogeneous. Mutations in the *MPLKIP* gene are found in several of the analysed cases, whereas mutations in *RNF113A* and *GTF2E2* genes have been so far reported in two and five patients, respectively. The observation that *GTF2E2* encodes the β-subunit of the general transcription factor IIE (TFIIE) unequivocally demonstrates the strong contribution of transcriptional impairments to TTD etiopathogenesis.

10.2 Clinical Features

All TTD patients exhibit sparse, dry and easily broken hair characterized by low sulphur and cysteine content (10–50% of normal) and alternating light and dark bands, called tiger tail pattern, under polarized microscopy (Fig. 10.1). These alterations, which affect hair from the scalp, eyebrows and eyelashes, are considered the key factors in the recognition of TTD and provide a diagnostic test for the disorder. Hair anomalies are associated with a wide spectrum of abnormalities affecting organs of ectodermal and neuroectodermal origin (such as the skin, hair follicles, nail, tooth enamel and brain) with varying degrees of severity, ranging from patients with severe neuroectodermal symptoms to (rare) minimally affected cases [2–4]. A relevant cysteine reduction is observed also in the nail clippings of the patients. Nails are short and broadened and may show longitudinal ridging as well as horizontal splits (nail dysplasia). Other clinical manifestations may affect the skin of TTD patients [4]. Frequently, a smooth, shiny membrane encases the affected newborns at birth (collodion baby), and ichthyosiform erythroderma is present during the first weeks of life. Notably, about 50% of patients exhibit cutaneous photosensitivity but no precancerous and cancerous skin lesions.

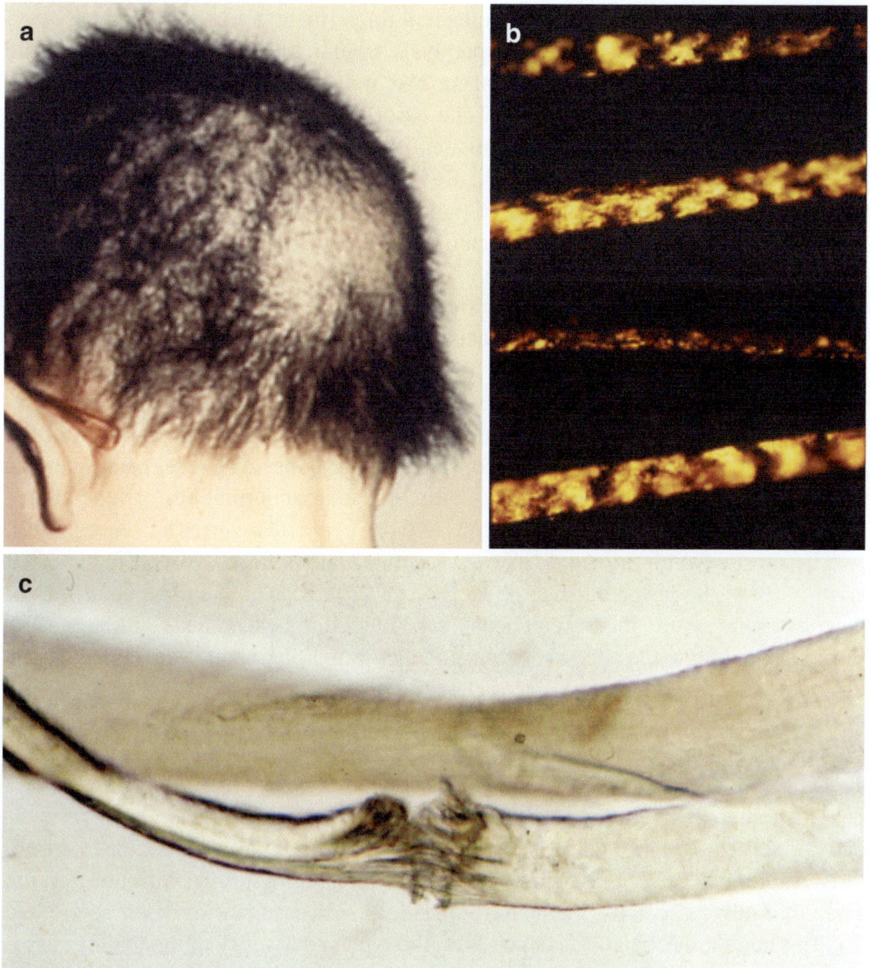

Fig. 10.1 Hair abnormalities in TTD patients. (**a**) Scalp hair is short, sparse and broken. (**b**) Hair shafts show alternating light and dark bands that confer a tiger-tail pattern under polarized microscopy. (**c**) A sharp transverse fracture characteristic of trichoschisis under light microscopy (from [4])

Patients are frequently born prematurely with a low birth weight for gestational age that was attributed to placental insufficiency. Indeed, mothers of TTD patients report gestational complications only when the pregnancy resulted in an affected child and not in their unaffected kids. They experienced pre-eclampsia (pregnancy-induced high blood pressure), HELLP (haemolysis, elevated liver enzymes and low platelet count) syndrome and/or reduced foetal movements [5–8].

Many TTD patients exhibit neurological and developmental defects of varying clinical severity with no clear evidence of progressive degeneration. Mental retardation, speech delays, reduced learning ability and impaired motor control (ataxia) are frequently described. Despite the intellectual impairment, a social and friendly personality is often reported. Spasticity, paralysis, tremor, hypotonia, decreased muscle tone, seizures and sensorineural hearing loss may also be observed. In addition, some patients present calcification of the basal ganglia, and some additional cases reveal hypomyelination of the white matter of the cerebrum likely caused by a delay in development (dysmyelination) [9]. Developmental defects also include microcephaly, congenital cataracts, delayed puberty and short stature. Delayed menarche and hypoplasia of female genitalia can be observed in affected females, whereas undescended testes and bilateral or unilateral cryptorchidism have been reported in male patients. Post-pubertal patients frequently present a delayed development of secondary sexual characters and reduced fertility. Skeletal deformations have been described in many cases and include genu valgum, coxa valga, pes valgus, cubital and tibial valgus deformity, ulnar deviation of the fingers, zygodactyly, clinodactyly, scoliosis, thoracic kyphosis, lumbosacral lordosis and metacarpal bones of the thumb reduced in size. Marked osteopenia in long bones and axial osteosclerosis have been also described (Fig. 10.2). Common ocular abnormalities include microcornea, nystagmus, cataracts, strabismus and dry eye/ocular surface disease [3, 10].

The most severely affected cases lack subcutaneous fatty tissue, a feature that contributes to the progeroid-like phenotype in TTD. Indeed, the absence of subcutaneous tissue results in sunken cheeks, receding chin, beaked nose and large ears, thus contributing to the aged appearance. Other progeroid features are dental caries, osteoporosis, hearing loss and cataracts.

The haematological features of β-thalassemia, including reduced mean red cell volume and mean corpuscular haemoglobin, increased haemoglobin A2 levels [3, 11, 12] and reduced synthesis of β-globin [13], have been reported in several patients, whereas neutropenia, anaemia and hypereosinophilia have been sporadically observed. Notably, TTD is also associated with recurrent infections, particularly respiratory infections, which can be life-threatening thus contributing to the early mortality: affected patients have a 20-fold increased risk of death before age 10 [3]. Though, the prognosis depends on the type and severity of the features associated with the hair alterations.

In the past, the acronyms BIDS (brittle hair, impaired intelligence, decreased fertility, short stature), IBIDS (BIDS with ichthyosis), PIBIDS (IBIDS with photosensitivity) and SIBIDS (IBIDS with osteosclerosis) have been used to describe subgroups of TTD patients. These acronyms, as well as the eponyms sulphur-deficient brittle hair syndrome, Tay's syndrome, Pollitt syndrome, Amish hair-brain syndrome and Sabinas syndrome (reviewed in [4]), do not adequately describe either the spectrum of clinical involvement in most patients or clinically relevant symptoms including the eye abnormalities and the recurrent infections reported in many cases. Nowadays, TTD patients are classified in two major groups depending on the presence or absence of cutaneous photosensitivity.

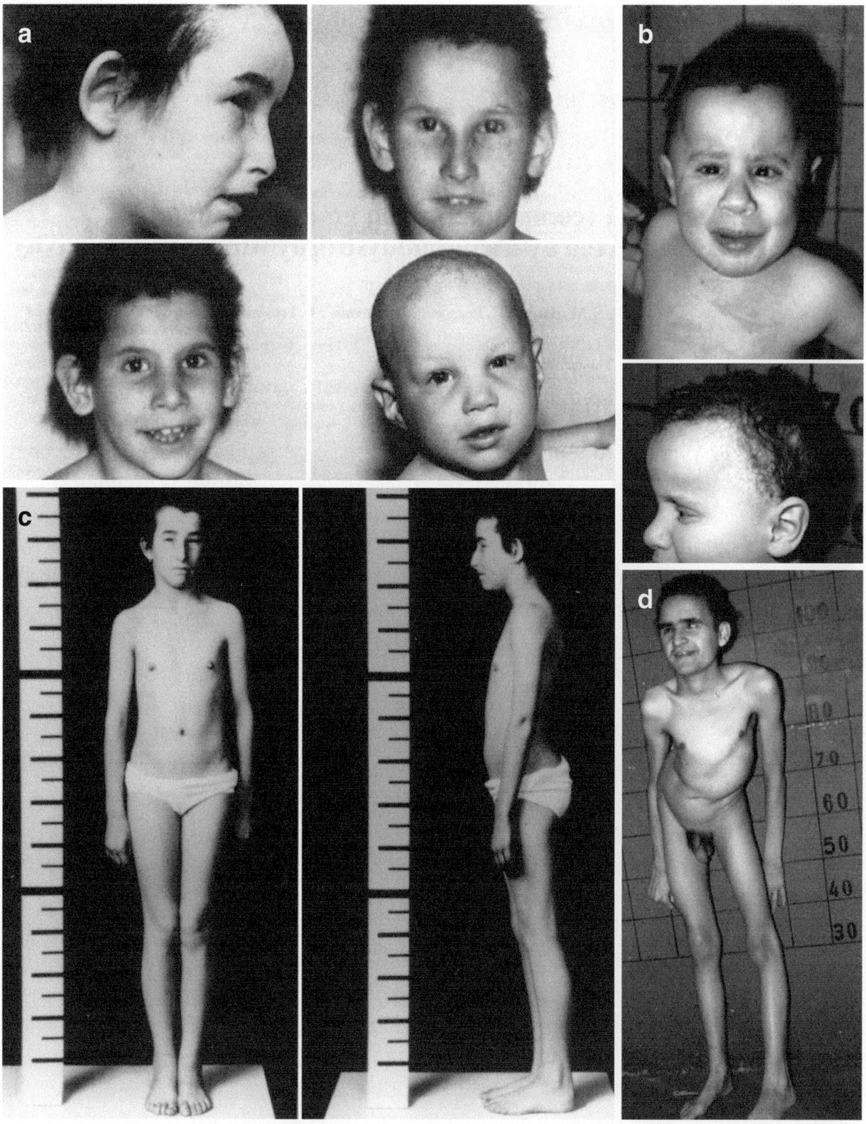

Fig. 10.2 TTD affected children. (**a**) The four Italian TTD patients firstly described as NER-defective and classified into the XP-D group [29]. (**b**) The face of a male patient. (**c**) A female photosensitive TTD patient aged 18 years. (**d**) A male TTD patient with complex skeletal deformities including severe scoliosis (from [4])

10.2.1 Photosensitive and Non-photosensitive Forms of TTD

In addition to hair abnormalities, both the photosensitive and the non-photosensitive forms of TTD show similar clinical outcome and include patients who differ in type and severity of symptoms. The increasing number of patients characterized at the cellular, genetic and molecular level has opened the possibility to investigate the correlation between clinical and genetic features.

The majority of photosensitive TTD patients are mutated in the *ERCC2/XPD* gene and display different degrees of clinical severity, ranging from moderately affected cases (showing mental and physical retardation, proneness to infections, reduced motor coordination and survival beyond early childhood) to severely affected cases (displaying very poor mental and motor performances, speech impairment, marked proneness to infections, failure to thrive and death during early childhood) [14].

The only two patients mutated in the *ERCC3/XPB* gene are two French siblings from first-cousin marriage. They were born at term with a similar presentation as a collodion baby and exhibited typical hair abnormalities associated with mild ichthyosis and mild cutaneous photosensitivity but no physical and mental impairment [15]. Finally, the six patients (from five families) mutated in the *GTF2H5* gene exhibit similar clinical symptoms of moderate severity [16–18]. They are all mildly affected by mental retardation and developmental delay. Cutaneous photosensitivity is present in all except one patient who developed a moderately differentiated squamous cell carcinoma (SCC) at the age of 42 years [18]. Ichthyosis, cataracts and collodion baby appearance at birth are reported in the affected member(s) of at least two families.

Among the non-photosensitive TTD cases, those mutated in *MPLKIP* gene represent the largest group. They display the characteristic features of TTD, including recurrent infections, intellectual impairment, delayed physical development, ichthyosis, nail dysplasia, hypotonia and osteopenia, which are all reported with varying severity [19, 20]. A recent study focused on five TTD patients with mutation in *MPLKIP* revealed that delayed bone age and seizure disorders are overrepresented, whereas pregnancy complications, low birth weight, collodion membrane, reduced height and weight, cataracts, β-thalassemia-like changes and brain dysmyelination are underrepresented or absent in this group of patients. In addition, three of the patients display autistic behaviours in contrast to the characteristic friendly, socially interactive personality of most TTD patients [21].

The two male cousins with non-photosensitive TTD due to mutations in the *RNF113A* gene are both affected by a severe phenotype. They had intrauterine growth restriction, progressive microcephaly with profound intellectual disability, aged appearance, short stature, facial dysmorphism, seizures, immunoglobulin deficiency, multiple endocrine abnormalities, genital anomalies, cerebellar hypoplasia and partial absence of the corpus callosum. The extended phenotype within this family includes panhypopituitarism, cutis marmorata and congenital short oesophagus [22]. In contrast, the first two reported unrelated patients with mutations in the *GTF2E2* gene revealed a moderate phenotype. In addition to short brittle hair with tiger tail pattern, both children present ichthyosis, short stature, microcephaly and developmental delay but no collodion membrane, recurrent infections, congenital cataracts, osteoporosis or dental caries [12]. Similar clinical features have been found in three additional patients from two families of Moroccan origin [23].

10.2.2 Incidence and Inheritance

TTD is a rare inherited genetic disorder with approximately 200 patients reported in the literature. Affected individuals have been described worldwide, and photosensitivity is present in about half of published cases. The incidence of the photosensitive form of TTD has been established at 1.2 per million livebirths in Western Europe and at 1.1 per million livebirths in the autochthonic western European population [24]. In the United States, the incidence of TTD is thought to be about one in one million newborns, whereas in Asian countries, the incidence is too low to obtain an accurate assessment [17]. The reviewing of the published cases indicates that males and females are similarly affected according to the autosomal recessive inheritance pattern inferred by the family pedigree of photosensitive and non-photosensitive TTD patients characterised so far at the cellular and molecular level. Notably, an X-linked form of non-photosensitive TTD has been recently reported in two Australian cousins mutated in the *RNF113A* gene [22].

10.3 Cellular Defects

The clinical photosensitivity reported in about half of TTD patients is paralleled by cellular alterations due to defects in NER, the only mechanism by which human cells remove the major DNA lesions induced by UV light. In addition, NER acts on a wide range of bulky DNA adducts and helix-distorting lesions caused by chemical mutagens and on some free radical-induced endogenous lesions. By the coordinated action of more than 30 gene products, NER operates through two subpathways, namely, global genome NER (GG-NER) and transcription-coupled NER (TC-NER), which differ in the DNA damage recognition step (reviewed in [25]). In GG-NER, helix-distorting lesions are detected throughout the genome by either the heterotrimeric XPC/RAD23B/Centrin2 complex or the UV-damaged DNA-binding protein complex that binds UV-induced lesions and facilitates XPC loading. In TC-NER, a lesion on the transcribed strand of active genes is recognized by the stalling of elongating RNA polymerase (RNApol) II [26]. After damage recognition, both GG-NER and TC-NER converge into a common pathway relying on the same repair factors to remove the lesion. The TFIIH complex is recruited to damaged sites to verify the lesion and locally unwind the DNA double helix [27]. The open complex formation is stabilized by XPA and RPA, and the damaged strand is incised on either side of the lesion by the endonucleases XPG and ERCC1-XPF, generating a lesion-containing oligomer of 24–32 nucleotides. Its replacement by DNA repair synthesis is followed by ligation of the repair patch to the contiguous parental DNA strand.

Specific functional assays on *in vitro* cultured fibroblasts are available to evaluate the cellular response to UV light in terms of UV-induced DNA repair synthesis (unscheduled DNA synthesis, UDS), recovery of RNA synthesis (RRS) at late times after UV irradiation and sensitivity to the lethal effects of UV. The most commonly used test relies on UDS analysis, which is performed by measuring the incorporation of labelled nucleotides into the genomic DNA of UV-irradiated cell samples either by autoradiography or liquid scintillation counting or using a fluorescence assay ([28] and references therein). Reduced levels of UDS confirm the diagnosis of photosensitive form of TTD, which is also characterized by reduced RRS and survival after UV irradiation. In addition, all NER-defective TTD patients typically show reduced cellular amount of the TFIIH complex (reviewed in [14]).

Further tests can be applied to identify the defective gene. The classical complementation assay for NER alterations is based on UDS analysis in heterodikaryons obtained following fusion of fibroblasts of the patient under study with cells representative of each of the various NER-defective groups [29, 30]. The restoration of normal UDS levels in heterodikaryons indicates that the patients are defective in different genes, whereas the maintenance of impaired UDS levels results from defects in the same gene and allows the classification of patients into the same complementation group (*e.g.* see [28, 31]). In parallel to the classical complementation test, genetic analysis can be carried out also by analysing UDS or RRS levels in patient cells after microinjection with wild-type cloned NER genes or after infection with recombinant lentiviruses expressing one of the cloned NER genes [32]. An alternative approach makes use of a host-cell reactivation assay by co-transfecting patient cells with a UV-treated plasmid and a plasmid expressing a cloned NER gene [33].

In at-risk pregnancies, the DNA repair tests can be performed on amniocytes and chorionic villus-derived cells, whereas molecular analysis can be carried out if the inactivating mutations have been previously identified in an affected family member.

Notably, non-photosensitive TTD patients show a normal cellular response to UV light in terms of UDS, RRS and survival levels, thus excluding NER deficiencies. No cellular/biochemical markers have been so far identified for the confirmation of the diagnosis in these cases. Hypersensitivity to the alkylating agent methyl methanesulphonate (MMS) has been found in cells from the two cousins mutated in the *RNF113A* gene [34]. Evidence for the use of MMS sensitivity as a routine test is still lacking. Thus, in NER-proficient TTD patients, only the sequencing of the genes responsible for the non-photosensitive form of TTD may be informative.

10.4 Genes Mutated in TTD: Properties and Functional Activities

Both forms of TTD are genetically heterogeneous (Table 10.1). All the three genes mutated in the photosensitive form encode components of the ten subunit complex TFIIH, namely, XPB, XPD and TTDA, also called p8 (reviewed in [35, 36]). TFIIH

Table 10.1 Photosensitive and non-photosensitive forms of trichothiodystrophy: causative genes and inheritance pattern

Phenotype	OMIM number	Mutated gene (synonyms)	OMIM number	Location	Inheritance
Photosensitive TTD	601675	*ERCC2 (XPD)*	126340	19q13.32	Autosomal recessive
	616390	*ERCC3 (XPB)*	133510	2q14.3	Autosomal recessive
	616395	*GTF2H5 (TTDA)*	608780	6q25.3	Autosomal recessive
Non-photosensitive TTD	234050	*MPLKIP (TTDN1)*	609188	7p14.1	Autosomal recessive
	616943	*GTF2E2*	189964	8p12	Autosomal recessive
	300953	*RNF113A*	300951	Xq24	X-linked

is organized in two main functional subcomplexes: the CORE, containing the six subunits XPB, p62, p52, p44, p34 and p8, and the cyclin-dependent kinase (CDK)-activating kinase (CAK) made of the CDK7, cyclin H and MAT1 subunits. The two subcomplexes are bridged by the XPD subunit that interacts with p44 and MAT1 (Fig. 10.3). A three-dimensional model derived from electron microscopy indicates that the human TFIIH complex is organized into a ring-like structure (CORE), with a hole suitably sized to accommodate a DNA molecule, and a roughly spherical bulge (CAK) appended to the ring-like structure. The subunits p34 and p44 are central for the CORE structural integrity [37].

TFIIH was initially identified as one out of the six auxiliary proteins, collectively known as general transcription factors (GTFs), essential for promoter recognition and transcription initiation by RNApol II. In the transcription reaction, RNApol II and GTFs converge on promoters in a highly ordered manner to form the preinitiation complex (PIC). PIC assembly is followed by promoter opening, first phosphodiester bond formation, promoter clearance, transcript elongation and termination. Later, TFIIH was shown to play a key role in NER and RNApol II transcription regulation and to participate in transcription of ribosomal RNA (rRNA) by RNApol I ([36] and references therein). The composition of TFIIH is dynamic to accomplish its distinct cellular functions. The entire TFIIH complex (holo-TFIIH) is engaged in transcription and recruited to DNA-damaged sites. Then, NER is driven by the dissociation of CAK from TFIIH. Following damage removal, the holo-TFIIH reappears on the chromatin, together with the resumption of transcription formerly inhibited by UV. In addition, CAK alone is involved in cell cycle regulation, during which it influences the activity of other CDKs, whereas CAK associated with XPD is required for the progression of the mitotic divisions during the late nuclear division steps. XPD also belongs to MMXD and CGX complexes, which are both implicated in chromosome segregation [38, 39].

Consistent with its key roles in fundamental cellular processes, TFIIH is an evolutionarily well-conserved complex. All TFIIH subunits are found in various eukaryotic species. XPB and XPD subunits have been also identified in different archaea and have been used as models for structural investigations.

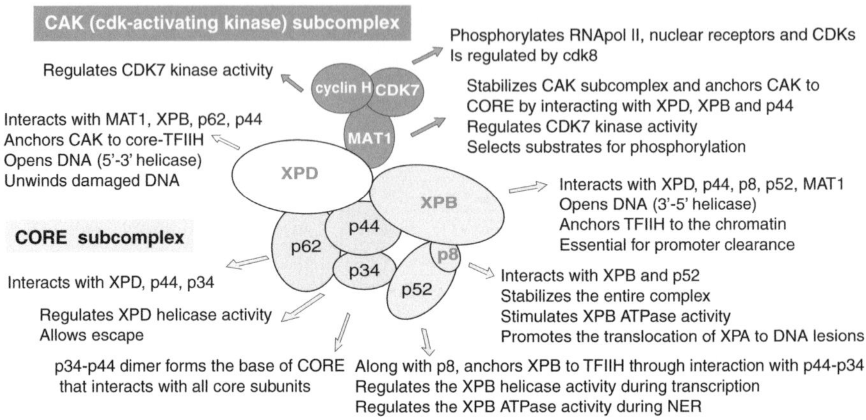

Fig. 10.3 Human TFIIH: organization and functions of its subunits. The six subunits composing the CORE TFIIH subcomplex are in light grey, the three subunits of the CDK-activating kinase (CAK) subcomplex are in grey and the XPD subunit that bridges the two TFIIH subcomplexes is in white. The designation of the subunits altered in the photosensitive form of TTD is in green. Organization adapted from [36, 41, 91]

10.4.1 Role of XPB, XPD and p8/TTDA Within the TFIIH Complex

TFIIH possesses several enzymatic activities: the protein kinase activity of the cyclin-dependent kinase CDK7, a master player in transcription and cell cycle regulation, and the two ATPase/DNA helicase activities with opposite polarity of XPB and XPD, which participate in DNA unwinding both in NER and basal transcription. Recent studies in yeast suggest that XPB/Ssl2 promotes DNA opening independently of its helicase activity by acting as a double-stranded DNA translocase [40]. The TFIIH enzymatic functions are modulated by protein-protein contacts within the complex [41] as well as by interactions with distinct components of the transcriptional machinery, including regulatory transcription factors.

During RNApol II transcription, TFIIH plays a key role in promoter opening through the helicase activity of XPB and the ATPase activity of XPD and in promoter escape through the CDK7 kinase activity that phosphorylates the carboxy-terminal domain (CTD) of the largest subunit of RNApol II. In addition, CDK7 phosphorylates several transcriptional factors, including the general transcription factor TFIIE, nuclear receptors (NRs) containing a functional A/B domain and transcriptional activators, such as Ets1 that when phosphorylated by CDK7 promotes the binding of liganded vitamin D receptor to its responsive promoter elements and triggers the subsequent recruitment of the transcription machinery. Phosphorylation of NRs by CDK7 may influence their partnerships with other proteins, thereby modulating NRs function. TFIIH also acts as a coactivator to stabilize NRs to their DNA-responsive elements in a phosphorylation-independent manner. Furthermore, it may interact with cofactors and influence their activities [36, 42]. TFIIH contributes also to RNApol I transcription elongation through the ATPase activity of XPB following TFIIH recruitment on ribosomal DNA in an active transcription-dependent manner [43].

The main role of TFIIH in NER is to open the double-stranded DNA around the lesion through the helicase activity of XPD and the ATPase activity of XPB, which contributes to anchoring TFIIH to the chromatin. The XPD helicase is regulated by the TFIIH subunit p44, whereas the XPB ATPase is regulated by the p52 subunit and the damage recognition factor XPC. Once damaged DNA has been unwound by TFIIH, the CORE subcomplex associates with NER-specific factors, including XPA that catalyses the detachment of the CAK from the CORE thereby triggering the incision/excision of the damaged oligonucleotide and its replacement by a new DNA fragment.

Because of its small size, the p8/TTDA subunit is the latest identified component of TFIIH (reviewed in [44]). Through a direct interaction with the C-terminal (C-ter) end of p52, p8/TTDA acts as a stabilizer of the entire complex and is essential to sustain normal cellular TFIIH concentration. p8/TTDA plays a relevant role in NER by further strengthening TFIIH anchoring to DNA after a genotoxic attack, by contributing to the regulation of the XPB ATPase activity in a DNA-dependent manner and by promoting the translocation of XPA to UV-damaged sites. Remarkably, p8/TTDA is the only TFIIH subunit for which a complete absence is compatible with life, implying that NER requires higher TFIIH levels than transcription, which in turn may be affected only when the reduced amount of TFIIH is unable to sustain an elevated transcriptional demand.

10.4.2 Role of the Genes Mutated in Non-photosensitive TTD Patients

The first disease gene responsible for the non-photosensitive form of TTD was initially identified as the uncharacterised chromosome 7 open reading frame 11 (*C7orf11*), and the disease locus was designated *TTDN1* (TTD non-photosensitive 1) [19]. Thereafter, it was termed *MPLKIP* (M-phase-specific PLK1-interacting protein) because during mitosis, the corresponding protein was shown to interact with polo-like kinase 1 (PLK1) and to be phosphorylated by CDK1 [45]. MPLKIP contains a glycine/proline-rich region but no putative conserved domains; it localizes to the nucleus and is expressed in foetal hair follicles. Its overexpression in HeLa cells causes nuclear fragmentation, whereas its knockdown results in multiple nuclei or multiple-polar mitotic spindles, suggesting that TTDN1 might be involved in mitosis regulation and cytokinesis [45]. Loss of *TTDN1* interferes with the recruitment of the activating signal cointegrator complex (ASCC) involved in the response to alkylation damage [34]. Defects in *MPLKIP* account for about 20% of the non-photosensitive TTD cases.

In 2015, a novel X-linked form of non-photosensitive TTD was identified in two cousins with a nonsense mutation in the *RNF113A* gene [22]. This intronless gene encodes a protein that contains a C3H1-type zinc finger domain and a C3HC4-type Ring (Really Interesting New Gene-type) finger domain. The former is a motif known to bind RNA, whereas the latter is probably involved in mediating protein-protein interactions. Strong evidence has been provided that RNF113A is the E3 ligase responsible for upstream ubiquitin signalling in the ASCC pathway induced specifically to repair alkylation damage [34].

Recently, two unrelated children with TTD clinical features have been reported to carry different homozygous missense mutations in *GTF2E2*, the gene encoding the β-subunit of the general transcription factor TFIIE [12]. TFIIE consists of two subunits, α and β, both of which contain several structural motifs that are essential for DNA binding and/or protein-protein interactions [46]. TFIIE is essential for the assembly and stabilization of the PIC at the promoter transcription start site. It enters the PIC after RNApol II, recruits TFIIH to the PIC, stimulates RNApol II CTD phosphorylation by CDK7 and regulates the helicase activity of XPB to catalyse the open complex formation, leading to promoter clearance. The three genes mutated in the non-photosensitive form account for a small proportion of patients, indicating that still unidentified, additional genes play a role in pathways whose alteration is responsible for pathological features detected in TTD patients.

10.5 Mutation Pattern and Genotype–Phenotype Relationships

Clinical and cellular features of photosensitive and non-photosensitive TTD patients investigated by gene sequencing are summarized in Table 10.2. The majority of NER-defective TTD patients are mutated in the *XPD* gene, and only a few cases are mutated in either *XPB* or *TTDA* (Fig. 10.4). Thus, *XPD* is very tolerant of changes, consistent with the dispensable role of its ATP-dependent helicase activity in basal transcription. Conversely, XPB is essential for transcription and tolerates only rare alterations. The two TTD siblings mutated in the *XPB* gene are homozygous for a mutation resulting in the p.Thr119Pro substitution [15], which slightly interferes with basal transcription activity *in vitro* [47].

The results of mutation analysis in 44 TTD patients mutated in the *XPD* gene (TTD/XP-D) [11, 13, 48–61] do not delineate any specific domain, being the mutations distributed across the gene with a prevalence in the last 1/3 of the gene (Fig. 10.4). Several changes are predicted to result in the substitution of single arginine residues at positions 112, 378, 487, 592, 636, 658 or 722, 4 of which (Arg112, Arg592, Arg658 and Arg722) were observed in the homozygous state in 11 patients. Interestingly, the alteration p.Arg658Cys specifically results in a temperature-sensitive

Table 10.2 Clinical and cellular features of photosensitive and non-photosensitive TTD patients

Mutated gene	Case no. (families)	Clinical symptoms[a] Hair alterations	Physical impairment	Neurological impairment	Skin photosens	NER efficiency[b] UDS % normal	UV survival	TFIIH level[c] % normal
Photosensitive form[d]								
TTDA	6 (5)	+	+	+	+	15–25	−−	30
XPB	2 (1)	+	−	−	+	40	−	40
XPD	44 (41)	+	+/++	+/++	+	10–50	−−/−	35–65
Non-photosensitive form[e]								
TTDN1	49 (18)	+	+/++	+/++	−	Normal	Normal	Normal
GTF2E2	5 (4)	+	++	++	−	Normal	Normal	Normal
RNF113A	2 (1)	+	+	+	−	Normal	Normal	Normal

NER nucleotide excision repair, *UDS* unscheduled DNA synthesis, *photos* photosensitivity

[a]Physical impairment: + moderate (survival beyond early childhood, delayed puberty and short stature); ++ severe (death during childhood and/or failure to thrive/dystrophy). Neurological impairment: + moderate severity (mental development at either preschool level or primary school level, axial hypotonia and reduced motor coordination); ++ severe (very poor mental and motor performances and speech)

[b]NER efficiency refers to the cellular response to UV. UV sensitivity: survival partially (−) or drastically (−−) reduced compared with normal. UDS: the ability to perform UV-induced DNA repair synthesis is expressed as a percentage of that in normal cells

[c]TFIIH level refers to the mean steady state of the subunits CDK7, p44 and p62 in patient cells expressed as percentage of that in normal cells analysed in parallel

[d]Lack of photosensitivity has been reported in a TTD-A patient who developed one SCC at the age of 42 years [18]

[e]Cutaneous burning associated with normal NER efficiency has been reported in the patient TTD343BE mutated in *MPLKIP* [21]

phenotype [62]. The remaining 33 patients are compound heterozygotes with various combinations of mutated *XPD* alleles. Mutations resulting in changes of the C-ter 30 amino acids (aa) were found in the functional hemizygous state in four patients. This finding implies that alterations in the C-ter region of XPD are still compatible with the function of TFIIH in NER and transcription, in agreement with the observation that this is the only part of the *XPD* gene that is poorly conserved. Other frequently observed mutations (the substitution of Arg616 and the Leu461Val change which is associated in the same allele with the deletion of the aa region 716–730) are likely to be non-functional [51], as previously discussed [14].

Remarkably, TTD patients mutated in *XPD* display different degrees of clinical severity, repair defect and reduction in the cellular amount of TFIIH (Table 10.2). The degree of impairment in the cellular response to UV is mutation-specific and does not correlate with either TFIIH reduction level or severity of clinical symptoms. Evaluation of the genotype-phenotype relationships suggests that the main determinant of the severity of the clinical features might be the effective *XPD*-gene dosage [11, 14, 52]. It is worthwhile mentioning that mutations in *XPB* and *XPD*

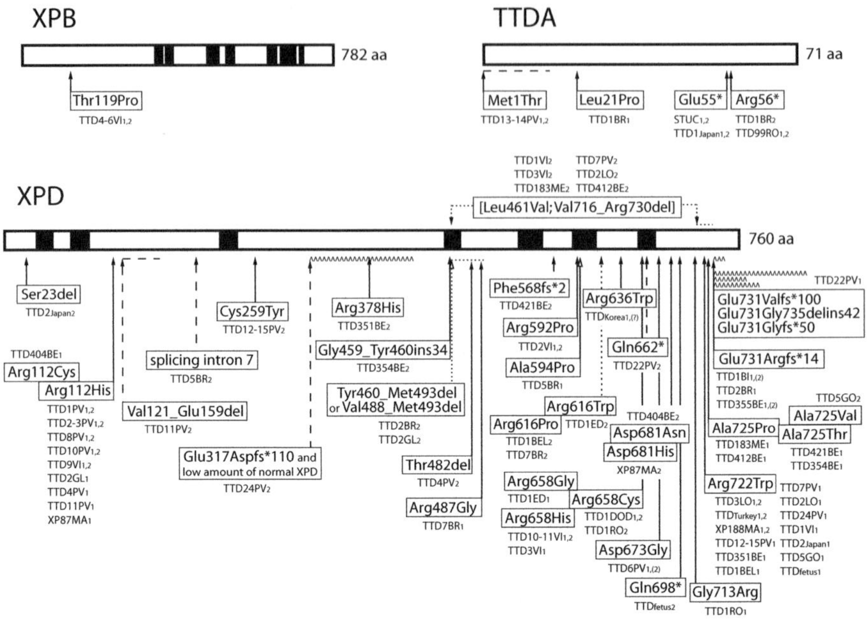

Fig. 10.4 Amino acid changes in the XPB, TTDA and XPD proteins from mutations described in photosensitive TTD patients. The XPB and XPD proteins are shown with the helicase domains in black. All the patients are designated according to nomenclature recommendations [92–94] except for STUC [17], TTD1Japan [18], TTD2Japan [55], TTDKorea [59], TTDTurkey [56], TTDfetus [61], XP87MA and XP188MA clinically diagnosed as TTD [58]. The numbers 1 and 2 after the patient designation denote the different alleles; the number 2 in brackets indicates an unexpressed allele. In XPD, the changes responsible for the pathological phenotype, mutations described as lethal [51] and those resulting in deletions likely to affect cellular viability are indicated by solid, dotted and dashed arrows, respectively. Mutation nomenclature follows the format indicated at www.hgvs.org/mutnomen. TTDA, XPB and XPD protein sequence refers to GenBank NP_997001.1, NP_000113.1, NP_000391.1, respectively

have been found in a variety of inherited diseases, which include TTD, the cancer-prone disorder xeroderma pigmentosum (XP) and rare clinical entities characterized by the association of XP with Cockayne syndrome (CS) and/or TTD features. The heterogeneity of the pathological phenotype in XP-B and XP-D patients as well as the wide range of severity in TTD/XP-D patients may be related to mutations in different sites of the genes, which differentially interfere with the stability and the conformation of the TFIIH complex. In turn, they may affect in slightly different ways TFIIH functional activities in NER, transcription and other cellular processes in which the complex is engaged (reviewed in [36]). Data in TTD ($Xpd^{Arg722Trp}$) and XP/CS ($Xpd^{Gly602Asp}$) mouse model support the notion that interactions between different recessive disease alleles, or biallelic effects, represent a potent source of disease heterogeneity in compound heterozygous patients [63]. The complexity of genotype-phenotype relationships in XP-D patients is underlined by the recent observation that the p.[Leu461Val;Val716_Arg730del]-encoding *XPD* allele expresses also p.[Leu461Val;Ala717Gly] as a result of authentic splicing [64]. In contrast to the p.[Leu461Val;Val716_Arg730del], which completely abolishes basal transcription *in vitro* [65] and behaves as null allele in *S. pombe* [51], the [Leu461Val;Ala717Gly] XPD protein retains a low ability to form TFIIH complex and a partial NER activity, thus explaining its association with a mild phenotype in XP/CS patients [64].

The resolution of the crystal structure of archaeal XPDs revealed that the mutations associated with TTD cause framework defects impacting TFIIH integrity, according to the reduced TFIIH amount typically found in TTD cells (reviewed in [66]). Conversely, the mutations found in XP map close to the ATP-binding pocket or at sites predicted to interact with DNA, thus drastically reducing the XPD helicase activity. Also the mutations found in XP/CS patients are clustered around the ATP-binding site and, in addition, are predicted to produce or prevent important conformational changes [67–69].

Mutation analysis in six patients mutated in the *GTF2H5/TTDA* gene [16–18] indicates that five patients are homozygous for mutations resulting in non-functional truncated peptides (p.Glu55* in two Japanese cases and p.Arg56* in the patient TTD99RO) or for a mutation in the start codon that results in the loss of protein synthesis or in the production of an N-terminally truncated polypeptide lacking the first and most conserved 15 aa, when a downstream AUG at codon 16 is used (the brothers TTD13PV and TTD14PV). The sixth patient (TTD1BR) is heterozygous for a nonsense mutation resulting in p.Arg56* and a missense mutation (p.Leu21Pro) localized in one of the hydrophobic patches on the protein surface that might be involved in protein-protein recognition (Fig. 10.4). It has been suggested that mutations in *TTDA* disrupt the p8-p52-XPB interaction network, thus impacting TFIIH function by destabilizing XPB and in turn the entire complex and/or by reducing the XPB ATPase activity [41].

As far as the non-photosensitive TTD form is concerned, mutations in *MPLKIP* have been reported in 49 NER-proficient TTD patients belonging to 18 families (Fig. 10.5) [19–21, 70–72]. Two distinct missense mutations were found in a consanguineous Amish kindred (p.Met144Val) and in a consanguineous Arab-Muslim kindred (p.Thr169Asn). Large genomic deletions (from 5 to at least 150 kb) including the complete gene or part of the coding region were detected in five cases (four families) and two cases (TTD9PV and TTD343BE), respectively. Two missense mutations were identified in the start codon in one patient (TTD402BE), likely resulting in the loss of translation initiation and no protein production. Furthermore, 6 mutations found in 23 cases (18 homozygotes from 8 families and 5 compound heterozygotes from 4 families) result in frameshifts predicted to produce severely truncated proteins, including 1 founder mutation identified in 10 patients from three Pakistani families. This mutation pattern indicates that the *MPLKIP* gene is not

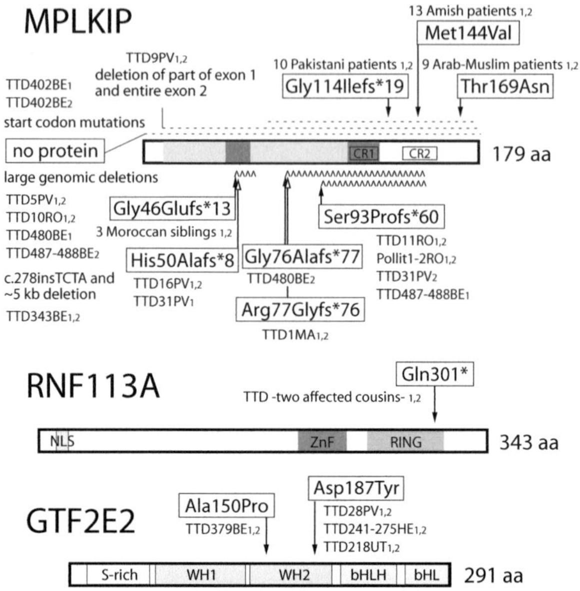

Fig. 10.5 Amino acid changes in the MPLKIP (TTDN1), RNF113A and GTF2E2 proteins from mutations described in non-photosensitive TTD patients. The MPLKIP protein is shown with the glycine-/proline-rich region (grey box), the low complexity regions detected by the BLASTP programme (dark grey boxes) and the two highly conserved C-terminal regions (CR1 and CR2) present among the candidate orthologs. The RNF113A protein is shown with a putative bipartite nuclear localisation sequence (NLS), the C3H1-type zinc finger (ZnF) and the C3HC4-type ring finger (RING) domains [22]. The GTF2E2 protein is shown with a Ser-rich domain (white box, S-rich), two winged-helix motifs (grey boxes, WH1 and WH2), a basic region-helix-loop-helix motif (bHLH) and a basic region-helix-loop domain (bHL) [46]. All the patients are designated according to nomenclature recommendations except for the 2 TTD affected cousins [22], the 10 Pakistani patients [72], the 13 Amish patients [19], the 9 Arab-Muslim patients [71], the 3 Moroccan siblings [19] and the 2 siblings with Pollitt syndrome [70]. The numbers 1 and 2 after the patient designation indicate the different alleles. Mutation nomenclature follows the format indicated at www.hgvs.org/mutnomen. MPLKIP, GTF2E2 and RNF113A protein sequence refers to GenBank NP_619646.1, NP_002086.1, NP_008909.1, respectively

essential for cell proliferation and viability. Evaluation of genotype-phenotype relationships in the *MPLKIP*-mutated TTD patients indicates that the severity of the clinical features does not correlate with the molecular defect. This observation implies that other genetic and/or environmental factors, besides *MPLKIP* mutations, are likely to be involved in determining the severity of the pathological phenotype.

The *RNF113A* gene was found mutated in two cousins carrying a nonsense mutation resulting in a truncated protein of 300 aa [22]. The mutation does not affect the nuclear localisation of the protein but markedly reduces RNF113A protein expression in patient lymphoblastoid cell lines [22] and confers increased sensitivity to MMS by interfering with its E3 ligase activity [34]. The spectrum of clinical symptoms is similar to that reported in severely affected TTD/XP-D cases.

Mutations in *GTF2E2* were found in a 10-year-old Asian boy (TTD379BE) and in a 16-year-old Moroccan girl (TTD28PV), who are homozygous for distinct missense mutations resulting in either p.Ala150Pro (TTD379BE) or p.Asp187Tyr (TTD28PV). The latter alteration has been reported in three other patients from two non-related families of Moroccan origin [23], suggesting the presence of a founder mutation. Both the aa changes alter the wing helix 2 domain of TFIIEβ and the p.Ala150Pro change is expected to destabilize the long α helix, causing its bending or local unfolding. Repair of UV-induced DNA damage was normal in the *GTF2E2*-mutated cells, indicating that TFIIE is not involved in NER. Decreased protein levels of the two TFIIE subunits as well as decreased phosphorylation of TFIIEα were found in cells from both children. Interestingly, decreased phosphorylation of TFIIEα was also seen in non-proliferating, confluent cells of TTD/XP-D but not XP/XP-D patients [12]. These findings support the notion that TTD clinical hallmarks are mainly due to interference with the role of TFIIH in transcription, whereas NER deficiency plays a major role in the XP pathological phenotype.

10.6 Altered Processes/Pathways Underlying TTD Clinical Symptoms

Mutations in *XPB*, *XPD* and *TTDA*, the three genes responsible for the photosensitive form of TTD, alter the structure and stability of the TFIIH complex, thus impairing its functionality in NER and transcription. Overlapping clinical features among photosensitive and non-photosensitive TTD patients indicate that the DNA repair defect is not the main determinant of the wide spectrum of clinical manifestations in TTD except for clinical and cellular photosensitivity. Furthermore, the finding that alterations in the β-subunit of the basal transcription factor TFIIE result in the non-photosensitive TTD pathological phenotype without affecting NER efficiency unequivocally demonstrates the strong contribution of transcriptional impairments to TTD etiopathogenesis [12]. It is not unlikely, however, that the accumulation of unrepaired DNA lesions in cells of photosensitive TTD patients may further exacerbate the consequences of transcriptional defects. Until recently, no functional role was identified for the other two disease genes (*RNF113A* and *MPLKIP*) of the non-photosensitive form of TTD. The most reasonable prediction was that they may encode proteins involved in transcription regulation or RNA processing of specific subset of genes. Recent data indicate that both proteins are involved in an alkylation-specific damage response. After MMS treatment, the alkylation repair complex

ASCC relocalizes to distinct nuclear foci that associate with alkylated nucleotides and coincide spatially with elongating RNApol II and splicing components. RNF113A is the E3 ligase responsible for upstream ubiquitin signalling in the ASCC pathway [34].

Since TTD transcriptional failures are compatible with life, it is likely that they do not ensure adequate gene expression in tissue-specific cell types, in late stages of tissue differentiation and/or in the orchestration of cellular responses to a wide variety of stimuli during normal tissue/organ development and homeostasis.

In reviewing recent experimental evidence supporting this notion, we will examine how distinct biochemical alterations may result in specific TTD clinical manifestations. Most of our knowledge on the mechanistic and functional defects underlying the TTD phenotype originates from studies on patients with mutations in *XPD*, the firstly identified [29] and most common TTD-associated defect, and on the corresponding Xpd^{TTD} mouse model. TTD mice largely recapitulate the clinical manifestations of the human disorder, including brittle hair, developmental abnormalities, reduced life span, UV sensitivity and skin abnormalities, thus demonstrating that the *XPD* mutation resulting in the single aa change Arg722Trp is responsible for the TTD phenotypic spectrum [73].

10.6.1 DNA Repair Defects

Skin photosensitivity is observed in all TTD patients with functional and structural alterations in the TFIIH complex. It is rather clear that the NER deficiency is the key determinant of the sun hypersensitivity and pigmentation anomalies reported in these patients who, however, do not exhibit an increased risk of sunlight-induced skin cancer, in striking contrast to XP and XP/CS.

No specific cellular/biochemical markers of the different cancer susceptibility in XP and TTD have been identified so far. The prevailing explanation for this paradox is that the transcriptional impairment and the limiting amounts of TFIIH typically detected in TTD cells do not support critical steps of skin tumour initiation/promotion by interfering with the expression of genes involved in carcinogenic pathways. In this respect, it is worthwhile to recall that the Xpd^{TTD} mice exhibit a lower spontaneous tumour incidence than wild-type mice [74] but are mildly cancer prone following exposure to a relatively high dose of UV-B. All tumours were histopathologically identified as SCCs, and no eye lesions were registered [63, 75, 76]. We still do not know whether this discrepancy between human and mouse phenotypes represents a significant difference in the complex functional network orbiting TFIIH or it just reflects interspecies differences in DNA damage responses [77] and in the outcome of the multifaceted interplay among DNA repair, transcription, differentiation and carcinogenesis.

The accumulation of DNA damage is one of the factors postulated to drive the ageing process [78]. Due to the intimate connection between the organism ageing and the cell senescence, it is worthwhile mentioning that cell senescence may lead to different outcomes–suppression of malignancy or acceleration of ageing–depending on the context and persistence of senescent cells *in vivo* ([79] and references therein). Precocious ageing in TTD may be favoured by unrepaired DNA damage that further compromises transcription, leading to functional inactivation of critical genes and enhanced apoptosis. Accordingly, alterations in the transcription profiles of Xpd^{TTD} mouse liver were associated to an increased apoptosis rate and increased but not sufficient cell proliferation, suggesting that the imbalance between cell loss and cell renewal may underline premature ageing [80]. The DNA damage accumulation is also a relevant risk factor for neurodegenerative diseases, which raises the spectre of NER involvement in these pathologies. The information available at present indicates that NER is essential for neuronal function and maintenance, but neurological abnormalities in TTD can be fully explained by perturbed transcriptional regulation affecting specific signalling pathways.

10.6.2 Transcriptional Deregulations

In vitro reconstituted TFIIH complex carrying aa changes in the XPD subunit revealed that mutations responsible for TTD, but not for XP, impair the basal transcriptional activity of TFIIH [65], thus demonstrating the different impact of *XPD* mutations on the distinct TFIIH functions and their relevance in TTD or XP clinical features. Further studies in patient cells have been fundamental to demonstrate the significance of transcriptional impairments in TTD etiopathogenesis. At first, the *HBB* gene encoding β-globin was found transcriptionally downregulated in reticulocytes from several TTD but not XP cases with mutations in *XPD* [13]. This finding allowed the identification of haematological features resembling β-thalassaemia in TTD patients. More recently, the expression of two other genes (*COL6A1* and *MMP-1*) was discovered to be deregulated in patient cells, leading to alterations at the level of the extracellular matrix (ECM). The ECM is a complex structural network of macromolecules that surrounds and supports cells within connective tissues. Metabolic disturbance as well as inherited or acquired structural alterations of the ECM gives rise to specific connective tissue disorders in which the bone, cartilage, skin, muscle, brain, eye and cardiovascular systems are differentially affected [81]. Due to malfunctioning TFIIH (Fig. 10.6), TTD dermal fibroblasts at high cell density fail to upregulate the expression of *COL6A1* leading to a reduced content of collagen (COL) type VI that may explain the muscle weakness, joint contractures, skin defects, distal hyper-extensibility and kyphosis in TTD [42]. Moreover, the overexpression and secretion of MMP-1, a zymogen that acts by degrading the interstitial collagens (type I, II and III), lead to reduced levels of collagen type I (COLI) in TTD fibroblast cultures as well as in the dermal layer of the patient skin. This alteration impacts on the migration and wound-healing properties of TTD, but not XP, dermal fibroblasts mutated in TFIIH [82]. As COLI is an important protein building block in the bone, tendons and ligaments, its reduction or alteration in the osteogenesis imperfecta (OI type I) disorder causes fragile bones and reduced bone density frequently associated with slight spinal curvature, loose joints, muscle weakness, lax ligaments, easy bruising, blue sclerae and early loss of hearing [83]. Notably, skeletal abnormalities that include kyphosis, osteopenia and osteosclerosis, sensorineural hearing loss, altered muscle tone, joints and deep tendon reflex are reported in several TTD cases [3]. ECM alterations might also explain other TTD-related features, including pregnancy complications when the foetus is affected by TTD [6–8, 84]. The placenta, whose function is to facilitate nutrient exchanges between the embryo and the mother, contains abundant ECM components and well-preserved endogenous growth factors. Since the placenta consists of both

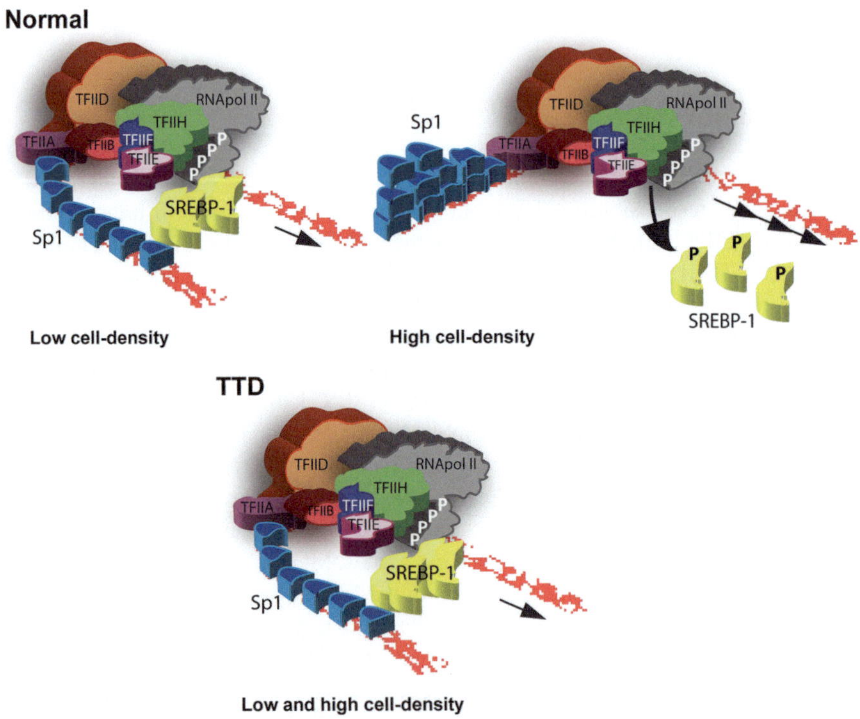

Fig. 10.6 Faulty TFIIH-dependent transcriptional derepression of *COL6A1* gene in high-cell density TTD fibroblasts. In low-cell density cultures of dermal fibroblasts, the RNApol II preinitiation complex and the transcription factors Sp1 and sterol regulatory element-binding protein 1 (SREBP-1) are assembled on *COL6A1* promoter where SREBP-1 downregulates *COL6A1* expression. In response to high cell density, TFIIH triggers the removal of SREBP-1 from *COL6A1* promoter, thus leading to *COL6A1* transcriptional upregulation. The persistent occupancy of SREBP-1 in high-cell density TTD fibroblasts maintains *COL6A1* expression at a low rate [42]

foetal and maternal tissues, pregnancy complications could be due to ECM alterations in the foetal side, which affect the structure and functionality of the placenta [82].

Further understanding of how transcriptional deregulations strongly contribute to TTD pathogenesis derived from studies on the TTD mouse model. In the skin of these mice, the expression of the *SPRR2* gene encoding the small proline-rich 2 protein was found reduced, thus linking hair loss and skin abnormalities with transcriptional impairments. In addition, several target genes of the peroxisome proliferator-activated receptors (PPARs) were found transcriptionally deregulated in Xpd^{TTD} mice, due to transactivation defects associated with a weak phosphorylation of PPAR by the Cdk7 subunit of TFIIH. Bearing in mind the involvement of PPARs in lipid metabolism and differentiation/survival of adipocytes, these alterations may account for the hypoplasia of adipose tissues that characterizes TTD [85]. Furthermore, a transcriptional deregulation of gluconeogenic genes and consequent abnormal amount of the corresponding enzymes were observed in the hepatic parenchyma of Xpd^{TTD} mice subjected to various periods of fasting, a process that requires a physiological adaptation and accurate gene expression regulation in liver cells. The dynamic interaction of TFIIH with the PPARγ coactivator-1α (PGC-1α) and the deacetylase Sirtuin 1 (SIRT1) was altered in Xpd^{TTD} liver cells. The former event is essential to promote the PGC-1α deacetylation by SIRT1, the correct recruitment of PGC-1α on the promoters of gluconeogenic genes and consequently their appropriate transcription [86]. Notably, comparative studies on the liver of young and old Xpd^{TTD} female mice revealed transcription profiles pointing towards a metabolic response involving reduced energy metabolism and reduced insulin growth factor-1 (IGF-1) signalling [80]. Moreover, a spatial and selective transcriptional deregulation of thyroid hormone target genes was shown in the brain of Xpd^{TTD} mouse embryos. Besides revealing an unexpected coactivator function of TFIIH, these results may explain some of TTD neurological manifestations as the thyroid hormone contributes to the regulation of several processes taking place during perinatal brain development, including myelinogenesis and neuronal cell migration [87]. Finally, in the brain of Xpd^{TTD} mice, an altered transcription by RNApol I was identified, indicating that a defective ribosomal protein gene expression likely contributes to the faulty mechanisms implicated in TTD pathogenesis [43].

10.7 Conclusions and Perspectives

Investigations on TTD patients greatly contributed to reveal the role of TFIIH complex in the crosstalk between DNA repair and transcription, to clarify the functional properties of the XPB, XPD and p8/TTDA subunits in these two fundamental processes and to disclose the structural function of XPD in TFIIH assembly, which is pivotal in regulating and coordinating the distinct functions of CORE and CAK subcomplexes in transcription, DNA repair and cell cycle progression [88]. Besides regulating CAK activity during mitosis, XPD has been shown to be part of the MMXD and

CGX complexes, which are both implicated in chromosome segregation [38, 39]. Recent observations suggest a role of XPD also in mitochondrial response to stress [89]. Future studies will clarify whether alterations in these processes (reviewed in [90]) might have a causal link with some of the TTD clinical manifestations. Relevant issues to be further explored include the functional role of *MPLKIP* and *RNF113A* genes as well as the cloning and characterization of the other still unidentified genes responsible for the non-photosensitive form of TTD. A deep knowledge of the molecular mechanisms affected by defects in these genes and their involvement in tumorigenesis, physical and mental development and/or ageing represents a key step in the identification of targets for effective therapeutic approaches.

Acknowledgements We acknowledge patients and their families as well as referring clinicians for their fundamental contribution to our knowledge of TTD. We thank all the members of our team at the IGM CNR Pavia for their contribution to the research over the years. Our studies mentioned in the text have been supported by grants from the Associazione Italiana per la Ricerca sul Cancro (AIRC IG 13537 and 17710), the European Community, the Cariplo Foundation and the Italian Ministry of University and Research.

References

1. Price VH, Odom RB, Ward WH, Jones FT. Trichothiodystrophy: sulfur-deficient brittle hair as a marker for a neuroectodermal symptom complex. Arch Dermatol. 1980;116:1375–84.
2. Itin PH, Sarasin A, Pittelkow MR. Trichothiodystrophy: update on the sulfur-deficient brittle hair syndromes. J Am Acad Dermatol. 2001;44:891–920; quiz 921–4. https://doi.org/10.1067/mjd.2001.114294.
3. Faghri S, Tamura D, Kraemer KH, Digiovanna JJ. Trichothiodystrophy: a systematic review of 112 published cases characterises a wide spectrum of clinical manifestations. J Med Genet. 2008;45:609–21. https://doi.org/10.1136/jmg.2008.058743.
4. Stefanini M, Ruggieri M. Trichothiodystrophy. New York: Springer; 2008. p. 821–45.
5. Tamura D, Merideth M, DiGiovanna JJ, et al. High-risk pregnancy and neonatal complications in the DNA repair and transcription disorder trichothiodystrophy: report of 27 affected pregnancies. Prenat Diagn. 2011;31:1046–53. https://doi.org/10.1002/pd.2829.
6. Tamura D, Khan SG, Merideth M, et al. Effect of mutations in XPD(ERCC2) on pregnancy and prenatal development in mothers of patients with trichothiodystrophy or xeroderma pigmentosum. Eur J Hum Genet. 2012;20:1308–10. https://doi.org/10.1038/ejhg.2012.90.
7. Moslehi R, Signore C, Tamura D, et al. Adverse effects of trichothiodystrophy DNA repair and transcription gene disorder on human fetal development. Clin Genet. 2010;77:365–73. https://doi.org/10.1111/j.1399-0004.2009.01336.x.
8. Moslehi R, Kumar A, Mills JL, et al. Phenotype-specific adverse effects of XPD mutations on human prenatal development implicate impairment of TFIIH-mediated functions in placenta. Eur J Hum Genet. 2012;20:626–31. https://doi.org/10.1038/ejhg.2011.249.
9. Kraemer KH, Patronas NJ, Schiffmann R, et al. Xeroderma pigmentosum, trichothiodystrophy and Cockayne syndrome: a complex genotype-phenotype relationship. Neuroscience. 2007;145:1388–96. https://doi.org/10.1016/j.neuroscience.2006.12.020.
10. Brooks BP, Thompson AH, Clayton JA, et al. Ocular manifestations of trichothiodystrophy. Ophthalmology. 2011;118:2335–42. https://doi.org/10.1016/j.ophtha.2011.05.036.

11. Botta E, Nardo T, Orioli D, et al. Genotype-phenotype relationships in trichothiodystrophy patients with novel splicing mutations in the XPD gene. Hum Mutat. 2009;30:438–45. https://doi.org/10.1002/humu.20912.
12. Kuschal C, Botta E, Orioli D, et al. GTF2E2 mutations destabilize the general transcription factor complex TFIIE in individuals with DNA repair-proficient trichothiodystrophy. Am J Hum Genet. 2016;98:627–42. https://doi.org/10.1016/j.ajhg.2016.02.008.
13. Viprakasit V, Gibbons RJ, Broughton BC, et al. Mutations in the general transcription factor TFIIH result in beta-thalassaemia in individuals with trichothiodystrophy. Hum Mol Genet. 2001;10:2797–802.
14. Stefanini M, Botta E, Lanzafame M, Orioli D. Trichothiodystrophy: from basic mechanisms to clinical implications. DNA Repair (Amst). 2010;9:2–10. https://doi.org/10.1016/j.dnarep.2009.10.005.
15. Weeda G, Rossignol M, Fraser RA, et al. The XPB subunit of repair/transcription factor TFIIH directly interacts with SUG1, a subunit of the 26S proteasome and putative transcription factor. Nucleic Acids Res. 1997;25:2274–83.
16. Giglia-Mari G, Coin F, Ranish JA, et al. A new, tenth subunit of TFIIH is responsible for the DNA repair syndrome trichothiodystrophy group A. Nat Genet. 2004;36:714–9. https://doi.org/10.1038/ng1387.
17. Moriwaki S, Saruwatari H, Kanzaki T, et al. Trichothiodystrophy group A: a first Japanese patient with a novel homozygous nonsense mutation in the GTF2H5 gene. J Dermatol. 2014;41:705–8. https://doi.org/10.1111/1346-8138.12549.
18. Takeichi T, Tomimura S, Okuno Y, et al. Trichothiodystrophy, complementation group A complicated with squamous cell carcinoma. J Eur Acad Dermatol Venereol. 2017;32:e75–7. https://doi.org/10.1111/jdv.14531.
19. Nakabayashi K, Amann D, Ren Y, et al. Identification of C7orf11 (TTDN1) gene mutations and genetic heterogeneity in nonphotosensitive trichothiodystrophy. Am J Hum Genet. 2005;76:510–6. https://doi.org/10.1086/428141.
20. Botta E, Offman J, Nardo T, et al. Mutations in the C7orf11 (TTDN1) gene in six nonphotosensitive trichothiodystrophy patients: no obvious genotype-phenotype relationships. Hum Mutat. 2007;28:92–6. https://doi.org/10.1002/humu.20419.
21. Heller ER, Khan SG, Kuschal C, et al. Mutations in the TTDN1 gene are associated with a distinct trichothiodystrophy phenotype. J Invest Dermatol. 2015;135:734–41. https://doi.org/10.1038/jid.2014.440.
22. Corbett MA, Dudding-Byth T, Crock PA, et al. A novel X-linked trichothiodystrophy associated with a nonsense mutation in RNF113A. J Med Genet. 2015;52:269–74. https://doi.org/10.1136/jmedgenet-2014-102418.
23. Theil AF, Mandemaker IK, van den Akker E, et al. Trichothiodystrophy causative TFIIEβ mutation affects transcription in highly differentiated tissue. Hum Mol Genet. 2017;26:4689–98. https://doi.org/10.1093/hmg/ddx351.
24. Kleijer WJ, Laugel V, Berneburg M, et al. Incidence of DNA repair deficiency disorders in western Europe: Xeroderma pigmentosum, Cockayne syndrome and trichothiodystrophy. DNA Repair (Amst). 2008;7:744–50. https://doi.org/10.1016/j.dnarep.2008.01.014.
25. Marteijn JA, Lans H, Vermeulen W, Hoeijmakers JHJ. Understanding nucleotide excision repair and its roles in cancer and ageing. Nat Rev Mol Cell Biol. 2014;15:465–81. https://doi.org/10.1038/nrm3822.
26. Spivak G. Transcription-coupled repair: an update. Arch Toxicol. 2016;90:2583–94. https://doi.org/10.1007/s00204-016-1820-x.
27. Lainé J-P, Egly JM. When transcription and repair meet: a complex system. Trends Genet. 2006;22:430–6. https://doi.org/10.1016/j.tig.2006.06.006.
28. Lehmann AR, McGibbon D, Stefanini M. Xeroderma pigmentosum. Orphanet J Rare Dis. 2011;6:70. https://doi.org/10.1186/1750-1172-6-70.

29. Stefanini M, Lagomarsini P, Arlett CF, et al. Xeroderma pigmentosum (complementation group D) mutation is present in patients affected by trichothiodystrophy with photosensitivity. Hum Genet. 1986;74:107–12.
30. Stefanini M, Vermeulen W, Weeda G, et al. A new nucleotide-excision-repair gene associated with the disorder trichothiodystrophy. Am J Hum Genet. 1993;53:817–21.
31. Stefanini M, Kraemer KH. Xeroderma pigmentosum. New York: Springer; 2008. p. 771–92.
32. Jia N, Nakazawa Y, Guo C, et al. A rapid, comprehensive system for assaying DNA repair activity and cytotoxic effects of DNA-damaging reagents. Nat Protoc. 2015;10:12–24. https://doi.org/10.1038/nprot.2014.194.
33. Emmert S, Slor H, Busch DB, et al. Relationship of neurologic degeneration to genotype in three xeroderma pigmentosum group G patients. J Invest Dermatol. 2002;118:972–82. https://doi.org/10.1046/j.1523-1747.2002.01782.x.
34. Brickner JR, Soll JM, Lombardi PM, et al. A ubiquitin-dependent signalling axis specific for ALKBH-mediated DNA dealkylation repair. Nature. 2017;551:389–93. https://doi.org/10.1038/nature24484.
35. Compe E, Egly JM. TFIIH: when transcription met DNA repair. Nat Rev Mol Cell Biol. 2012;13:343–54. https://doi.org/10.1038/nrm3350.
36. Compe E, Egly JM. Nucleotide excision repair and transcriptional regulation: TFIIH and beyond. Annu Rev Biochem. 2016;85:265–90. https://doi.org/10.1146/annurev-biochem-060815-014857.
37. Radu L, Schoenwetter E, Braun C, et al. The intricate network between the p34 and p44 subunits is central to the activity of the transcription/DNA repair factor TFIIH. Nucleic Acids Res. 2017;45:10872–83. https://doi.org/10.1093/nar/gkx743.
38. Li X, Urwyler O, Suter B. Drosophila Xpd regulates Cdk7 localization, mitotic kinase activity, spindle dynamics, and chromosome segregation. PLoS Genet. 2010;6:e1000876. https://doi.org/10.1371/journal.pgen.1000876.
39. Yeom E, Hong S-T, Choi K-W. Crumbs interacts with Xpd for nuclear division control in Drosophila. Oncogene. 2015;34:2777–89. https://doi.org/10.1038/onc.2014.202.
40. Fishburn J, Tomko E, Galburt E, Hahn S. Double-stranded DNA translocase activity of transcription factor TFIIH and the mechanism of RNA polymerase II open complex formation. Proc Natl Acad Sci U S A. 2015;112:3961–6. https://doi.org/10.1073/pnas.1417709112.
41. Luo J, Cimermancic P, Viswanath S, et al. Architecture of the human and yeast general transcription and DNA repair factor TFIIH. Mol Cell. 2015;59:794–806. https://doi.org/10.1016/j.molcel.2015.07.016.
42. Orioli D, Compe E, Nardo T, et al. XPD mutations in trichothiodystrophy hamper collagen VI expression and reveal a role of TFIIH in transcription derepression. Hum Mol Genet. 2013;22:1061–73. https://doi.org/10.1093/hmg/dds508.
43. Nonnekens J, Perez-Fernandez J, Theil AF, et al. Mutations in TFIIH causing trichothiodystrophy are responsible for defects in ribosomal RNA production and processing. Hum Mol Genet. 2013;22:2881–93. https://doi.org/10.1093/hmg/ddt143.
44. Theil AF, Hoeijmakers JHJ, Vermeulen W. TTDA: big impact of a small protein. Exp Cell Res. 2014;329:61–8. https://doi.org/10.1016/j.yexcr.2014.07.008.
45. Zhang Y, Tian Y, Chen Q, et al. TTDN1 is a Plk1-interacting protein involved in maintenance of cell cycle integrity. Cell Mol Life Sci. 2007;64:632–40. https://doi.org/10.1007/s00018-007-6501-8.
46. Tanaka A, Akimoto Y, Kobayashi S, et al. Association of the winged helix motif of the TFIIEα subunit of TFIIE with either the TFIIEβ subunit or TFIIB distinguishes its functions in transcription. Genes Cells. 2015;20:203–16. https://doi.org/10.1111/gtc.12212.
47. Bradsher J, Coin F, Egly JM. Distinct roles for the helicases of TFIIH in transcript initiation and promoter escape. J Biol Chem. 2000;275:2532–8.
48. Broughton BC, Steingrimsdottir H, Weber CA, Lehmann AR. Mutations in the xeroderma pigmentosum group D DNA repair/transcription gene in patients with trichothiodystrophy. Nat Genet. 1994;7:189–94. https://doi.org/10.1038/ng0694-189.

49. Takayama K, Salazar EP, Broughton BC, et al. Defects in the DNA repair and transcription gene ERCC2(XPD) in trichothiodystrophy. Am J Hum Genet. 1996;58:263–70.
50. Takayama K, Danks DM, Salazar EP, et al. DNA repair characteristics and mutations in the ERCC2 DNA repair and transcription gene in a trichothiodystrophy patient. Hum Mutat. 1997;9:519–25. https://doi.org/10.1002/(SICI)1098-1004(1997)9:6<519::AID-HUMU4>3.0.CO;2-X.
51. Taylor EM, Broughton BC, Botta E, et al. Xeroderma pigmentosum and trichothiodystrophy are associated with different mutations in the XPD (ERCC2) repair/transcription gene. Proc Natl Acad Sci U S A. 1997;94:8658–63.
52. Botta E, Nardo T, Broughton BC, et al. Analysis of mutations in the XPD gene in Italian patients with trichothiodystrophy: site of mutation correlates with repair deficiency, but gene dosage appears to determine clinical severity. Am J Hum Genet. 1998;63:1036–48. https://doi.org/10.1086/302063.
53. Boyle J, Ueda T, Oh K-S, et al. Persistence of repair proteins at unrepaired DNA damage distinguishes diseases with ERCC2 (XPD) mutations: cancer-prone xeroderma pigmentosum vs. non-cancer-prone trichothiodystrophy. Hum Mutat. 2008;29:1194–208. https://doi.org/10.1002/humu.20768.
54. Zhou X, Khan SG, Tamura D, et al. Brittle hair, developmental delay, neurologic abnormalities, and photosensitivity in a 4-year-old girl. J Am Acad Dermatol. 2010;63:323–8. https://doi.org/10.1016/j.jaad.2010.03.041.
55. Usuda T, Saijo M, Tanaka K, et al. A Japanese trichothiodystrophy patient with XPD mutations. J Hum Genet. 2011;56:77–9. https://doi.org/10.1038/jhg.2010.123.
56. Pehlivan D, Cefle K, Raams A, et al. A Turkish trichothiodystrophy patient with homozygous XPD mutation and genotype-phenotype relationship. J Dermatol. 2012;39:1016–21. https://doi.org/10.1111/j.1346-8138.2012.01662.x.
57. Zhou X, Khan SG, Tamura D, et al. Abnormal XPD-induced nuclear receptor transactivation in DNA repair disorders: trichothiodystrophy and xeroderma pigmentosum. Eur J Hum Genet. 2013;21:831–7. https://doi.org/10.1038/ejhg.2012.246.
58. Schäfer A, Gratchev A, Seebode C, et al. Functional and molecular genetic analyses of nine newly identified XPD-deficient patients reveal a novel mutation resulting in TTD as well as in XP/CS complex phenotypes. Exp Dermatol. 2013;22:486–9. https://doi.org/10.1111/exd.12166.
59. Shin S, Kim J, Kim Y, et al. Analysis of mutations in the XPD gene in a patient with brittle hair. Ann Clin Lab Sci. 2013;43:323–7.
60. Brauns B, Schubert S, Lehmann J, et al. Photosensitive form of trichothiodystrophy associated with a novel mutation in the XPD gene. Photodermatol Photoimmunol Photomed. 2016;32:110–2. https://doi.org/10.1111/phpp.12225.
61. Miguet M, Thevenon J, Laugel V, et al. Mutations in the ERCC2 (XPD) gene associated with severe fetal ichthyosis and dysmorphic features. Prenat Diagn. 2016;36:1276–9. https://doi.org/10.1002/pd.4965.
62. Vermeulen W, Rademakers S, Jaspers NG, et al. A temperature-sensitive disorder in basal transcription and DNA repair in humans. Nat Genet. 2001;27:299–303. https://doi.org/10.1038/85864.
63. van de Ven M, Andressoo JO, van der Horst GTJ, et al. Effects of compound heterozygosity at the Xpd locus on cancer and ageing in mouse models. DNA Repair (Amst). 2012;11:874–83. https://doi.org/10.1016/j.dnarep.2012.08.003.
64. Horibata K, Kono S, Ishigami C, et al. Constructive rescue of TFIIH instability by an alternative isoform of XPD derived from a mutated XPD allele in mild but not severe XP-D/CS. J Hum Genet. 2015;60:259–65. https://doi.org/10.1038/jhg.2015.18.
65. Dubaele S, Proietti De Santis L, Bienstock RJ, et al. Basal transcription defect discriminates between xeroderma pigmentosum and trichothiodystrophy in XPD patients. Mol Cell. 2003;11:1635–46.
66. Lehmann AR. XPD structure reveals its secrets. DNA Repair (Amst). 2008;7:1912–5. https://doi.org/10.1016/j.dnarep.2008.07.008.

67. Fan L, Fuss JO, Cheng QJ, et al. XPD helicase structures and activities: insights into the cancer and aging phenotypes from XPD mutations. Cell. 2008;133:789–800. https://doi.org/10.1016/j.cell.2008.04.030.
68. Liu H, Rudolf J, Johnson KA, et al. Structure of the DNA repair helicase XPD. Cell. 2008;133:801–12. https://doi.org/10.1016/j.cell.2008.04.029.
69. Wolski SC, Kuper J, Hänzelmann P, et al. Crystal structure of the FeS cluster-containing nucleotide excision repair helicase XPD. PLoS Biol. 2008;6:e149. https://doi.org/10.1371/journal.pbio.0060149.
70. Swagemakers SMA, Jaspers NGJ, Raams A, et al. Pollitt syndrome patients carry mutation in TTDN1. Meta Gene. 2014;2:616–8. https://doi.org/10.1016/j.mgene.2014.08.001.
71. Pode-Shakked B, Marek-Yagel D, Greenberger S, et al. A novel mutation in the C7orf11 gene causes nonphotosensitive trichothiodystrophy in a multiplex highly consanguineous kindred. Eur J Med Genet. 2015;58:685–8. https://doi.org/10.1016/j.ejmg.2015.10.012.
72. Shah K, Ali RH, Ansar M, et al. Mitral regurgitation as a phenotypic manifestation of nonphotosensitive trichothiodystrophy due to a splice variant in MPLKIP. BMC Med Genet. 2016;17:13. https://doi.org/10.1186/s12881-016-0275-5.
73. de Boer J, de Wit J, van Steeg H, et al. A mouse model for the basal transcription/DNA repair syndrome trichothiodystrophy. Mol Cell. 1998;1:981–90.
74. Wijnhoven SWP, Beems RB, Roodbergen M, et al. Accelerated aging pathology in ad libitum fed Xpd(TTD) mice is accompanied by features suggestive of caloric restriction. DNA Repair (Amst). 2005;4:1314–24. https://doi.org/10.1016/j.dnarep.2005.07.002.
75. de Boer J, van Steeg H, Berg RJ, et al. Mouse model for the DNA repair/basal transcription disorder trichothiodystrophy reveals cancer predisposition. Cancer Res. 1999;59:3489–94.
76. Andressoo JO, Mitchell JR, de Wit J, et al. An Xpd mouse model for the combined xeroderma pigmentosum/Cockayne syndrome exhibiting both cancer predisposition and segmental progeria. Cancer Cell. 2006;10:121–32. https://doi.org/10.1016/j.ccr.2006.05.027.
77. D'Errico M, Teson M, Calcagnile A, et al. Differential role of transcription-coupled repair in UVB-induced response of human fibroblasts and keratinocytes. Cancer Res. 2005;65:432–8.
78. Lenart P, Krejci L. DNA, the central molecule of aging. Mutat Res. 2016;786:1–7. https://doi.org/10.1016/j.mrfmmm.2016.01.007.
79. Ribezzo F, Shiloh Y, Schumacher B. Systemic DNA damage responses in aging and diseases. Semin Cancer Biol. 2016;37–38:26–35. https://doi.org/10.1016/j.semcancer.2015.12.005.
80. Park JY, Cho M-O, Leonard S, et al. Homeostatic imbalance between apoptosis and cell renewal in the liver of premature aging Xpd mice. PLoS One. 2008;3:e2346. https://doi.org/10.1371/journal.pone.0002346.
81. Bateman JF, Boot-Handford RP, Lamandé SR. Genetic diseases of connective tissues: cellular and extracellular effects of ECM mutations. Nat Rev Genet. 2009;10:173–83. https://doi.org/10.1038/nrg2520.
82. Arseni L, Lanzafame M, Compe E, et al. TFIIH-dependent MMP-1 overexpression in trichothiodystrophy leads to extracellular matrix alterations in patient skin. Proc Natl Acad Sci U S A. 2015;112:1499–504. https://doi.org/10.1073/pnas.1416181112.
83. Byers PH, Pyott SM. Recessively inherited forms of osteogenesis imperfecta. Annu Rev Genet. 2012;46:475–97. https://doi.org/10.1146/annurev-genet-110711-155608.
84. Moslehi R, Ambroggio X, Nagarajan V, et al. Nucleotide excision repair/transcription gene defects in the fetus and impaired TFIIH-mediated function in transcription in placenta leading to preeclampsia. BMC Genomics. 2014;15:373. https://doi.org/10.1186/1471-2164-15-373.
85. Compe E, Drané P, Laurent C, et al. Dysregulation of the peroxisome proliferator-activated receptor target genes by XPD mutations. Mol Cell Biol. 2005;25:6065–76. https://doi.org/10.1128/MCB.25.14.6065-6076.2005.
86. Traboulsi H, Davoli S, Catez P, et al. Dynamic partnership between TFIIH, PGC-1α and SIRT1 is impaired in trichothiodystrophy. PLoS Genet. 2014;10:e1004732. https://doi.org/10.1371/journal.pgen.1004732.
87. Compe E, Malerba M, Soler L, et al. Neurological defects in trichothiodystrophy reveal a coactivator function of TFIIH. Nat Neurosci. 2007;10:1414–22. https://doi.org/10.1038/nn1990.

88. Cameroni E, Stettler K, Suter B. On the traces of XPD: cell cycle matters – untangling the genotype-phenotype relationship of XPD mutations. Cell Div. 2010;5:24. https://doi.org/10.1186/1747-1028-5-24.
89. Liu J, Fang H, Chi Z, et al. XPD localizes in mitochondria and protects the mitochondrial genome from oxidative DNA damage. Nucleic Acids Res. 2015;43:5476–88. https://doi.org/10.1093/nar/gkv472.
90. Houten BV, Kuper J, Kisker C. Role of XPD in cellular functions: to TFIIH and beyond. DNA Repair (Amst). 2016;44:136–42. https://doi.org/10.1016/j.dnarep.2016.05.019.
91. Schultz P, Fribourg S, Poterszman A, et al. Molecular structure of human TFIIH. Cell. 2000;102:599–607.
92. Kleijer WJ, de Weerd-Kastelein EA, Sluyter ML, et al. UV-induced DNA repair synthesis in cells of patients with different forms of xeroderma pigmentosum and of heterozygotes. Mutat Res. 1973;20:417–28.
93. Cleaver JE, Bootsma D, Friedberg E. Human diseases with genetically altered DNA repair processes. Genetics. 1975;79(Suppl):215–25.
94. Lehmann AR, Bootsma D, Clarkson SG, et al. Nomenclature of human DNA repair genes. Mutat Res. 1994;315:41–2.

Chapter 11
Rothmund–Thomson Syndrome

Hideo Kaneko

Abstract Rothmund–Thomson syndrome is an autosomal recessive genetic disorder which is characterized by poikiloderma of the face, small stature, sparse scalp hair, juvenile cataracts, radial aplasia, and predisposition to cancers. Facial redness is particularly characteristic of this syndrome with redness gradually spreading over the four limbs. The redness appears within a year of birth and then progresses to poikiloderma. The causative gene for Rothmund–Thomson syndrome is *RECQL4*, which is essential for genetic replication and repair. *RECQL4* mutations are found in approximately 60% of all patients with Rothmund–Thomson syndrome. Some researchers classify Rothmund–Thomson syndrome with *RECQL4* mutations as type II and that without *RECQL4* mutations type I. Rothmund–Thomson type I is characterized by poikiloderma, ectodermal malformation, and juvenile cataracts, whereas Rothmund–Thomson type II is characterized by poikiloderma, congenital bone defects, the complication of osteosarcoma in infancy, and the complication of skin cancer with aging.

Keywords Rothmund–Thomson syndrome · RAPADILINO syndrome · Baller–Gerold syndrome · RECQL4 · Poikiloderma · Osteosarcoma

11.1 Introduction

There are five human RecQ-like proteins (RECQL1, BLM, WRN, RECQL4, and RECQ5), each having 3′ to 5′ DNA helicase activity but little sequence similarity outside the helicase motifs. Three of these helicases (BLM, WRN, and Rothmund–Thomson) show genomic instability and cancer susceptibility, but each also has distinctive features. RECQL4 is the causative gene for Rothmund–Thomson

H. Kaneko
Department of Clinical Research, National Hospital Organization, Nagara Medical Center, Gifu, Japan
e-mail: hkaneko@nagara-lan.hosp.go.jp

C. Nishigori, K. Sugasawa (eds.), *DNA Repair Disorders*,
https://doi.org/10.1007/978-981-10-6722-8_11

syndrome (OMIM 266280) characterized by poikiloderma and skeletal defects [1, 2]. Homozygous or compound heterozygous mutations of RECQL4 gene causes Rothmund–Thomson syndrome [3].

11.2 Epidemiology

Approximately 300 patients are reported to have Rothmund–Thomson syndrome worldwide [4]. On the basis of a questionnaire survey conducted by the departments of pediatrics and dermatology, and core hospitals for cancer, as part of the Research on Measures for Intractable Diseases supported by the Ministry of Health, Labour and Welfare in Japan, ten patients were identified to have Rothmund–Thomson syndrome in Japan [5]. Of these ten, eight were males, among whom seven had small stature (Table 11.1). The birth weights of two patients were low. Eight patients had poikiloderma, which was the basis for the diagnosis in many of them. Regarding cancer development, two patients had osteosarcoma. Four of the ten patients had undergone *RECQL4* screening, but none were found to carry *RECQL4* mutations.

11.3 Cause of Disease

Rothmund–Thomson type II is caused by mutations of the RECQL4 protein, which belongs to the RecQ helicase family (Fig. 11.1) [6]. Other proteins that also belong to the RecQ helicase family are WRN and BLM, mutations of which cause Werner syndrome and Bloom syndrome, respectively. Werner, Bloom, and Rothmund–Thomson syndromes have chromosomal instability and predisposition to cancers as their common characteristics. *RECQL4* is located on chromosome 8q24.3 and encodes a protein with 1208 amino acids and a molecular weight of 133 kDa. Helicase is a protein that unwinds the double-stranded DNA into single strands and plays an important role in genetic replication and repair. Rothmund–Thomson type I comprises a variety of pathological conditions. The pathogenesis of Rothmund–Thomson type I has not yet been clarified despite intensive research efforts. There are two diseases that are related to Rothmund–Thomson syndrome, namely, RAPADILINO syndrome and Baller–Gerold syndrome. The name RAPADILINO is an acronym for the following features commonly observed in affected patients: radial hypoplasia/aplasia, patella hypoplasia, palate hypoplasia/cleft palate, diarrhea, dislocated joints, little size, limb malformation, nose slender, and normal intelligence. Poikiloderma is not observed in patients with RAPADILINO syndrome [7, 8]. Baller–Gerold syndrome is characterized by brachycephaly due to the premature fusion of coronal sutures, prominent forehead, bulging eyes, low-set ears,

Table 11.1 Clinical feature of Rothmund–Thomson in Japan

No	Sex	Present (age)	Height (age)	Body weight (age)	Skin lesion (age)	Malignancy (age)	Others (age)
1	M	8 (death)	103 cm (7)	14 kg (7)	Poikiloderma (10 M) Pigmentation (10 M)	Osteosarcoma (7)	Death caused by lung metastasis of osteosarcoma (8)
2	M	10 (alive)	135 cm (10)	33 kg (10)	Sun-sensitive erythema (7) Pigmentation (2)	None	Syndactyly
3	M	10 (alive)	84 cm (3) 130 cm (10)	2126 g (at birth) 26 kg (10)	Sun-sensitive erythema (6 M) Poikiloderma (6 M) Sparse hair (6 M)	None	Tooth dysplasia (6) Pulmonary valve stenosis ventricular aneurysm cretinism (supplementation of thyroid hormone)
4	F	25 (alive)	150 cm (18)	34 kg (18)	Poikiloderma telangiectasia pigmentation sparse hair (5 M) Decreased sweating (14)	Osteosarcoma (20)	Scoliosis (14) Atrophic gastritis (16) Insufficiency of pancreatic function (fatty degeneration) (16)
5	M	37 (alive)	168 cm (37)	55 kg (37)	Erythema nodosum (34)	None	Syndactyly
6	M	25 (death) Younger brother of case 5	150 cm (25)	31 kg (25)	Recurrent skin ulcer and infectious granuloma	None	Acute glomerulonephritis, renal insufficiency, respiratory insufficiency (25), psychomotor retardation
7	M	6 (alive)	100 cm (6)	2000 g (at birth) 13 kg (6)	Reticulated pigmentation on the face and lower limbs and poikiloderma (1)	None	
8	M	27 (alive)	144.7 cm (20)	43 kg (20)	Poikiloderma (13)	None	Micrognathia
9	M	1 (alive)	72 cm (14 M)	7600 g (14 M)	Poikiloderma, reticulated erythema, and blister	None	
10	F	5 (alive)	106.4 cm (5)	19.5 kg (5)	Poikiloderma and reticulated depigmentation	None	Hearing loss and bilateral cataract

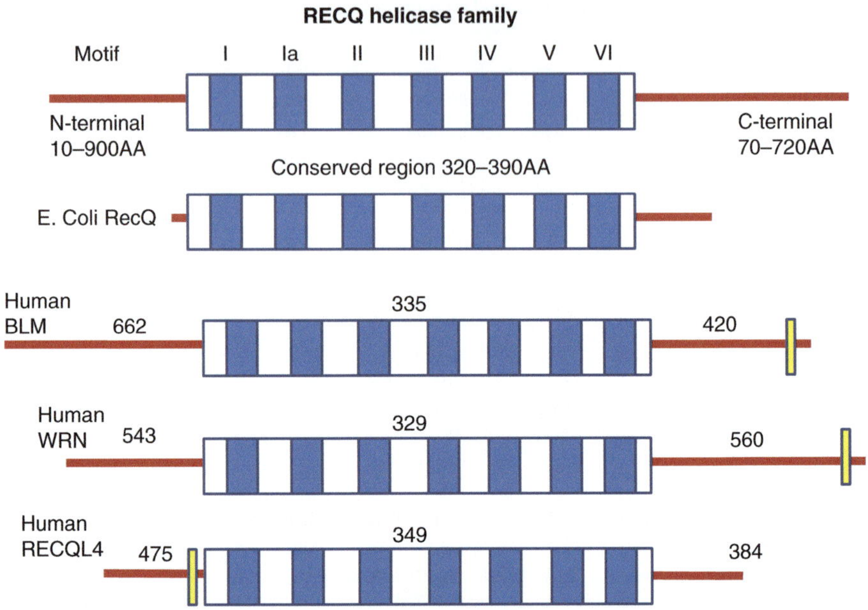

Fig. 11.1 Features of RECQ helicase family proteins and functional motif [6]. I–VI are helicase motif. AA: amino acids. Yellow boxes indicated nuclear localization signal. The number indicated the number of amino acids for N-terminal, C-terminal, and helicase motif, respectively

radial aplasia (Fig. 11.2), absence of a thumb, poikiloderma, and the complications of osteosarcoma, skin cancer, and malignant lymphoma. Cataracts are characteristic of Rothmund–Thomson syndrome, whereas dislocated joints and patella hypoplasia are characteristic of RAPADILINO syndrome. Moreover, craniostenosis is characteristic of Baller–Gerold syndrome.

To understand Baller- Gerold syndrome patients in Japan, a nationwide survey was conducted, which identified two families and three patients affected by the syndrome [9]. All the three patients showed radial defects and craniosynostosis. In one patient who showed a dislocated joint of the hip and flexion contracture of both the elbow joints and wrists at birth, a homozygous large deletion in the RECQL4 gene was identified (Fig. 11.3).

Because these diseases often exhibit common phenotypic features, in the future it would be more beneficial to diagnose these syndromes differently on the basis of causative genes, classifying them into those with *RECQL4* mutations and those without *RECQL4* mutations [10].

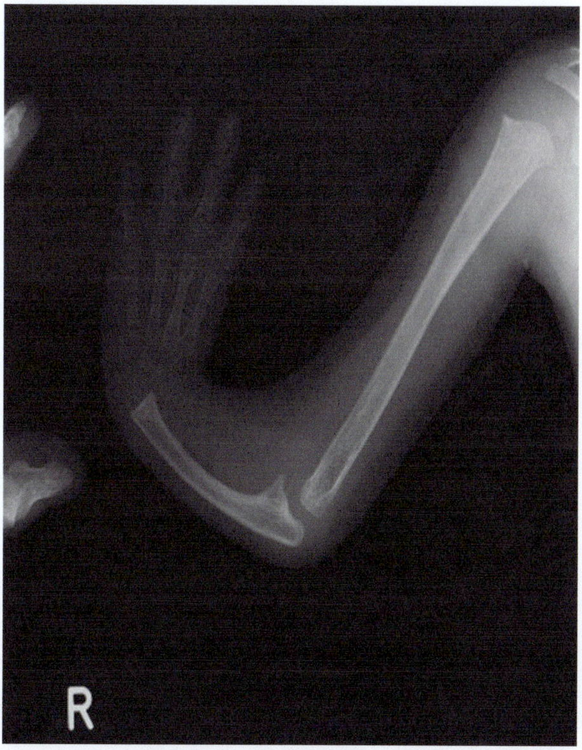

Fig. 11.2 Radial ray defects of Baller–Gerold syndrome revealed by X-ray examination

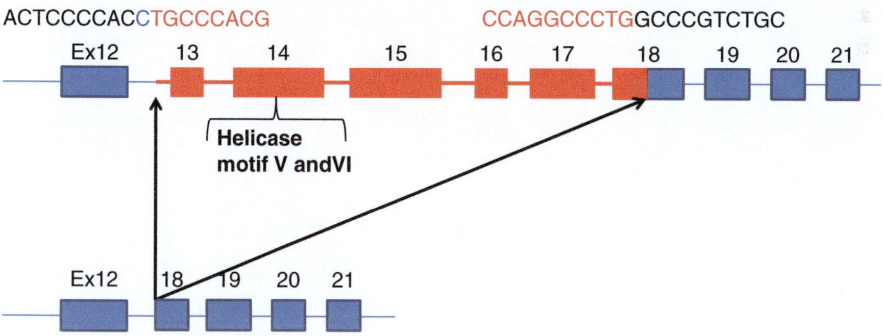

Fig. 11.3 Japanese Baller–Gerold syndrome detected homozygous deletion of the RECQL4 gene from intron 12 to the former part of exon 18, resulting in the deletion of amino acids after 687th. Red boxes indicated the deleted exons. Exon 14 encodes the helicase motif V and VI

11.4 Pathological Conditions

RECQL4 dysfunction leads to abnormal gene replication, increased sensitivity to oxidants, and abnormal DNA repair. It is considered that *RECQL4* dysfunction causes the characteristic skin findings and osteosarcoma, because RECQL4 plays a particularly significant role in the replication and repair of bone and skin tissue.

11.5 Diagnosis and Differential Diagnosis

Poikiloderma is a pathognomonic symptom for the diagnosis of Rothmund–Thomson syndrome. In addition Rothmund–Thomson syndrome is suspected when patients have small stature, skeletal abnormalities (dental hypoplasia and reduced bone density), sparse scalp hair, juvenile cataracts, intractable diarrhea in infancy, and nail abnormalities (Fig. 11.4). All patients with osteosarcoma should be examined for Rothmund–Thomson syndrome. Those patients with a family history of Rothmund–Thomson syndrome should undergo *RECQL4* screening; and those who found to carry *RECQL4* mutations should be diagnosed as having Rothmund–Thomson syndrome. Because *RECQL4* mutations are not found in approximately

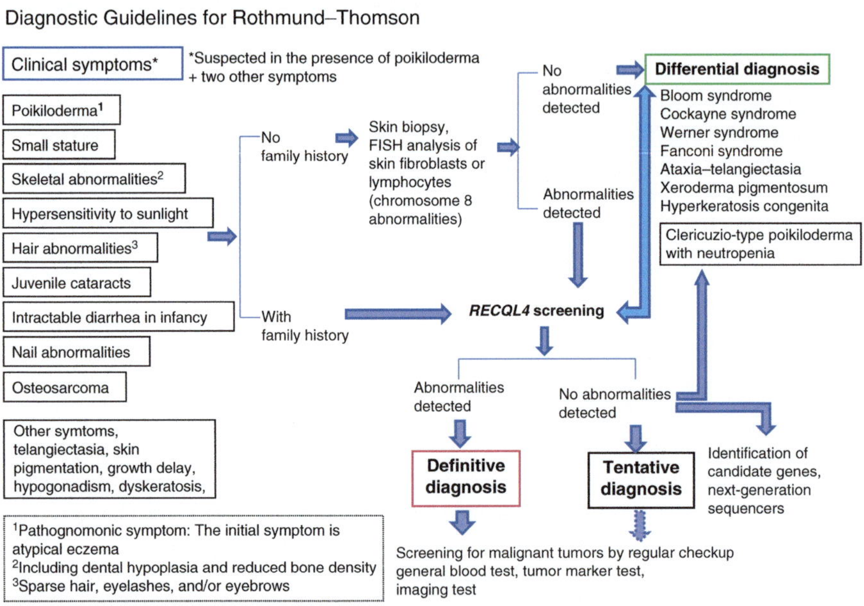

Fig. 11.4 Diagnostic guidelines for Rothmund–Thomson syndrome

40% of patients with Rothmund–Thomson syndrome, the diagnosis should be tentative, and at some time later, these patients should be screened for candidate genes using next-generation sequencers. When chromosome 8 abnormalities are detected by fluorescence in situ hybridization (FISH) analysis in patients without a family history of Rothmund–Thomson syndrome, they should undergo *RECQL4* screening. Skeletal abnormalities are frequent in persons with Rothmund-Thomson syndrome with *RECQL4* mutations compared with persons without *RECQL4* mutations. If they do not have *RECQL4* mutations, a diagnosis which differentiates from other diseases is required. The suspected diseases are discussed below.

The components of the causative genes for Bloom syndrome and Werner syndrome are helicases, which are the same as that for Rothmund–Thomson syndrome. Bloom syndrome and Werner syndrome show similar phenotypic features to Rothmund–Thomson syndrome [11]. Bloom syndrome is characterized by small stature, redness of the skin due to hypersensitivity to sunlight, and immune deficiency. In a chromosome test, the frequency of sister chromatid exchamges is high in patients with Bloom syndrome. Werner syndrome is typically associated with premature aging that is characterized by cataracts in both eyes, premature graying of the hair, and the calcification of subcutaneous tissue. Ataxia-telangiectasia is characterized by telangiectasia of the eyes, progressive ataxia, and immune deficiency. Xeroderma pigmentosum is characterized by hypersensitivity to sunlight from early postnatal life. The incidence of skin cancer is high in patients with xeroderma pigmentosum. Hyperkeratosis congenita is characterized by abnormal skin pigmentation, nail abnormalities, and leukoplakia and is associated with abnormal myeloid differentiation. Clericuzio-type poikiloderma with neutropenia, which is caused by the mutation of USB1 gene, is often diagnosed as Rothmund-Thomson syndrome.

11.6 Treatment and Prognosis

Patients with Rothmund–Thomson syndrome require regular follow-ups owing to the risk of cancer development. The life expectancy of those patients who do not have cancers is not poor. The 5-year survival rate for patients with osteosarcoma that developed as a complication of Rothmund–Thomson syndrome and that for patients with osteosarcoma that is not a complication of Rothmund–Thomson syndrome is similar (60–70%). Cataract and skeletal abnormalities are treated mainly by supportive measures. Genetic counseling should be provided to patients and their families.

References

1. Rothmund A. Uber cataracte in Verbindung mit einer eigenthuemlichen Hautdegeneration. Albrecht von Graefes Arch Klin Exp Ophthalmol. 1868;14:159–82.

2. Thomson MS. Poikiloderma congenitale. Br J Dermatol. 1936;48:221–34.
3. Kitao S, Shimamoto A, Goto M, Miller RW, Smithson WA, Lindor NM, Furuichi Y. Mutations in RECQL4 cause a subset of cases of Rothmund–Thomson syndrome. Nat Genet. 1999;22:82–4.
4. Larizza L, Roversi G, Volpi L. Rothmund–Thomson syndrome. Orphanet J Rare Dis. 2010;5:2.
5. Kaneko H (research representative). Survey of genetic repair defects (Bloom syndrome, Rothmund–Thomson syndrome, RAPADILINO syndrome, and Baller-Gerold syndrome) and research on early diagnosis. Research Grants for Research on Measures for Intractable Diseases supported by the Ministry of Health, Labour and Welfare of Japan. Annual Report 2011 (in Japanese).
6. Nakayama H. RecQ family helicases: roles as tumor suppressor proteins. Oncogene. 2002;21:9008–21.
7. Siitonen HA, Sotkasiira J, Biervliet M, Benmansour A, Capri Y, Cormier-Daire V, Crandall B, Hannula-Jouppi K, Hennekam R, Herzog D, Keymolen K, Lipsanen-Nyman M, Miny P, Plon SE, Riedl S, Sarkar A, Vargas FR, Verloes A, Wang LL, Kääriäinen H, Kestilä M. The mutation spectrum in RECQL4 disease. Eur J Hum Genet. 2009;17:151–8.
8. Siitonen HA, Kopra O, Kääriäinen H, Haravuori H, Winter RM, Säämänen AM, Peltonen L, Kestilä M. Molecular defect of RAPADILINO syndrome expands the phenotype spectrum of RECQL diseases. Hum Mol Genet. 2003;12:2837–44.
9. Kaneko H, Izumi R, Oda H, Ohara O, Sameshima K, Ohnishi H, Fukao T, Michinori Funato M. Nationwide survey of Baller-Gerold syndrome in Japanese population. Mol Med Rep. 2017;15(5):3222–4.
10. Kellermayer R, Siitonen HA, Hadzsiev K, Kestilä M, Kosztolanyl G. A patient with Rothmund–Thomson syndrome and all features of RAPADILINO. Arch Dermatol. 2005;141:617–20.
11. Kaneko H, Kondo N. Clinical features of Bloom syndrome and function of the causative gene, BLM helicase. Expert Rev Mol Diagn. 2004;4:393–401.

Chapter 12
Translesion DNA Synthesis

Chikahide Masutani and Fumio Hanaoka

Abstract Human DNA polymerase η (pol η) is the gene product that is altered in the variant form of xeroderma pigmentosum. Pol η has a structure that can accommodate the cyclobutane pyrimidine dimer, the most prominent ultraviolet-induced DNA lesion. Pol η catalyzes efficient and accurate translesion DNA synthesis (TLS) under the fine control of systems involving interactions with mono-ubiquitinated proliferating cell nuclear antigen. Pol η can also catalyze TLS past cisplatin lesions, which might contribute to the resistance of tumors to chemotherapy. Other Y-family polymerases, pol ι, pol κ, and REV1, and a B-family polymerase pol ζ can contribute to erroneous TLS past ultraviolet-induced lesions. However, these polymerases also contribute to the maintenance of genomic stability in the presence of their cognate DNA lesions. A-family polymerases, pol θ and pol ν, also have TLS abilities, and pol θ has an important role in an alternative end-joining repair pathway for DNA double-strand breaks, protecting against genomic instability. PrimPol is a protein with DNA polymerase and primase activities that is capable of initiating *de novo* DNA/RNA synthesis and that also has the capacity to bypass modifications that stall the replisome, by TLS or origin-independent re-priming. This chapter summarizes our current knowledge relating to DNA polymerases that are capable of catalyzing TLS.

Keywords Xeroderma pigmentosum variant · Translesion DNA synthesis · DNA damage tolerance · DNA polymerase eta

---

C. Masutani (✉)
Department of Genome Dynamics, Research Institute of Environmental Medicine, Nagoya University, Nagoya, Japan
e-mail: masutani@riem.nagoya-u.ac.jp

F. Hanaoka
Department of Life Science, Graduate School of Science, Gakushuin University, Tokyo, Japan

Life Science Center for Survival Dynamics, Tsukuba Advanced Research Alliance, University of Tsukuba, Tsukuba-shi, Ibaraki, Japan

C. Nishigori, K. Sugasawa (eds.), *DNA Repair Disorders*,
https://doi.org/10.1007/978-981-10-6722-8_12

12.1 Introduction

Cells have DNA damage tolerance mechanisms to overcome replication blockage caused by lesions on DNA template strands. These mechanisms are classified into two distinct pathways: translesion DNA synthesis (TLS) and homology-directed repair (or template switch, about which little is currently known). In TLS, specialized DNA polymerases catalyze DNA synthesis with damaged DNA as the template. To date, 15 template DNA-dependent DNA polymerases have been identified in human cells (Table 12.1). They are classified in five families by the similarities of their amino acid sequences. Pol α, pol δ, and pol ε are B-family polymerases with important roles in maintaining genomic stability, especially during DNA replication. Pol β is essential for base excision repair, and other X-family polymerases are thought to be involved in nonhomologous end-joining repair. Pol γ belongs to the A family and replicates mitochondrial DNA. Y-family DNA polymerases have central roles in TLS [1]. In addition to Y-family polymerases, DNA polymerases belonging to the A, B, X, and archaeo-eukaryotic primase (AEP) families can catalyze TLS. In this chapter, we summarize the properties of Y-family polymerases and other polymerases with TLS activities.

12.2 Pol η

12.2.1 Biological Properties

RAD30 in *Saccharomyces cerevisiae* was the first gene that was shown to encode a DNA polymerase that can replicate efficiently past a *cis-syn* cyclobutane pyrimidine dimer (CPD) [2]; this enzyme was named DNA polymerase η (pol η). In the same year, human pol η was discovered by purification of a protein from HeLa cell nuclear extracts that corrected defective TLS for CPD lesions of xeroderma pigmentosum variant (XP-V) cell extracts [3, 4]. The human protein (which is encoded by *POLH*) was also independently identified on the basis of its homology to the yeast *RAD30* gene product [5]. Introduction of *POLH* cDNA corrects the ultraviolet sensitivity of XP-V cells [6, 7].

Table 12.1 DNA-dependent human DNA polymerases

Family	Name	Function
A	Pol γ	Mitochondrial DNA replication and repair
	Pol ν, θ	TLS, micro homology-directed end-joining repair
B	Pol α, δ, ε	Nuclear DNA replication and repair
	Pol ζ	TLS, mutagenesis, Fanconi anemia pathway
X	Pol β, λ, μ	Base excision repair, nonhomologous end-joining repair
Y	Pol η, ι, κ, Rev1	TLS, mutagenesis
AEP	PrimPol	TLS, re-priming DNA synthesis

Pol η efficiently and accurately inserts two A nucleotides opposite template T-T CPD lesions at a level similar to that with undamaged TT [3, 8, 9]. Human pol η has a higher affinity for a template with a CPD lesion than for an undamaged TT template, and binding is stabilized by incorporation of dAMP opposite the 3′ T of the CPD, which enables pol η to incorporate a nucleotide opposite the 5′ T of the CPD. Pol η then preferentially incorporates two additional nucleotides beyond the TT dimer position, which makes the TLS patch resistant to the exonuclease activity of pol δ [10, 11]. Crystal structures of human pol η with template–primer DNA structures that can be produced during the bypass reaction through *cis-syn* cyclobutane thymine dimers show that pol η acts like a molecular splint to stabilize damaged DNA in a normal B-form conformation, and its large active site accommodates the thymine dimer with excellent stereochemistry [12]. In addition, catalysis of the formation of the phosphodiester bond by human pol η has been visualized by time-resolved X-ray crystallography [13]. Evidence indicates that the binding of dATP to the human pol η-DNA complex is thermodynamically favored for cyclobutane T-T dimer-containing DNA over undamaged DNA [14]. These observations support the suggested importance of pol η in TLS related to CPDs, the most prominent DNA lesions induced by ultraviolet irradiation. In contrast to CPDs, pol η rarely bypasses 6-4-photoproduct (6-4PP) lesions, which are also induced by ultraviolet irradiation (Fig. 12.1). Although pol η is able to insert a G nucleotide opposite 6-4PP lesions, it is unable to extend from the inserted nucleotide [9, 15]. However, mutagenic TLS of 6-4PPs by pol η is observed in human cells, implying that other polymerase(s) accomplish this extension [16]. More importantly, nucleotide excision repair (NER), which is missing in cells from patients with xeroderma pigmentosum types A–G,

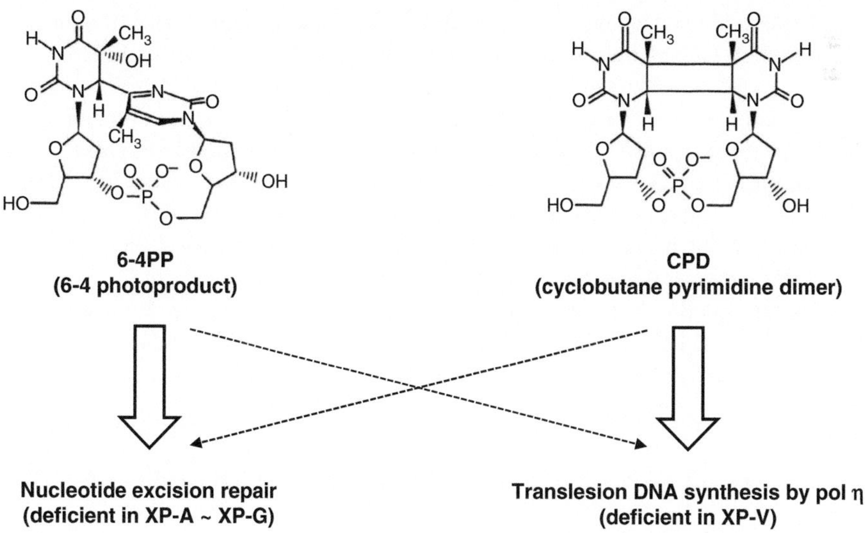

Fig. 12.1 Pathways for repair or bypass of DNA lesions induced by ultraviolet irradiation. Nucleotide excision repair and pol η-catalyzed translesion DNA synthesis have principal roles in the prevention of genomic instability by 6-4PPs and CPDs, respectively

removes 6-4PP efficiently and prevents skin cancer but removes CPDs inefficiently throughout the genome [17]. Thus, TLS and NER represent complementary systems that enable human cells to tolerate CPDs and 6-4PPs, respectively (Fig. 12.1).

Polh-knockout mice are viable and fertile and exhibit a high incidence of ultraviolet-induced epithelial skin tumors [18, 19]. Ultraviolet-induced mutation frequencies are significantly higher in epidermal cells of pol η-deficient mice than in wild-type mice [20, 21]. In addition, mouse embryonic fibroblasts (MEFs) from *Polh*-knockout mice display ultraviolet sensitivity and hypermutability [22]. These results indicate that pol η is the enzyme that is responsible for the prevention of ultraviolet-induced mutations.

Pol η mis-incorporates nucleotides with a frequency of 10^{-2} to 10^{-3} during the copying of undamaged templates [8, 23, 24]. Despite this low fidelity, overexpression of pol η in human fibroblasts rarely induces mutations [25], indicating that control mechanisms are in place to prevent untargeted mutagenesis. The low processivity of pol η suggests that mismatches could be removed by other exonuclease activities [26].

For some kinds of DNA lesions (other than those induced by ultraviolet), pol η is capable of catalyzing TLS, often mutagenically. Human pol η replication of 7,8-dihydro-8-oxoguanine (8-oxoG) lesions is efficient but error-prone [27–30]. It also efficiently incorporates correct and incorrect C and T bases opposite O^6-methyl-2′-deoxyguanosine (O^6-meG) [31]. Apurinic/apyrimidinic (AP) sites, benzo[*a*]pyrene diol epoxide-guanine (BPDE-G), 2-acetylaminofluorene-guanine (AAF-G), and thymine glycol lesions also result in inefficient and error-prone DNA synthesis by pol η [9, 30, 32–37]. Generally, it seems that Watson–Crick hydrogen bonding of base pairs is required for pol η lesion bypass reactions [33].

Human pol η can also bypass *cis*-diamminedichloroplatinum (cisplatin)-adducted GG (cisplatin-GG) by insertion of two C nucleotides [9, 38, 39]. Structural analyses of human pol η complexed with intrastrand cisplatin-1,2-cross-linked DNA demonstrate that nucleotidyl transfer requires the DNA to rotate into an active conformation [40]. Pol η undergoes a specific backbone rearrangement to accommodate the base dimer and minimize the DNA distortion around the lesion. The enzyme is able to incorporate deoxycytidine opposite cisplatin-cross-linked guanines but (as with 6-4PPs) is inefficient at extending primers after cisplatin lesions, which necessitates a second translesion DNA polymerase to complete bypass synthesis [39]. Human pol ζ can extend DNA synthesis after the pol η-mediated nucleotide insertion opposite cisplatin lesions both *in vitro* and in cells [41, 42]. Pol η is required for cellular tolerance to cisplatin [43, 44], suggesting that this enzyme contributes to chemotherapy resistance in tumors.

12.2.2 Domain Compositions and Mutations in Patients with XP-V

Although Y-family polymerases do not contain some sequences that are conserved among replicative DNA polymerases, they do have conventional right hand-like catalytic cores consisting of palm, finger, and thumb domains [45–47]. However,

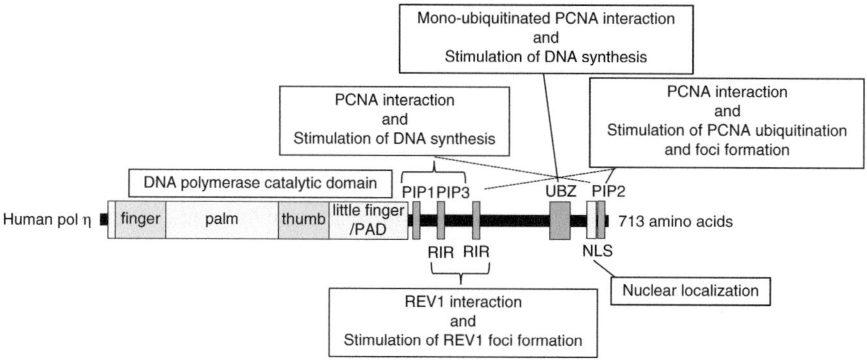

Fig. 12.2 Schematic representation of structural and functional domains of human pol η. The DNA polymerase catalytic domain consists of palm, finger, thumb, and little finger (PAD) domains. Positions of PIP (PCNA-interacting protein), RIR (REV1-interaction region), UBZ (ubiquitin-binding zinc finger), and NLS (nuclear localization signal) and their roles are indicated

Y-family polymerases have spacious active sites that facilitate lesion bypass, compensating for their low fidelity. The finger and thumb domains of Y-family polymerases are smaller and stubbier than those of other polymerase families. In addition, Y-family polymerases have a unique domain called the little finger or polymerase-associated domain. This structure weakens the interactions between polymerases, DNA, and incoming nucleotides, contributing to the low processivity and fidelity of Y-family polymerases.

Human pol η consists of 713 amino acids and contains residues for catalytic domains at its N terminus (Fig. 12.2). The originally identified protein that corrected replication defects of XP-V cell-free extracts consisted of the N-terminal 511 amino acids and was truncated at its C-terminus [3, 4]. The N-terminal 432 residues of pol η exhibit basal DNA polymerase activity [48]. These observations indicate that the C-terminal residues are dispensable for DNA polymerase activity.

The C terminus of pol η contains regulatory elements for cellular functions. A sequence that is located close to the C terminus is important for nuclear localization of pol η in human cells [6]. Human pol η also has three proliferating cell nuclear antigen (PCNA)-interacting protein (PIP) motifs, one ubiquitin-binding zinc finger (UBZ) motif, and two REV1-interacting region (RIR) motifs, one of which overlaps with a PIP.

In response to ultraviolet irradiation, pol η forms nuclear foci that co-localize with PCNA. Formation of pol η foci is almost completely abolished by mutations in the C-terminal PIP (PIP2) sequence [6]. However, such mutations only partially compromise the ability to rescue the ultraviolet sensitivity of XP-V cells. The UBZ domain is required for the interaction of pol η with mono-ubiquitinated PCNA, and point mutations in the UBZ domain severely affect the ability of XP-V cells to cope with ultraviolet-induced DNA damage [49, 50], probably by disturbing the chromatin loading of pol η [51]. Another PIP-like domain (PIP1) is located close to the DNA polymerase domain [52], and an allele encoding pol η lacking the C terminus (including the UBZ sequence) but retaining an intact PIP1 can promote cellular survival in response to ultraviolet irradiation [48]. The roles of the different regions in the functions of pol η were unclear until the identification of the third PIP (PIP3), which overlaps with one of the RIR motifs [51]. The three PIPs contribute differentially to two distinct functions: stimulation of DNA synthesis and promotion of PCNA ubiquitination (Fig. 12.2). The latter function is strongly associated with formation of nuclear pol η foci. Importantly, the three PIP boxes and UBZ exert redundant and additive effects in cells, enabling pol η activity to be finely regulated by interaction with mono-ubiquitinated PCNA. Mutations in these motifs are rarely found in patients because of their functional redundancy in cellular survival of ultraviolet irradiation.

Mutations in the *POLH* gene have been extensively characterized [4, 53–57]. Many of the mutations found in patients result in severe truncations of the protein and are effectively null alleles. Some patients have truncations at the C terminus such that the catalytic domains are intact, and extracts from cells of these patients are competent for TLS. However, these truncated proteins do not localize in the nucleus and fail to correct the defects in XP-V cells [6]. Missense mutations located in the conserved catalytic domain of the protein have also been identified. These mutations often correlate with protein stability determined by structural modeling of the pol η polymerase domain. A patient was identified with two missense mutations; one mutation translated to the nuclear localization signal and the second to the stop codon, leading to a protein with eight extra amino acids. This longer protein bypasses CPD lesions *in vitro* but is extremely unstable and nearly undetectable in cells [58].

Human cells have Y-family polymerases pol ι, pol κ, and REV1 in addition to pol η, but, unlike in *POLH*, no mutations have been identified in *POLI*, *POLK*, and *REV1* in patients with XP-V to date.

12.3 Other TLS Polymerases

12.3.1 Pol ι

The gene encoding pol ι (*POLI*) was identified as a paralog of *POLH* [59]. Pol ι is a highly error-prone enzyme that preferentially mis-incorporates G opposite a template T [60, 61]. The fidelity and efficiency of pol ι differ greatly depending on the template nucleotides. Biochemical and structural analyses revealed that human pol ι is specialized for Hoogsteen base pairing, whereby the templating base is driven to the *syn* conformation [62–64]. The narrow active site prevents Watson–Crick bond formation with an incoming dATP, with a three- to tenfold preference for the wobble base G over A [60, 61, 65, 66].

Pol ι rarely replicates through CPD lesions, but pol ι-dependent bypass of thymine-uracil (T-U) CPD does occur, with mis-incorporation of G opposite the 3′ U at a tenfold lower frequency than correct incorporation of A [67]. Because C in an ultraviolet-induced photoproduct undergoes deamination to produce U, mis-incorporation of G opposite U can decrease the mutagenic potential of T-C CPD lesions [68]. Pol ι preferentially incorporates A opposite the 3′ T of 6-4PPs [8, 69, 70].

Pol ι efficiently and accurately replicates small N^2-guanine adducts that inhibit Watson–Crick hydrogen bonding by forming Hoogsteen hydrogen bonds [71–73]. Pol ι can also replicate through 8-oxoG lesions, with predominant insertion of a correct C base via Hoogsteen bonding [68, 70, 74]. Pol ι is also reported to contribute to tolerance for cross-linker-induced replication stress, together with p53 [75].

Because pol ι protein expression is naturally deficient in the strains derived from 129 mice [76], the allele from these strains has been commonly utilized for analysis of the *Poli* gene. However, 129-derived mouse strains also have a small amount of a splice variant form of pol ι protein that is catalytically active but unstable [77]. Several reports suggest that hypermutation in XP-V cells is attributable to inaccurate TLS by pol ι [16, 78–81]. In addition, the induction of mutations by ultraviolet irradiation in epidermal cells of pol η-deficient mice is reduced by the addition of pol ι deficiency [21]. On the other hand, *Poli* deficiency, but not *Polh* deficiency, promotes the formation of mesenchymal tumors induced by ultraviolet irradiation [19]. Although induction of mutagenesis by ultraviolet irradiation in *Polh*-deficient MEFs is not reduced in *Polh–Poli*-deficient cells, it is lower in *Polh–Poli–Polk*-deficient MEFs [22], suggesting that pol κ has a larger contribution than pol ι to mutation in a pol η-deficient background. During somatic hypermutation of antibody genes, pol ι contributes to the generation of tandem mutations together with pol ζ [82], although pol ι deficiency does not alter the mutational spectra [83].

12.3.2 Pol κ

Pol κ proteins are orthologs of the *E. coli* DinB protein [84–87]. Pol κ misincorporates nucleotides with a frequency of about 10^{-2} to 10^{-4} [85, 88]. Unlike pol η [25], ectopic overexpression of pol κ induces chromosomal instability in human cells [89]. In mice, ultraviolet irradiation and treatment with 3-methylcholanthrene induces *Polk* expression [90, 91].

Pol κ is unable to bypass CPD, 6-4PP, and cisplatin-GG lesions [85, 87, 92–94]. Pol κ bypassing of AP sites depends on the surrounding sequence context and occurs via a single-nucleotide deletion [87, 94]. Pol κ can catalyze error-prone bypass past AAF-G and 8-oxoG lesions predominantly via insertions of T and A, respectively [87, 94, 95].

Pol κ can accurately replicate thymine glycols and bulky N^2-guanine lesions, such as BPDE-G, N^2-[3-methoxyestra-1,3,5(10)-trien-6-yl]-2′-guanine (an estrogen-derived adduct), bulky N^2-alkyl-G adducts, and N^2-furfuryl-G [37, 71, 94, 96–99]. Pol κ stabilizes BPDE-G-adducted DNA in the same active conformation as regular B-form DNA substrates and maintains a Watson–Crick base pair of BPDE-G:C, ensuring correct nucleotide insertion opposite the bulky adduct [100].

In addition to the nucleotide insertion activity, pol κ is a promiscuous extender of mispaired primer termini [93], producing deletion and insertion mutations at high rates [101, 102]. Pol κ elongates from G opposite a 3′ T of a CPD, A opposite 8-oxoG, T opposite O^6-meG, and G and T opposite BPDE-G [37, 93, 103].

Polk-deficient mice are viable, fertile, and able to mount a normal immune response [104]. *Polk*-knockout mouse embryonic stem cells and MEFs exhibit significant and moderate sensitivity to BPDE and ultraviolet irradiation, respectively [104–107]. Notably, *Polk*-knockout embryonic stem cells show increased numbers of BPDE-induced mutations with a characteristic spectrum [106, 108]. Ultraviolet sensitivity of pol κ-depleted human cells and *Polk*-deficient MEFs seems to be additive with pol η depletion and *Polh* deficiency, respectively, implying distinct roles for the two polymerases for the bypass of lesions induced by ultraviolet irradiation. Pol κ acts together with pol ζ and REV1 to bypass ultraviolet-induced lesions and BPDE-G lesions [22, 42, 81, 109, 110].

12.3.3 REV1

The mammalian genes encoding REV1 were identified as orthologs of yeast *REV1* [111–113]. REV1 preferentially inserts C efficiently opposite template G, U, O^6-meG, AP sites, and N^2-adducted guanines but inefficiently opposite 8-oxoG [113–115].

Downregulation of *REV1* in human cells enhances sensitivity to cisplatin and ionizing radiation and reduces induction of mutations by ultraviolet irradiation and cisplatin [111, 116, 117]. However, REV1 is unable to insert any nucleotides opposite template 6-4PP and CPD lesions [115]. The non-catalytic BRCA1 C terminus (BRCT) domain of REV1 seems to be required to bypass 6-4PP lesions. Analyses of *Rev1*-knockout mice and mice genetically engineered to express BRCT domain-truncated or catalytically dead Rev1 support the idea that Rev1 has non-catalytic functions in TLS via the BRCT domain [118–120]. *Rev1*-knockout mice were generated from a 129/OLA-derived embryonic stem cell line, and, although subsequent backcrosses to 129/OLA produced mice with reduced body size, no *Rev1*-knockout mice were obtained beyond the F2 backcross into C57BL/6 mice, indicating strain dependence of the phenotype. Notably, despite a reduction in ultraviolet-induced mutagenesis in *Rev1*-deficient cells, ultraviolet-induced skin carcinogenesis is accelerated in *Rev1*-knockout mice, which is associated with the induction of inflammatory hyperplasia [121].

Mammalian REV1 proteins interact with pol η, pol ι, pol κ, REV7, and the REV3–REV7 complex through the REV1 C-terminal region [122–129]. These interactions are mutually exclusive [122, 128]. Two RIRs have been identified in human pol η, and one RIR has been found in each of pol ι and pol κ [127, 130]. The REV1–pol η interaction enhances the accumulation of endogenous REV1 at sites of damage by ultraviolet irradiation and suppresses spontaneous mutagenesis in human cells [131], although exogenously expressed REV1 is able to form foci in a pol η-independent manner [129]. In addition, the REV1–pol κ interaction seems to be essential for the TLS functions of pol κ in MEFs [127], and the REV1–pol η interaction also seems to be required for the TLS function of pol κ [22]. RIR motifs of Y-family polymerases overlap with PIP motifs, contributing to fine regulation of TLS polymerases [51, 132]. Although REV1 is indispensable for TLS mediated by pol η, pol ι, and pol κ, it is not required for TLS by pol ζ and promotes predominantly error-free TLS opposite ultraviolet-induced lesions in humans [133]. These observations imply a hierarchy for the recruitment of Y-family polymerases for TLS reactions in mammals.

12.3.4 Pol ζ

Pol ζ, which contributes to TLS, belongs to the B family that includes the replicative DNA polymerases, pol α, pol δ, and pol ε. At a minimum, pol ζ consists of a catalytic subunit REV3 and an accessory subunit REV7. Mammalian REV3 (DNA polymerase ζ catalytic subunit, encoded by *REV3L*) and REV7 (mitotic arrest-deficient 2-like protein 2, encoded by *MAD2L2*) were identified on the basis of homology and from the results of a yeast two-hybrid screen, respectively [134–136]. Expression of human *REV3L* is damage inducible [137]. In addition to these core subunits, as in yeast, human pol ζ can contain POLD2 and POLD3, which are also subunits of pol δ [41].

Downregulation of *REV3L* in human cells does not affect sensitivity to ultraviolet irradiation, BPDE, and cisplatin but reduces mutations induced by them [134, 138–140]. *REV3L*-knockout human cells show enhanced UV sensitivity and reduced UV mutagenesis compared with wild-type cells [79]. Thus, pol ζ is responsible for mutagenic TLS past these lesions, apparently with the participation of pol κ [42, 81, 109, 110]. *Rev3l*-knockout mice are embryonically lethal [141–145]. MEFs established from *Rev3l*-knockout mice with a *Trp53*-deficient background exhibit sensitivity to ultraviolet radiation, methyl methanesulfonate, mitomycin C (MMC), and ionizing radiation [146, 147]. Analysis of conditional *Rev3l*-knockout mice demonstrates that *Rev3l* deficiency enhances spontaneous tumorigenesis [148].

Depletion of REV7 in human cells enhances sensitivity to cisplatin and ultraviolet irradiation and results in a reduced frequency of induced mutations [149, 150]. *Mad2l2*-knockout mice exhibit growth retardation and partial embryonic lethality, and mice that survive to adulthood are infertile because of progressive loss of primordial germ cells in the embryo [151]. Interaction of REV7 with REV3 is necessary to prevent spontaneous chromosome breaks and confer resistance to UV radiation and cisplatin, suggesting that REV7 contributes to damage tolerance by pol ζ [152]. Notably, REV7 has an important role as a barrier to homologous recombination, suppressing end resection at DNA double-strand break sites [153, 154]. In addition, *MAD2L2* has been identified as a Fanconi anemia gene, mutated in a patient with severe bone marrow failure [155], suggesting the contribution of pol ζ to interstrand cross-link repair.

12.3.5 Pol θ and Pol ν

Human pol θ and pol ν are encoded by *POLQ* and *POLN*, respectively; homologs of these genes are exclusively detected in the genomes of multicellular eukaryotes [156–158]. Although these polymerases belong to the A family, the conserved exonuclease domain is mutated in pol θ and pol ν, and they can catalyze low-fidelity DNA synthesis [156, 157, 159, 160].

Pol θ, but not pol ν, is able to replicate AP sites by the insertion of A [159, 160]. Pol ν can mediate error-free replication of templates with psoralen DNA interstrand cross-links [161]. Pol θ and pol ν are able to replicate DNA with thymine glycol lesions but are unable to incorporate any base opposite template CPD and 6-4PP lesions. However, pol θ is able to extend from an A opposite a 3′ T of a T-T 6-4PP lesion, following insertion by another polymerase [162].

A mouse strain with the *Polq* mutation *Polq*^{chaos1} (where "chaos" stands for "chromosome aberration occurring spontaneously") was isolated by the screening of a library for the phenotype of elevated levels of spontaneous and radiation-induced micronuclei in peripheral blood cells [163]. *Polq*-knockout mice are viable and fertile and display the "chaos" phenotype [164]. *Polq*-knockout and *Polq*^{chaos1} cells display high sensitivity to ionizing radiation, bleomycin, and etoposide but only slightly enhanced or normal sensitivity to lesions induced by ultraviolet irradiation, cisplatin, and MMC [164–166]. Intriguingly, pol θ has an important role in an alternative end-joining repair pathway for DNA double-strand breaks and protects against genomic instability [167, 168]. Pol θ accounts for most repairs associated with microhomologies (sequences with ≥2 bp of homology); the efficiency of this process results from the coupling of microhomology searching to removal of nonhomologous tails and microhomology-primed synthesis across broken ends. In contrast to nonhomologous end joining, pol θ-mediated end joining efficiently repairs end structures that are expected to occur after aborted homology-directed repair (5′–3′ resected ends) or replication fork collapse [169]. Structural analysis of the DNA polymerase domain of pol θ demonstrated that it uses a specialized thumb subdomain to establish unique upstream contacts to the primer DNA strand, including an interaction with the 3′-terminal phosphate of the primer by one of five distinctive insertion loops of the protein. These contacts enable pol θ to grasp the primer and bypass DNA lesions or extend poorly annealed DNA termini, to mediate end joining [170]. Pol θ has an N-terminal helicase domain, although no helicase activity has been detected [157].

Pol ν-depleted cells display sensitivity to MMC and a reduction in the frequency of homologous recombination [161, 171]. Structural analyses revealed that, instead of a simple open-to-closed movement of the O helix upon binding of a correct incoming nucleotide, pol ν has a different open state and requires the finger domain to swing sideways and undergo both opening and closing motions to accommodate the nascent base pair. A unique cavity and the flexibility of the thumb domain enable pol ν to generate and accommodate a looped-out primer strand. Primer loop-out may be a mechanism for DNA trinucleotide repeat expansion [172].

12.3.6 PrimPol

The DNA-directed primase–polymerase, PrimPol, is classified in the AEP superfamily. This enzyme is capable of initiating *de novo* DNA/RNA synthesis and DNA chain elongation and has the capacity to bypass modifications that stall the replisome by TLS or origin-independent re-priming, enabling discontinuous synthesis of the leading strand.

Human PrimPol has the ability to initiate DNA chains with deoxynucleotides, unlike regular primases, which exclusively use ribonucleotides [173]. PrimPol uses its primase activity to mediate uninterrupted fork progression after ultraviolet irradiation and to reinitiate DNA synthesis after dNTP depletion [174]. Although PrimPol is unable to directly replicate G quadruplexes, it can bind and re-prime downstream of these structures [175]. PrimPol, pol η, and pol ζ redundantly prevent cell death and facilitate unperturbed cell cycle progression in chicken cells [176]. PrimPol-deleted cells also exhibit increased sensitivity to chain-terminating nucleoside analogs, compared with wild-type cells [176].

In addition to re-priming, PrimPol is required for replication fork progression on DNA templates damaged by ultraviolet irradiation through its ability to catalyze TLS [177]. PrimPol is also a DNA polymerase that can bypass the most common oxidative lesions in DNA, such as abasic sites and 8-oxoG [173, 178]. PrimPol is able to extend RNA primers with ribonucleotides, even when bypassing 8-oxoG lesions, suggesting a possible role as a TLS polymerase assisting transcription. PrimPol displays a high degree of versatility to accept or induce distortions of both primer and template strands, creating alternative alignments based on microhomology that could serve to skip unreadable lesions and connect separate strands [179]. PrimPol is a mutagenic polymerase with a unique error specificity that is highly biased toward insertion–deletion errors [180]. Human PrimPol interacts directly with replication factor-a protein 1 (RPA1) and is recruited to sites of DNA damage and stalled replication forks in an RPA1-dependent manner in cells; the fidelity of *in vitro* DNA synthesis by PrimPol is also regulated through this interaction [180, 181].

By exome-sequencing analysis of a family with high myopia, a novel missense variant of the gene that encodes PrimPol has been identified [182]. This mutation decreases the activity of PrimPol *in vitro* and *in vivo*, suggesting that a DNA replication defect associated with this PrimPol mutant contributes to the onset of high myopia [183]. However, the mutation seems to occur randomly in the general population, and further studies are required to establish whether a general link exists between the PrimPol mutation and high myopia.

12.4 Concluding Remarks

Pol η has a crucial role in the prevention of skin cancers induced by ultraviolet irradiation, and deficiency of pol η results in a cancer-prone syndrome. In addition to pol η, cells have multiple DNA polymerases that vary in their abilities to bypass different kinds of DNA lesions. In the cells of patients with XP-V, in which pol η is inactivated, other TLS polymerases catalyze erroneous TLS instead of pol η, resulting in genomic instability that contributes to cancer formation. However, in addition to TLS pathways, human cells have another DNA damage tolerance mechanism called template switch or homology-directed repair, in which arrested DNA synthesis is restored by replication with a newly synthesized daughter strand as a template; this mechanism is, therefore, considered to be error-free. To determine why this error-free mechanism is not adopted in XP-V cells, it will be necessary to take a holistic view of the picture of DNA damage tolerance, to understand its perturbation in patients with XP-V, and to constitute a new strategy for therapy.

Acknowledgments This work was supported by Grants-in-Aid from the Ministry of Education, Culture, Sports, Science, and Technology of Japan and by Takeda Science Foundation.

References

1. Ohmori H, Friedberg EC, Fuchs RP, Goodman MF, Hanaoka F, Hinkle D, Kunkel TA, Lawrence CW, Livneh Z, Nohmi T, et al. The Y-family of DNA polymerases. Mol Cell. 2001;8:7–8.
2. Johnson RE, Prakash S, Prakash L. Efficient bypass of a thymine-thymine dimer by yeast DNA polymerase poleta. Science. 1999;283:1001–4.
3. Masutani C, Araki M, Yamada A, Kusumoto R, Nogimori T, Maekawa T, Iwai S, Hanaoka F. Xeroderma pigmentosum variant (XP-V) correcting protein from HeLa cells has a thymine dimer bypass DNA polymerase activity. EMBO J. 1999;18:3491–501.
4. Masutani C, Kusumoto R, Yamada A, Dohmae N, Yokoi M, Yuasa M, Araki M, Iwai S, Takio K, Hanaoka F. The *XPV* (xeroderma pigmentosum variant) gene encodes human DNA polymerase eta. Nature. 1999;399:700–4.
5. Johnson RE, Kondratick CM, Prakash S, Prakash L. hRAD30 mutations in the variant form of xeroderma pigmentosum. Science. 1999;285:263–5.
6. Kannouche P, Broughton BC, Volker M, Hanaoka F, Mullenders LH, Lehmann AR. Domain structure, localization, and function of DNA polymerase eta, defective in xeroderma pigmentosum variant cells. Genes Dev. 2001;15:158–72.
7. Yamada A, Masutani C, Iwai S, Hanaoka F. Complementation of defective translesion synthesis and UV light sensitivity in xeroderma pigmentosum variant cells by human and mouse DNA polymerase eta. Nucleic Acids Res. 2000;28:2473–80.
8. Johnson RE, Washington MT, Prakash S, Prakash L. Fidelity of human DNA polymerase eta. J Biol Chem. 2000;275:7447–50.
9. Masutani C, Kusumoto R, Iwai S, Hanaoka F. Mechanisms of accurate translesion synthesis by human DNA polymerase eta. EMBO J. 2000;19:3100–9.
10. Kusumoto R, Masutani C, Shimmyo S, Iwai S, Hanaoka F. DNA binding properties of human DNA polymerase eta: implications for fidelity and polymerase switching of translesion synthesis. Genes Cells. 2004;9:1139–50.

11. McCulloch SD, Kokoska RJ, Masutani C, Iwai S, Hanaoka F, Kunkel TA. Preferential *cis-syn* thymine dimer bypass by DNA polymerase eta occurs with biased fidelity. Nature. 2004;428:97–100.
12. Biertumpfel C, Zhao Y, Kondo Y, Ramon-Maiques S, Gregory M, Lee JY, Masutani C, Lehmann AR, Hanaoka F, Yang W. Structure and mechanism of human DNA polymerase eta. Nature. 2010;465:1044–8.
13. Nakamura T, Zhao Y, Yamagata Y, Hua YJ, Yang W. Watching DNA polymerase eta make a phosphodiester bond. Nature. 2012;487:196–201.
14. Ucisik MN, Hammes-Schiffer S. Relative binding free energies of adenine and guanine to damaged and undamaged DNA in human DNA polymerase eta: clues for fidelity and overall efficiency. J Am Chem Soc. 2015;137:13240–3.
15. Johnson RE, Haracska L, Prakash S, Prakash L. Role of DNA polymerase eta in the bypass of a (6-4) TT photoproduct. Mol Cell Biol. 2001;21:3558–63.
16. Yoon JH, Prakash L, Prakash S. Error-free replicative bypass of (6-4) photoproducts by DNA polymerase zeta in mouse and human cells. Genes Dev. 2010;24:123–8.
17. Naegeli H, Sugasawa K. The xeroderma pigmentosum pathway: decision tree analysis of DNA quality. DNA Repair (Amst). 2011;10:673–83.
18. Lin Q, Clark AB, McCulloch SD, Yuan T, Bronson RT, Kunkel TA, Kucherlapati R. Increased susceptibility to UV-induced skin carcinogenesis in polymerase eta-deficient mice. Cancer Res. 2006;66:87–94.
19. Ohkumo T, Kondo Y, Yokoi M, Tsukamoto T, Yamada A, Sugimoto T, Kanao R, Higashi Y, Kondoh H, Tatematsu M, et al. UV-B radiation induces epithelial tumors in mice lacking DNA polymerase eta and mesenchymal tumors in mice deficient for DNA polymerase iota. Mol Cell Biol. 2006;26:7696–706.
20. Ikehata H, Chang Y, Yokoi M, Yamamoto M, Hanaoka F. Remarkable induction of UV-signature mutations at the 3′-cytosine of dipyrimidine sites except at 5′-TCG-3′ in the UVB-exposed skin epidermis of xeroderma pigmentosum variant model mice. DNA Repair (Amst). 2014;22:112–22.
21. Kanao R, Yokoi M, Ohkumo T, Sakurai Y, Dotsu K, Kura S, Nakatsu Y, Tsuzuki T, Masutani C, Hanaoka F. UV-induced mutations in epidermal cells of mice defective in DNA polymerase eta and/or iota. DNA Repair (Amst). 2015;29:139–46.
22. Ito W, Yokoi M, Sakayoshi N, Sakurai Y, Akagi J, Mitani H, Hanaoka F. Stalled Poleta at its cognate substrate initiates an alternative translesion synthesis pathway via interaction with REV1. Genes Cells. 2012;17:98–108.
23. Matsuda T, Bebenek K, Masutani C, Hanaoka F, Kunkel TA. Low fidelity DNA synthesis by human DNA polymerase-eta. Nature. 2000;404:1011–3.
24. Matsuda T, Bebenek K, Masutani C, Rogozin IB, Hanaoka F, Kunkel TA. Error rate and specificity of human and murine DNA polymerase eta. J Mol Biol. 2001;312:335–46.
25. King NM, Nikolaishvili-Feinberg N, Bryant MF, Luche DD, Heffernan TP, Simpson DA, Hanaoka F, Kaufmann WK, Cordeiro-Stone M. Overproduction of DNA polymerase eta does not raise the spontaneous mutation rate in diploid human fibroblasts. DNA Repair (Amst). 2005;4:714–24.
26. Bebenek K, Matsuda T, Masutani C, Hanaoka F, Kunkel TA. Proofreading of DNA polymerase eta-dependent replication errors. J Biol Chem. 2001;276:2317–20.
27. Avkin S, Livneh Z. Efficiency, specificity and DNA polymerase-dependence of translesion replication across the oxidative DNA lesion 8-oxoguanine in human cells. Mutat Res. 2002;510:81–90.
28. Haracska L, Yu SL, Johnson RE, Prakash L, Prakash S. Efficient and accurate replication in the presence of 7,8-dihydro-8-oxoguanine by DNA polymerase eta. Nat Genet. 2000;25:458–61.
29. Su Y, Patra A, Harp JM, Egli M, Guengerich FP. Roles of residues Arg-61 and Gln-38 of human DNA polymerase eta in bypass of deoxyguanosine and 7,8-Dihydro-8-oxo-2′-deoxyguanosine. J Biol Chem. 2015;290:15921–33.
30. Zhang Y, Yuan F, Wu X, Rechkoblit O, Taylor JS, Geacintov NE, Wang Z. Error-prone lesion bypass by human DNA polymerase eta. Nucleic Acids Res. 2000;28:4717–24.

31. Patra A, Zhang Q, Guengerich FP, Egli M. Mechanisms of insertion of dCTP and dTTP opposite the DNA lesion O^6-methyl-2′-deoxyguanosine by human DNA polymerase eta. J Biol Chem. 2016;291(46):24304–13.
32. Chiapperino D, Kroth H, Kramarczuk IH, Sayer JM, Masutani C, Hanaoka F, Jerina DM, Cheh AM. Preferential misincorporation of purine nucleotides by human DNA polymerase eta opposite benzo[a]pyrene 7,8-diol 9,10-epoxide deoxyguanosine adducts. J Biol Chem. 2002;277:11765–71.
33. Haracska L, Washington MT, Prakash S, Prakash L. Inefficient bypass of an abasic site by DNA polymerase eta. J Biol Chem. 2001;276:6861–6.
34. Kusumoto R, Masutani C, Iwai S, Hanaoka F. Translesion synthesis by human DNA polymerase eta across thymine glycol lesions. Biochemistry. 2002;41:6090–9.
35. Patra A, Zhang Q, Lei L, Su Y, Egli M, Guengerich FP. Structural and kinetic analysis of nucleoside triphosphate incorporation opposite an abasic site by human translesion DNA polymerase eta. J Biol Chem. 2015;290:8028–38.
36. Yasui M, Dong H, Bonala RR, Suzuki N, Ohmori H, Hanaoka F, Johnson F, Grollman AP, Shibutani S. Mutagenic properties of 3-(deoxyguanosin-N2-yl)-2-acetylaminofluorene, a persistent acetylaminofluorene-derived DNA adduct in mammalian cells. Biochemistry. 2004;43:15005–13.
37. Zhang Y, Wu X, Guo D, Rechkoblit O, Geacintov NE, Wang Z. Two-step error-prone bypass of the (+)- and (−)-trans-anti-BPDE-N^2-dG adducts by human DNA polymerases eta and kappa. Mutat Res. 2002;510:23–35.
38. Vaisman A, Masutani C, Hanaoka F, Chaney SG. Efficient translesion replication past oxaliplatin and cisplatin GpG adducts by human DNA polymerase eta. Biochemistry. 2000;39:4575–80.
39. Zhao Y, Biertumpfel C, Gregory MT, Hua YJ, Hanaoka F, Yang W. Structural basis of human DNA polymerase eta-mediated chemoresistance to cisplatin. Proc Natl Acad Sci U S A. 2012;109:7269–74.
40. Alt A, Lammens K, Chiocchini C, Lammens A, Pieck JC, Kuch D, Hopfner KP, Carell T. Bypass of DNA lesions generated during anticancer treatment with cisplatin by DNA polymerase eta. Science. 2007;318:967–70.
41. Lee YS, Gregory MT, Yang W. Human Pol zeta purified with accessory subunits is active in translesion DNA synthesis and complements Pol eta in cisplatin bypass. Proc Natl Acad Sci U S A. 2014;111:2954–9.
42. Shachar S, Ziv O, Avkin S, Adar S, Wittschieben J, Reissner T, Chaney S, Friedberg EC, Wang Z, Carell T, et al. Two-polymerase mechanisms dictate error-free and error-prone translesion DNA synthesis in mammals. EMBO J. 2009;28:383–93.
43. Albertella MR, Green CM, Lehmann AR, O'Connor MJ. A role for polymerase eta in the cellular tolerance to cisplatin-induced damage. Cancer Res. 2005;65:9799–806.
44. Bassett E, King NM, Bryant MF, Hector S, Pendyala L, Chaney SG, Cordeiro-Stone M. The role of DNA polymerase eta in translesion synthesis past platinum-DNA adducts in human fibroblasts. Cancer Res. 2004;64:6469–75.
45. Prakash S, Johnson RE, Prakash L. Eukaryotic translesion synthesis DNA polymerases: specificity of structure and function. Annu Rev Biochem. 2005;74:317–53.
46. Sale JE, Lehmann AR, Woodgate R. Y-family DNA polymerases and their role in tolerance of cellular DNA damage. Nat Rev Mol Cell Biol. 2012;13:141–52.
47. Yang W. Damage repair DNA polymerases Y. Curr Opin Struct Biol. 2003;13:23–30.
48. Acharya N, Yoon JH, Hurwitz J, Prakash L, Prakash S. DNA polymerase eta lacking the ubiquitin-binding domain promotes replicative lesion bypass in humans cells. Proc Natl Acad Sci U S A. 2010;107:10401–5.
49. Bienko M, Green CM, Crosetto N, Rudolf F, Zapart G, Coull B, Kannouche P, Wider G, Peter M, Lehmann AR, et al. Ubiquitin-binding domains in Y-family polymerases regulate translesion synthesis. Science. 2005;310:1821–4.

50. Kannouche PL, Wing J, Lehmann AR. Interaction of human DNA polymerase eta with monoubiquitinated PCNA: a possible mechanism for the polymerase switch in response to DNA damage. Mol Cell. 2004;14:491–500.
51. Masuda Y, Kanao R, Kaji K, Ohmori H, Hanaoka F, Masutani C. Different types of interaction between PCNA and PIP boxes contribute to distinct cellular functions of Y-family DNA polymerases. Nucleic Acids Res. 2015;43:7898–910.
52. Acharya N, Yoon JH, Gali H, Unk I, Haracska L, Johnson RE, Hurwitz J, Prakash L, Prakash S. Roles of PCNA-binding and ubiquitin-binding domains in human DNA polymerase eta in translesion DNA synthesis. Proc Natl Acad Sci U S A. 2008;105:17724–9.
53. Broughton BC, Cordonnier A, Kleijer WJ, Jaspers NG, Fawcett H, Raams A, Garritsen VH, Stary A, Avril MF, Boudsocq F, et al. Molecular analysis of mutations in DNA polymerase eta in xeroderma pigmentosum-variant patients. Proc Natl Acad Sci U S A. 2002;99:815–20.
54. Inui H, Oh KS, Nadem C, Ueda T, Khan SG, Metin A, Gozukara E, Emmert S, Slor H, Busch DB, et al. Xeroderma pigmentosum-variant patients from America, Europe, and Asia. J Invest Dermatol. 2008;128:2055–68.
55. Ono R, Masaki T, Takeuchi S, Shimizu A, Tanioka M, Kambe N, Matsue H, Kamide R, Nishigori C. Three school-age cases of xeroderma pigmentosum variant type. Photodermatol Photoimmunol Photomed. 2013;29:132–9.
56. Opletalova K, Bourillon A, Yang W, Pouvelle C, Armier J, Despras E, Ludovic M, Mateus C, Robert C, Kannouche P, et al. Correlation of phenotype/genotype in a cohort of 23 xeroderma pigmentosum-variant patients reveals 12 new disease-causing POLH mutations. Hum Mutat. 2014;35:117–28.
57. Tanioka M, Masaki T, Ono R, Nagano T, Otoshi-Honda E, Matsumura Y, Takigawa M, Inui H, Miyachi Y, Moriwaki S, et al. Molecular analysis of DNA polymerase eta gene in Japanese patients diagnosed as xeroderma pigmentosum variant type. J Invest Dermatol. 2007;127:1745–51.
58. Ahmed-Seghir S, Pouvelle C, Despras E, Cordonnier A, Sarasin A, Kannouche PL. Aberrant C-terminal domain of polymerase eta targets the functional enzyme to the proteosomal degradation pathway. DNA Repair (Amst). 2015;29:154–65.
59. McDonald JP, Rapic-Otrin V, Epstein JA, Broughton BC, Wang X, Lehmann AR, Wolgemuth DJ, Woodgate R. Novel human and mouse homologs of *Saccharomyces cerevisiae* DNA polymerase eta. Genomics. 1999;60:20–30.
60. Tissier A, McDonald JP, Frank EG, Woodgate R. Poliota, a remarkably error-prone human DNA polymerase. Genes Dev. 2000;14:1642–50.
61. Zhang Y, Yuan F, Wu X, Wang Z. Preferential incorporation of G opposite template T by the low-fidelity human DNA polymerase iota. Mol Cell Biol. 2000;20:7099–108.
62. Johnson RE, Prakash L, Prakash S. Biochemical evidence for the requirement of Hoogsteen base pairing for replication by human DNA polymerase iota. Proc Natl Acad Sci U S A. 2005;102:10466–71.
63. Nair DT, Johnson RE, Prakash L, Prakash S, Aggarwal AK. An incoming nucleotide imposes an anti to syn conformational change on the templating purine in the human DNA polymerase-iota active site. Structure. 2006;14:749–55.
64. Nair DT, Johnson RE, Prakash S, Prakash L, Aggarwal AK. Replication by human DNA polymerase-iota occurs by Hoogsteen base-pairing. Nature. 2004;430:377–80.
65. Choi JY, Lim S, Eoff RL, Guengerich FP. Kinetic analysis of base-pairing preference for nucleotide incorporation opposite template pyrimidines by human DNA polymerase iota. J Mol Biol. 2009;389:264–74.
66. Kirouac KN, Ling H. Structural basis of error-prone replication and stalling at a thymine base by human DNA polymerase iota. EMBO J. 2009;28:1644–54.
67. Vaisman A, Takasawa K, Iwai S, Woodgate R. DNA polymerase iota-dependent translesion replication of uracil containing cyclobutane pyrimidine dimers. DNA Repair (Amst). 2006;5:210–8.

68. Vaisman A, Woodgate R. Unique misinsertion specificity of poliota may decrease the mutagenic potential of deaminated cytosines. EMBO J. 2001;20:6520–9.
69. Tissier A, Frank EG, McDonald JP, Iwai S, Hanaoka F, Woodgate R. Misinsertion and bypass of thymine-thymine dimers by human DNA polymerase iota. EMBO J. 2000;19:5259–66.
70. Zhang Y, Yuan F, Wu X, Taylor JS, Wang Z. Response of human DNA polymerase iota to DNA lesions. Nucleic Acids Res. 2001;29:928–35.
71. Choi JY, Guengerich FP. Kinetic evidence for inefficient and error-prone bypass across bulky N^2-guanine DNA adducts by human DNA polymerase iota. J Biol Chem. 2006;281:12315–24.
72. Pence MG, Blans P, Zink CN, Hollis T, Fishbein JC, Perrino FW. Lesion bypass of N2-ethylguanine by human DNA polymerase iota. J Biol Chem. 2009;284:1732–40.
73. Washington MT, Minko IG, Johnson RE, Haracska L, Harris TM, Lloyd RS, Prakash S, Prakash L. Efficient and error-free replication past a minor-groove N^2-guanine adduct by the sequential action of yeast Rev1 and DNA polymerase zeta. Mol Cell Biol. 2004;24:6900–6.
74. Kirouac KN, Ling H. Unique active site promotes error-free replication opposite an 8-oxoguanine lesion by human DNA polymerase iota. Proc Natl Acad Sci U S A. 2011;108:3210–5.
75. Hampp S, Kiessling T, Buechle K, Mansilla SF, Thomale J, Rall M, Ahn J, Pospiech H, Gottifredi V, Wiesmuller L. DNA damage tolerance pathway involving DNA polymerase iota and the tumor suppressor p53 regulates DNA replication fork progression. Proc Natl Acad Sci U S A. 2016;113:E4311–9.
76. McDonald JP, Frank EG, Plosky BS, Rogozin IB, Masutani C, Hanaoka F, Woodgate R, Gearhart PJ. 129-derived strains of mice are deficient in DNA polymerase iota and have normal immunoglobulin hypermutation. J Exp Med. 2003;198:635–43.
77. Aoufouchi S, De Smet A, Delbos F, Gelot C, Guerrera IC, Weill JC, Reynaud CA. 129-derived mouse strains express an unstable but catalytically active DNA polymerase iota variant. Mol Cell Biol. 2015;35:3059–70.
78. Dumstorf CA, Clark AB, Lin Q, Kissling GE, Yuan T, Kucherlapati R, McGregor WG, Kunkel TA. Participation of mouse DNA polymerase iota in strand-biased mutagenic bypass of UV photoproducts and suppression of skin cancer. Proc Natl Acad Sci U S A. 2006;103:18083–8.
79. Gueranger Q, Stary A, Aoufouchi S, Faili A, Sarasin A, Reynaud CA, Weill JC. Role of DNA polymerases eta, iota and zeta in UV resistance and UV-induced mutagenesis in a human cell line. DNA Repair (Amst). 2008;7:1551–62.
80. Wang Y, Woodgate R, McManus TP, Mead S, McCormick JJ, Maher VM. Evidence that in xeroderma pigmentosum variant cells, which lack DNA polymerase eta, DNA polymerase iota causes the very high frequency and unique spectrum of UV-induced mutations. Cancer Res. 2007;67:3018–26.
81. Ziv O, Geacintov N, Nakajima S, Yasui A, Livneh Z. DNA polymerase zeta cooperates with polymerases kappa and iota in translesion DNA synthesis across pyrimidine photodimers in cells from XPV patients. Proc Natl Acad Sci U S A. 2009;106:11552–7.
82. Maul RW, MacCarthy T, Frank EG, Donigan KA, McLenigan MP, Yang W, Saribasak H, Huston DE, Lange SS, Woodgate R, et al. DNA polymerase iota functions in the generation of tandem mutations during somatic hypermutation of antibody genes. J Exp Med. 2016;213:1675–83.
83. Martomo SA, Yang WW, Vaisman A, Maas A, Yokoi M, Hoeijmakers JH, Hanaoka F, Woodgate R, Gearhart PJ. Normal hypermutation in antibody genes from congenic mice defective for DNA polymerase iota. DNA Repair (Amst). 2006;5:392–8.
84. Gerlach VL, Aravind L, Gotway G, Schultz RA, Koonin EV, Friedberg EC. Human and mouse homologs of *Escherichia coli* DinB (DNA polymerase IV), members of the UmuC/DinB superfamily. Proc Natl Acad Sci U S A. 1999;96:11922–7.
85. Johnson RE, Prakash S, Prakash L. The human DINB1 gene encodes the DNA polymerase Poltheta. Proc Natl Acad Sci U S A. 2000;97:3838–43.
86. Ogi T, Kato T Jr, Kato T, Ohmori H. Mutation enhancement by DINB1, a mammalian homologue of the *Escherichia coli* mutagenesis protein dinB. Genes Cells. 1999;4:607–18.

87. Ohashi E, Ogi T, Kusumoto R, Iwai S, Masutani C, Hanaoka F, Ohmori H. Error-prone bypass of certain DNA lesions by the human DNA polymerase kappa. Genes Dev. 2000;14:1589–94.
88. Zhang Y, Yuan F, Xin H, Wu X, Rajpal DK, Yang D, Wang Z. Human DNA polymerase kappa synthesizes DNA with extraordinarily low fidelity. Nucleic Acids Res. 2000;28:4147–56.
89. Bavoux C, Leopoldino AM, Bergoglio V, O-Wang J, Ogi T, Bieth A, Judde JG, Pena SD, Poupon MF, Helleday T, et al. Up-regulation of the error-prone DNA polymerase kappa promotes pleiotropic genetic alterations and tumorigenesis. Cancer Res. 2005;65:325–30.
90. Ogi T, Mimura J, Hikida M, Fujimoto H, Fujii-Kuriyama Y, Ohmori H. Expression of human and mouse genes encoding polkappa: testis-specific developmental regulation and AhR-dependent inducible transcription. Genes Cells. 2001;6:943–53.
91. Velasco-Miguel S, Richardson JA, Gerlach VL, Lai WC, Gao T, Russell LD, Hladik CL, White CL, Friedberg EC. Constitutive and regulated expression of the mouse Dinb (Polkappa) gene encoding DNA polymerase kappa. DNA Repair (Amst). 2003;2:91–106.
92. Gerlach VL, Feaver WJ, Fischhaber PL, Friedberg EC. Purification and characterization of pol kappa, a DNA polymerase encoded by the human DINB1 gene. J Biol Chem. 2001;276:92–8.
93. Washington MT, Johnson RE, Prakash L, Prakash S. Human DINB1-encoded DNA polymerase kappa is a promiscuous extender of mispaired primer termini. Proc Natl Acad Sci U S A. 2002;99:1910–4.
94. Zhang Y, Yuan F, Wu X, Wang M, Rechkoblit O, Taylor JS, Geacintov NE, Wang Z. Error-free and error-prone lesion bypass by human DNA polymerase kappa in vitro. Nucleic Acids Res. 2000;28:4138–46.
95. Suzuki N, Ohashi E, Hayashi K, Ohmori H, Grollman AP, Shibutani S. Translesional synthesis past acetylaminofluorene-derived DNA adducts catalyzed by human DNA polymerase kappa and *Escherichia coli* DNA polymerase IV. Biochemistry. 2001;40:15176–83.
96. Fischhaber PL, Gerlach VL, Feaver WJ, Hatahet Z, Wallace SS, Friedberg EC. Human DNA polymerase kappa bypasses and extends beyond thymine glycols during translesion synthesis in vitro, preferentially incorporating correct nucleotides. J Biol Chem. 2002;277: 37604–11.
97. Rechkoblit O, Zhang Y, Guo D, Wang Z, Amin S, Krzeminsky J, Louneva N, Geacintov NE. Trans-lesion synthesis past bulky benzo[a]pyrene diol epoxide N^2-dG and N^6-dA lesions catalyzed by DNA bypass polymerases. J Biol Chem. 2002;277:30488–94.
98. Suzuki N, Itoh S, Poon K, Masutani C, Hanaoka F, Ohmori H, Yoshizawa I, Shibutani S. Translesion synthesis past estrogen-derived DNA adducts by human DNA polymerases eta and kappa. Biochemistry. 2004;43:6304–11.
99. Suzuki N, Ohashi E, Kolbanovskiy A, Geacintov NE, Grollman AP, Ohmori H, Shibutani S. Translesion synthesis by human DNA polymerase kappa on a DNA template containing a single stereoisomer of dG-(+)- or dG-(−)-anti-N(2)-BPDE (7,8-dihydroxy-anti-9,10-epoxy-7,8,9,10-tetrahydrobenzo[a]pyrene). Biochemistry. 2002;41:6100–6.
100. Jha V, Bian C, Xing G, Ling H. Structure and mechanism of error-free replication past the major benzo[a]pyrene adduct by human DNA polymerase kappa. Nucleic Acids Res. 2016;44:4957–67.
101. Ohashi E, Bebenek K, Matsuda T, Feaver WJ, Gerlach VL, Friedberg EC, Ohmori H, Kunkel TA. Fidelity and processivity of DNA synthesis by DNA polymerase kappa, the product of the human DINB1 gene. J Biol Chem. 2000;275:39678–84.
102. Wolfle WT, Washington MT, Prakash L, Prakash S. Human DNA polymerase kappa uses template-primer misalignment as a novel means for extending mispaired termini and for generating single-base deletions. Genes Dev. 2003;17:2191–9.
103. Haracska L, Prakash L, Prakash S. Role of human DNA polymerase kappa as an extender in translesion synthesis. Proc Natl Acad Sci U S A. 2002;99:16000–5.
104. Schenten D, Gerlach VL, Guo C, Velasco-Miguel S, Hladik CL, White CL, Friedberg EC, Rajewsky K, Esposito G. DNA polymerase kappa deficiency does not affect somatic hypermutation in mice. Eur J Immunol. 2002;32:3152–60.

105. Bi X, Slater DM, Ohmori H, Vaziri C. DNA polymerase kappa is specifically required for recovery from the benzo[a]pyrene-dihydrodiol epoxide (BPDE)-induced S-phase checkpoint. J Biol Chem. 2005;280:22343–55.
106. Ogi T, Shinkai Y, Tanaka K, Ohmori H. Polkappa protects mammalian cells against the lethal and mutagenic effects of benzo[a]pyrene. Proc Natl Acad Sci U S A. 2002;99:15548–53.
107. Stancel JN, McDaniel LD, Velasco S, Richardson J, Guo C, Friedberg EC. Polk mutant mice have a spontaneous mutator phenotype. DNA Repair (Amst). 2009;8:1355–62.
108. Avkin S, Goldsmith M, Velasco-Miguel S, Geacintov N, Friedberg EC, Livneh Z. Quantitative analysis of translesion DNA synthesis across a benzo[a]pyrene-guanine adduct in mammalian cells: the role of DNA polymerase kappa. J Biol Chem. 2004;279:53298–305.
109. Yoon JH, Prakash L, Prakash S. Highly error-free role of DNA polymerase eta in the replicative bypass of UV-induced pyrimidine dimers in mouse and human cells. Proc Natl Acad Sci U S A. 2009;106:18219–24.
110. Yoon JH, Prakash S, Prakash L. Genetic control of translesion synthesis on leading and lagging DNA strands in plasmids derived from Epstein-Barr virus in human cells. MBio. 2012;3:e00271–12.
111. Gibbs PE, Wang XD, Li Z, McManus TP, McGregor WG, Lawrence CW, Maher VM. The function of the human homolog of *Saccharomyces cerevisiae* REV1 is required for mutagenesis induced by UV light. Proc Natl Acad Sci U S A. 2000;97:4186–91.
112. Lin W, Xin H, Zhang Y, Wu X, Yuan F, Wang Z. The human REV1 gene codes for a DNA template-dependent dCMP transferase. Nucleic Acids Res. 1999;27:4468–75.
113. Masuda Y, Takahashi M, Fukuda S, Sumii M, Kamiya K. Mechanisms of dCMP transferase reactions catalyzed by mouse Rev1 protein. J Biol Chem. 2002;277:3040–6.
114. Masuda Y, Kamiya K. Biochemical properties of the human REV1 protein. FEBS Lett. 2002;520:88–92.
115. Zhang Y, Wu X, Rechkoblit O, Geacintov NE, Taylor JS, Wang Z. Response of human REV1 to different DNA damage: preferential dCMP insertion opposite the lesion. Nucleic Acids Res. 2002;30:1630–8.
116. Lin X, Okuda T, Trang J, Howell SB. Human REV1 modulates the cytotoxicity and mutagenicity of cisplatin in human ovarian carcinoma cells. Mol Pharmacol. 2006;69:1748–54.
117. Sharma S, Hicks JK, Chute CL, Brennan JR, Ahn JY, Glover TW, Canman CE. REV1 and polymerase zeta facilitate homologous recombination repair. Nucleic Acids Res. 2012;40:682–91.
118. Jansen JG, Langerak P, Tsaalbi-Shtylik A, van den Berk P, Jacobs H, de Wind N. Strand-biased defect in C/G transversions in hypermutating immunoglobulin genes in Rev1-deficient mice. J Exp Med. 2006;203:319–23.
119. Jansen JG, Tsaalbi-Shtylik A, Langerak P, Calleja F, Meijers CM, Jacobs H, de Wind N. The BRCT domain of mammalian Rev1 is involved in regulating DNA translesion synthesis. Nucleic Acids Res. 2005;33:356–65.
120. Masuda K, Ouchida R, Li Y, Gao X, Mori H, Wang JY. A critical role for REV1 in regulating the induction of C:G transitions and A:T mutations during Ig gene hypermutation. J Immunol. 2009;183:1846–50.
121. Tsaalbi-Shtylik A, Verspuy JW, Jansen JG, Rebel H, Carlee LM, van der Valk MA, Jonkers J, de Gruijl FR, de Wind N. Error-prone translesion replication of damaged DNA suppresses skin carcinogenesis by controlling inflammatory hyperplasia. Proc Natl Acad Sci U S A. 2009;106:21836–41.
122. Guo C, Fischhaber PL, Luk-Paszyc MJ, Masuda Y, Zhou J, Kamiya K, Kisker C, Friedberg EC. Mouse Rev1 protein interacts with multiple DNA polymerases involved in translesion DNA synthesis. EMBO J. 2003;22:6621–30.
123. Hara K, Hashimoto H, Murakumo Y, Kobayashi S, Kogame T, Unzai S, Akashi S, Takeda S, Shimizu T, Sato M. Crystal structure of human REV7 in complex with a human REV3 fragment and structural implication of the interaction between DNA polymerase zeta and REV1. J Biol Chem. 2010;285:12299–307.

124. Kikuchi S, Hara K, Shimizu T, Sato M, Hashimoto H. Structural basis of recruitment of DNA polymerase zeta by interaction between REV1 and REV7 proteins. J Biol Chem. 2012;287:33847–52.
125. Masuda Y, Ohmae M, Masuda K, Kamiya K. Structure and enzymatic properties of a stable complex of the human REV1 and REV7 proteins. J Biol Chem. 2003;278:12356–60.
126. Murakumo Y, Ogura Y, Ishii H, Numata S, Ichihara M, Croce CM, Fishel R, Takahashi M. Interactions in the error-prone postreplication repair proteins hREV1, hREV3, and hREV7. J Biol Chem. 2001;276:35644–51.
127. Ohashi E, Hanafusa T, Kamei K, Song I, Tomida J, Hashimoto H, Vaziri C, Ohmori H. Identification of a novel REV1-interacting motif necessary for DNA polymerase kappa function. Genes Cells. 2009;14:101–11.
128. Ohashi E, Murakumo Y, Kanjo N, Akagi J, Masutani C, Hanaoka F, Ohmori H. Interaction of hREV1 with three human Y-family DNA polymerases. Genes Cells. 2004;9:523–31.
129. Tissier A, Kannouche P, Reck MP, Lehmann AR, Fuchs RP, Cordonnier A. Co-localization in replication foci and interaction of human Y-family members, DNA polymerase pol eta and REVl protein. DNA Repair (Amst). 2004;3:1503–14.
130. Ohmori H, Hanafusa T, Ohashi E, Vaziri C. Separate roles of structured and unstructured regions of Y-family DNA polymerases. Adv Protein Chem Struct Biol. 2009;78:99–146.
131. Akagi J, Masutani C, Kataoka Y, Kan T, Ohashi E, Mori T, Ohmori H, Hanaoka F. Interaction with DNA polymerase eta is required for nuclear accumulation of REV1 and suppression of spontaneous mutations in human cells. DNA Repair (Amst). 2009;8:585–99.
132. Boehm EM, Powers KT, Kondratick CM, Spies M, Houtman JC, Washington MT. The proliferating cell nuclear antigen (PCNA)-interacting protein (PIP) motif of DNA polymerase eta mediates its interaction with the C-terminal domain of Rev1. J Biol Chem. 2016;291:8735–44.
133. Yoon JH, Park J, Conde J, Wakamiya M, Prakash L, Prakash S. Rev1 promotes replication through UV lesions in conjunction with DNA polymerases eta, iota, and kappa but not DNA polymerase zeta. Genes Dev. 2015;29:2588–602.
134. Gibbs PE, McGregor WG, Maher VM, Nisson P, Lawrence CW. A human homolog of the *Saccharomyces cerevisiae* REV3 gene, which encodes the catalytic subunit of DNA polymerase zeta. Proc Natl Acad Sci U S A. 1998;95:6876–80.
135. Murakumo Y, Roth T, Ishii H, Rasio D, Numata S, Croce CM, Fishel R. A human REV7 homolog that interacts with the polymerase zeta catalytic subunit hREV3 and the spindle assembly checkpoint protein hMAD2. J Biol Chem. 2000;275:4391–7.
136. Van Sloun PP, Romeijn RJ, Eeken JC. Molecular cloning, expression and chromosomal localisation of the mouse Rev3l gene, encoding the catalytic subunit of polymerase zeta. Mutat Res. 1999;433:109–16.
137. Yu Y, Yang J, Zhu F, Xu F. Response of REV3 promoter to N-methyl-N′-nitro-N-nitrosoguanidine. Mutat Res. 2004;550:49–58.
138. Diaz M, Watson NB, Turkington G, Verkoczy LK, Klinman NR, McGregor WG. Decreased frequency and highly aberrant spectrum of ultraviolet-induced mutations in the hprt gene of mouse fibroblasts expressing antisense RNA to DNA polymerase zeta. Mol Cancer Res. 2003;1:836–47.
139. Li Z, Zhang H, McManus TP, McCormick JJ, Lawrence CW, Maher VM. hREV3 is essential for error-prone translesion synthesis past UV or benzo[a]pyrene diol epoxide-induced DNA lesions in human fibroblasts. Mutat Res. 2002;510:71–80.
140. Wu F, Lin X, Okuda T, Howell SB. DNA polymerase zeta regulates cisplatin cytotoxicity, mutagenicity, and the rate of development of cisplatin resistance. Cancer Res. 2004;64:8029–35.
141. Bemark M, Khamlichi AA, Davies SL, Neuberger MS. Disruption of mouse polymerase zeta (Rev3) leads to embryonic lethality and impairs blastocyst development in vitro. Curr Biol. 2000;10:1213–6.
142. Esposito G, Godindagger I, Klein U, Yaspo ML, Cumano A, Rajewsky K. Disruption of the Rev3l-encoded catalytic subunit of polymerase zeta in mice results in early embryonic lethality. Curr Biol. 2000;10:1221–4.

143. O-Wang J, Kajiwara K, Kawamura K, Kimura M, Miyagishima H, Koseki H, Tagawa M. An essential role for REV3 in mammalian cell survival: absence of REV3 induces p53-independent embryonic death. Biochem Biophys Res Commun. 2002;293:1132–7.
144. Van Sloun PP, Varlet I, Sonneveld E, Boei JJ, Romeijn RJ, Eeken JC, De Wind N. Involvement of mouse Rev3 in tolerance of endogenous and exogenous DNA damage. Mol Cell Biol. 2002;22:2159–69.
145. Wittschieben J, Shivji MK, Lalani E, Jacobs MA, Marini F, Gearhart PJ, Rosewell I, Stamp G, Wood RD. Disruption of the developmentally regulated Rev3l gene causes embryonic lethality. Curr Biol. 2000;10:1217–20.
146. Jansen JG, Tsaalbi-Shtylik A, Hendriks G, Verspuy J, Gali H, Haracska L, de Wind N. Mammalian polymerase zeta is essential for post-replication repair of UV-induced DNA lesions. DNA Repair (Amst). 2009;8:1444–51.
147. Wittschieben JP, Reshmi SC, Gollin SM, Wood RD. Loss of DNA polymerase zeta causes chromosomal instability in mammalian cells. Cancer Res. 2006;66:134–42.
148. Wittschieben JP, Patil V, Glushets V, Robinson LJ, Kusewitt DF, Wood RD. Loss of DNA polymerase zeta enhances spontaneous tumorigenesis. Cancer Res. 2010;70:2770–8.
149. Cheung HW, Chun AC, Wang Q, Deng W, Hu L, Guan XY, Nicholls JM, Ling MT, Chuan Wong Y, Tsao SW, et al. Inactivation of human MAD2B in nasopharyngeal carcinoma cells leads to chemosensitization to DNA-damaging agents. Cancer Res. 2006;66:4357–67.
150. McNally K, Neal JA, McManus TP, McCormick JJ, Maher VM. hRev7, putative subunit of hPolzeta, plays a critical role in survival, induction of mutations, and progression through S-phase, of UV((254nm))-irradiated human fibroblasts. DNA Repair (Amst). 2008;7:597–604.
151. Watanabe N, Mii S, Asai N, Asai M, Niimi K, Ushida K, Kato T, Enomoto A, Ishii H, Takahashi M, et al. The REV7 subunit of DNA polymerase zeta is essential for primordial germ cell maintenance in the mouse. J Biol Chem. 2013;288:10459–71.
152. Tomida J, Takata K, Lange SS, Schibler AC, Yousefzadeh MJ, Bhetawal S, Dent SY, Wood RD. REV7 is essential for DNA damage tolerance via two REV3L binding sites in mammalian DNA polymerase zeta. Nucleic Acids Res. 2015;43:1000–11.
153. Boersma V, Moatti N, Segura-Bayona S, Peuscher MH, van der Torre J, Wevers BA, Orthwein A, Durocher D, Jacobs JJ. MAD2L2 controls DNA repair at telomeres and DNA breaks by inhibiting 5′ end resection. Nature. 2015;521:537–40.
154. Xu G, Chapman JR, Brandsma I, Yuan J, Mistrik M, Bouwman P, Bartkova J, Gogola E, Warmerdam D, Barazas M, et al. REV7 counteracts DNA double-strand break resection and affects PARP inhibition. Nature. 2015;521:541–4.
155. Bluteau D, Masliah-Planchon J, Clairmont C, Rousseau A, Ceccaldi R, Dubois d'Enghien C, Bluteau O, Cuccuini W, Gachet S, Peffault de Latour R, et al. Biallelic inactivation of REV7 is associated with Fanconi anemia. J Clin Invest. 2016;126:3580–4.
156. Marini F, Kim N, Schuffert A, Wood RD. POLN, a nuclear PolA family DNA polymerase homologous to the DNA cross-link sensitivity protein Mus308. J Biol Chem. 2003;278:32014–9.
157. Seki M, Marini F, Wood RD. POLQ (Pol theta), a DNA polymerase and DNA-dependent ATPase in human cells. Nucleic Acids Res. 2003;31:6117–26.
158. Sharief FS, Vojta PJ, Ropp PA, Copeland WC. Cloning and chromosomal mapping of the human DNA polymerase theta (POLQ), the eighth human DNA polymerase. Genomics. 1999;59:90–6.
159. Seki M, Masutani C, Yang LW, Schuffert A, Iwai S, Bahar I, Wood RD. High-efficiency bypass of DNA damage by human DNA polymerase Q. EMBO J. 2004;23:4484–94.
160. Takata K, Shimizu T, Iwai S, Wood RD. Human DNA polymerase N (POLN) is a low fidelity enzyme capable of error-free bypass of 5S-thymine glycol. J Biol Chem. 2006;281:23445–55.
161. Zietlow L, Smith LA, Bessho M, Bessho T. Evidence for the involvement of human DNA polymerase N in the repair of DNA interstrand cross-links. Biochemistry. 2009;48:11817–24.
162. Seki M, Wood RD. DNA polymerase theta (POLQ) can extend from mismatches and from bases opposite a (6-4) photoproduct. DNA Repair (Amst). 2008;7:119–27.
163. Shima N, Hartford SA, Duffy T, Wilson LA, Schimenti KJ, Schimenti JC. Phenotype-based identification of mouse chromosome instability mutants. Genetics. 2003;163:1031–40.

164. Shima N, Munroe RJ, Schimenti JC. The mouse genomic instability mutation chaos1 is an allele of Polq that exhibits genetic interaction with Atm. Mol Cell Biol. 2004;24:10381–9.
165. Goff JP, Shields DS, Seki M, Choi S, Epperly MW, Dixon T, Wang H, Bakkenist CJ, Dertinger SD, Torous DK, et al. Lack of DNA polymerase theta (POLQ) radiosensitizes bone marrow stromal cells in vitro and increases reticulocyte micronuclei after total-body irradiation. Radiat Res. 2009;172:165–74.
166. Li Y, Gao X, Wang JY. Comparison of two POLQ mutants reveals that a polymerase-inactive POLQ retains significant function in tolerance to etoposide and gamma-irradiation in mouse B cells. Genes Cells. 2011;16:973–83.
167. Ceccaldi R, Liu JC, Amunugama R, Hajdu I, Primack B, Petalcorin MI, O'Connor KW, Konstantinopoulos PA, Elledge SJ, Boulton SJ, et al. Homologous-recombination-deficient tumours are dependent on Poltheta-mediated repair. Nature. 2015;518:258–62.
168. Mateos-Gomez PA, Gong F, Nair N, Miller KM, Lazzerini-Denchi E, Sfeir A. Mammalian polymerase theta promotes alternative NHEJ and suppresses recombination. Nature. 2015;518:254–7.
169. Wyatt DW, Feng W, Conlin MP, Yousefzadeh MJ, Roberts SA, Mieczkowski P, Wood RD, Gupta GP, Ramsden DA. Essential roles for polymerase theta-mediated end joining in the repair of chromosome breaks. Mol Cell. 2016;63:662–73.
170. Zahn KE, Averill AM, Aller P, Wood RD, Doublie S. Human DNA polymerase theta grasps the primer terminus to mediate DNA repair. Nat Struct Mol Biol. 2015;22:304–11.
171. Moldovan GL, Madhavan MV, Mirchandani KD, McCaffrey RM, Vinciguerra P, D'Andrea AD. DNA polymerase POLN participates in cross-link repair and homologous recombination. Mol Cell Biol. 2010;30:1088–96.
172. Lee YS, Gao Y, Yang W. How a homolog of high-fidelity replicases conducts mutagenic DNA synthesis. Nat Struct Mol Biol. 2015;22:298–303.
173. Garcia-Gomez S, Reyes A, Martinez-Jimenez MI, Chocron ES, Mouron S, Terrados G, Powell C, Salido E, Mendez J, Holt IJ, et al. PrimPol, an archaic primase/polymerase operating in human cells. Mol Cell. 2013;52:541–53.
174. Mouron S, Rodriguez-Acebes S, Martinez-Jimenez MI, Garcia-Gomez S, Chocron S, Blanco L, Mendez J. Repriming of DNA synthesis at stalled replication forks by human PrimPol. Nat Struct Mol Biol. 2013;20:1383–9.
175. Schiavone D, Jozwiakowski SK, Romanello M, Guilbaud G, Guilliam TA, Bailey LJ, Sale JE, Doherty AJ. PrimPol is required for replicative tolerance of G Quadruplexes in vertebrate cells. Mol Cell. 2016;61:161–9.
176. Kobayashi K, Guilliam TA, Tsuda M, Yamamoto J, Bailey LJ, Iwai S, Takeda S, Doherty AJ, Hirota K. Repriming by PrimPol is critical for DNA replication restart downstream of lesions and chain-terminating nucleosides. Cell Cycle. 2016;15:1997–2008.
177. Bianchi J, Rudd SG, Jozwiakowski SK, Bailey LJ, Soura V, Taylor E, Stevanovic I, Green AJ, Stracker TH, Lindsay HD, et al. PrimPol bypasses UV photoproducts during eukaryotic chromosomal DNA replication. Mol Cell. 2013;52:566–73.
178. Zafar MK, Ketkar A, Lodeiro MF, Cameron CE, Eoff RL. Kinetic analysis of human PrimPol DNA polymerase activity reveals a generally error-prone enzyme capable of accurately bypassing 7,8-dihydro-8-oxo-2′-deoxyguanosine. Biochemistry. 2014;53:6584–94.
179. Martinez-Jimenez MI, Garcia-Gomez S, Bebenek K, Sastre-Moreno G, Calvo PA, Diaz-Talavera A, Kunkel TA, Blanco L. Alternative solutions and new scenarios for translesion DNA synthesis by human PrimPol. DNA Repair (Amst). 2015;29:127–38.
180. Guilliam TA, Jozwiakowski SK, Ehlinger A, Barnes RP, Rudd SG, Bailey LJ, Skehel JM, Eckert KA, Chazin WJ, Doherty AJ. Human PrimPol is a highly error-prone polymerase regulated by single-stranded DNA binding proteins. Nucleic Acids Res. 2015;43:1056–68.
181. Wan L, Lou J, Xia Y, Su B, Liu T, Cui J, Sun Y, Lou H, Huang J. hPrimpol1/CCDC111 is a human DNA primase-polymerase required for the maintenance of genome integrity. EMBO Rep. 2013;14:1104–12.
182. Zhao F, Wu J, Xue A, Su Y, Wang X, Lu X, Zhou Z, Qu J, Zhou X. Exome sequencing reveals CCDC111 mutation associated with high myopia. Hum Genet. 2013;132:913–21.
183. Keen BA, Jozwiakowski SK, Bailey LJ, Bianchi J, Doherty AJ. Molecular dissection of the domain architecture and catalytic activities of human PrimPol. Nucleic Acids Res. 2014;42:5830–45.

Chapter 13
Ataxia-Telangiectasia and Nijmegen Breakage Syndrome

Junya Kobayashi

Abstract Ataxia-telangiectasia (A-T) and Nijmegen breakage syndrome (NBS) are well-known single-gene disorders, which have similar cellular phenotypes, including chromosome instability, radioresistant DNA synthesis, and hypersensitivity to radiation. Such phenotypic similarity implies direct physical association and/or functional interaction between respective gene products. Indeed, the NBS1 protein responsible for NBS interacts with ATM kinase implicated in A-T and regulates ATM activation upon DNA damage; however, NBS1-mediated homologous recombination does not seem to require ATM. Moreover, ATM is activated by oxidative stress independently of NBS1. Thus, ATM and NBS1 are likely to have distinct functions in radiation-induced DNA damage responses or other cellular responses to genomic stresses such as oxidative stress, which should underlie different clinical manifestations of A-T and NBS.

13.1 Clinical Features of A-T, NBS, and ATLD

Ataxia-telangiectasia (A-T; OMIM#208900; Table 13.1) is a rare neurodegenerative disease with worldwide incidence of 1 in 40,000–200,000 live births characterized by poor coordination, hypersensitivity to radiation, and chromosomal instability, which enhances the risk of cancer. A-T was first described in 1958 [1]; later, Gotoff *et al.* reported that A-T patients showed hypersensitivity to radiotherapy [2]. Typical clinical manifestations of this disorder include severe neurodegeneration due to progressive cerebellar atrophy and telangiectasia (small dilated blood vessels), especially in the eyes and occasionally on the facial skin. In addition, patients have general immunodeficiency, predisposition to malignancies (particularly lymphoma), and hypersensitivity to ionizing radiation (IR); some of them also show growth retardation, premature aging, and insulin resistance [3–5]. The cellular phenotype in A-T is characterized by chromosomal instability, radioresistant DNA synthesis

J. Kobayashi
Radiation Biology Center, Graduate School of Biostudies, Kyoto University, Kyoto, Japan
e-mail: jkobayashi@house.rbc.kyoto-u.ac.jp

C. Nishigori, K. Sugasawa (eds.), *DNA Repair Disorders*,
https://doi.org/10.1007/978-981-10-6722-8_13

Table 13.1 Major characteristics of A-T, NBS, and ATLD

Characteristics	A-T	NBS	ATLD
OMIM	208900	251260	604391
Responsible gene	*ATM*	*NBS1*	*MRE11*
Function of gene product	Protein kinase	Regulator protein	Nuclease
Clinical features			
Progressive cerebellar	+	−	+
Oculomotor apraxia	+	−	+
Microcephaly	−	+	−
Birdlike face	−	+	−
Telangiectasia	+	−	−
Immunodeficiency	+	+	−
Decreases of IgA, IgG, and IgE	+	+	−
Cancer predisposition	+	+	−
Deficiency of glucose metabolism	+	−	ND
Growth retardation	+	+	ND
Ovarian dysgenesis	+	+	+
Increased radiosensitivity	+	+	+
Cellular phenotypes			
Increased radiosensitivity	+	+	+
Radio-resistant DNA synthesis	+	+	+
Chromosome instability	+	+	+

ND: not determined

(RDS), and hypersensitivity to DNA-damaging agents, especially those inducing double-strand breaks (DSBs) [3–5]; it has also been reported in other genetic conditions such as A-T variants and A-T-like disorder.

The most famous A-T variant is a rare autosomal recessive Nijmegen breakage syndrome (NBS; OMIN#251260; Table 13.1). NBS was first described in 1979 in a Dutch boy with microcephaly, growth retardation, IgA deficiency, and chromosomal rearrangements, which are also observed in A-T [6]. The boy's brother had similar clinical manifestations; therefore Weemaes *et al.* reported the condition as a genetic disease and named it Nijmegen breakage syndrome [7]. Typical clinical features of NBS are developmental microcephaly, which results in birdlike face, mild growth retardation, premature ovarian failure, immunodeficiency, and predisposition to cancer, most frequently in hematological organs [8, 9]. NBS is very similar to A-T in the cellular phenotype, namely, hypersensitivity to radiation, RDS, and genomic instability caused by chromosomal aberrations. The estimated worldwide incidence of NBS is 1 in 100,000 live births; however, NBS is more common in Eastern Europe, where its frequency can be as high as about 1:150 in some populations, because the origin of the founder mutation is in Poland.

A-T-like disorder (ATLD; OMIM#604391; Table 13.1) has been reported as an A-T-related disease [10, 11]. ATLD patients were first described by Hernandez *et al.* in 1993 [12]. A common feature of ATLD and A-T is progressive cerebellar ataxia; however, ATLD patients show normal levels of total IgG, IgA, and IgM and do not have telangiectasia. Although immunodeficiency and cancer predisposition typical for A-T and NBS are uncommon in ATLD, cellular phenotypes are similar. Therefore, interactions between the factors responsible for DNA damage responses in A-T and related syndromes have been expected before the discovery of the encoding genes.

13.2 Interaction Between ATM and NBS1 in DSB Damage Responses

Several critical studies have identified mutations responsible for A-T in the *ATM* gene located at 11q22-23 [13, 14]. ATM is a 350-kDa protein of 3056 amino acids belonging to the family of phosphatidylinositol 3-kinase-like protein kinases [15]. This family also includes the catalytic subunit of the DNA-dependent protein kinase (DNA-PKcs) involved in the nonhomologous end-joining (NHEJ) pathway of DSB repair [16] and the A-T and Rad3-related (ATR) protein, which responds primarily to stalled replication forks (replication stress) [17]. These protein kinases preferably phosphorylate serine or threonine residues followed by glutamine (S/TQ motif) and have redundant functions. The *ATM* gene is widely conserved in eukaryotes and has an ortholog (*TEL1/tel1*) in yeast [18]. *ATR* is also conserved in yeast (*MEC1* and *rad3* genes in budding and fission yeast, respectively), but *DNA-PKcs* is not [18].

The gene mutated in NBS is *NBS1*, which has been mapped to chromosome 8q21–24; it contains 16 exons [19, 20] and is widely conserved in eukaryotes. Human *NBS1* gene encodes 754-amino acid (aa) protein with weak homology to Xrs2 in budding yeast and stronger homology to yNbs1 in fission yeast. NBS1 has a fork-head-associated domain (aa 20–108) and two BRCA1 C-terminus (BRCT) domains (BRCT1, aa 111–197, and BRCT2, aa 219–327) at the C-terminal region (Fig. 13.1).

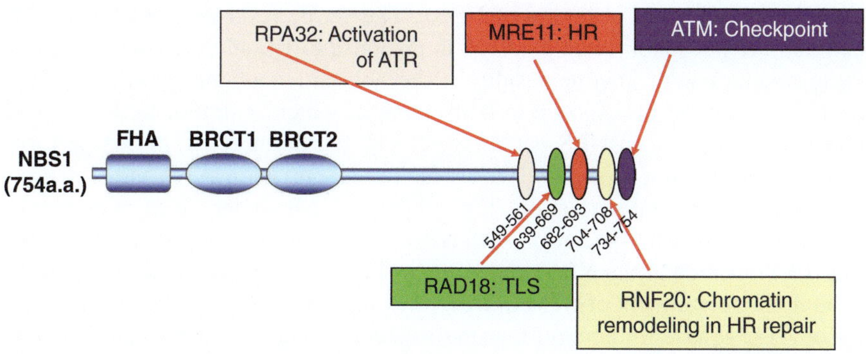

Fig. 13.1 Structure of human NBS1 protein. NBS1 contains several interaction motifs in the N- and C-termini. The N-terminal FHA/BRCT domains interact with DNA damage-responding proteins such as MDC1, γ-H2AX, and TopBP1. The C-terminal motifs are indispensable for binding to MRE11, ATM, RAD18, RNF20, and RPA32. These interactions might be important for the maintenance of genomic stability

ATLD is caused by mutations in the *MRE11* gene [10, 11], containing 20 exons and mapped to chromosome 11q21. The encoded 80-kDa MRE11 protein of 708 aa has a motif critical for nuclease activity (3′ to 5′ exo- and endonuclease) at the N-terminus and DNA-binding motifs at both central and C-terminal regions and is widely conserved in eukaryotes. It was shown that in budding yeast, yMre11 interacted with NBS1 ortholog Xrs2 [21]. In humans, MRE11 binds to NBS1 at the C-terminus (aa 682–693; Fig. 13.1) and to RAD50, forming the MRN complex [22]. Some patients with NBS-like disorder (NBSLD: OMIM#613078) have defective RAD50 and demonstrate cellular phenotypes characteristic for NBS, A-T, and ATLD, including radiation hypersensitivity, RDS, and chromosome instability [23]. The MRN complex functions in homologous recombination (HR) repair, which is, similar to NHEJ, a major DSB repair pathway. When DSBs are generated by IR, the MRN complex is rapidly recruited to DSB sites, and HR repair is initiated via resection of DSB ends, resulting in the formation of more than 30 single-stranded DNA (ssDNA) tails at both DSB ends. Then, the replication protein A (RPA) complex binds to ssDNA and is subsequently replaced with RAD51, which facilitates HR repair [24]. The MRN complex, particularly MRE11, is indispensable for the initial resection step in HR as evidenced by the findings in NBS1- or MRE11-deficient DT40 chicken and human cells [25, 26].

An intrinsic feature of A-T is RDS, which is a defect in the activation of the intra-S phase checkpoint. A-T cells are also known for failure to activate G1 and G2 checkpoints [27, 28]. A major pathway regulating the G1 checkpoint is the suppression of CDK/cyclin activity by cell cycle inhibitory protein p21, whose expression is induced by p53 activated after IR-induced DSB damage. ATM was found to participate in this pathway by phosphorylating p53 and promoting p21 transcription [29, 30]. ATM can also phosphorylate a number of other regulators of cell cycle checkpoints in response to DSB formation (Fig. 13.2). Most of ATM substrates have consensus sequences containing SQ/TQ motifs [31] that are found in the central region of NBS1. The serine residues at positions 278 and 343 are phosphorylated by ATM in response to IR, and the substitution of these residues with alanine results in the abrogation of the ATM-dependent intra-S checkpoint in response to DSB damage, which is similar to the cellular phenotype observed in NBS patients [32, 33]. Furthermore, ATM can perform autophosphorylation at serine 1981, which is essential for ATM monomerization and its sufficient activation (Fig. 13.2) [34]. Moreover, ATM is evidently activated by a direct association with NBS1 via a conserved motif (aa 734–754; Fig. 13.1) located at the NBS1 C-terminus (Fig. 13.2) [35]. NBS cells are deficient in the recruitment of ATM to DSB sites after IR and showed reduced activity of ATM kinase, indicating that the interaction between NBS1 and ATM is indispensable for sufficient ATM activation [35]. MRE11-defective ATLD cells and RAD50-defective NBSLD cells also show the reduction of ATM-dependent phosphorylation and activation of cell cycle checkpoint, suggesting that both MRE11 and RAD50 are also important for ATM activation via the MRN complex.

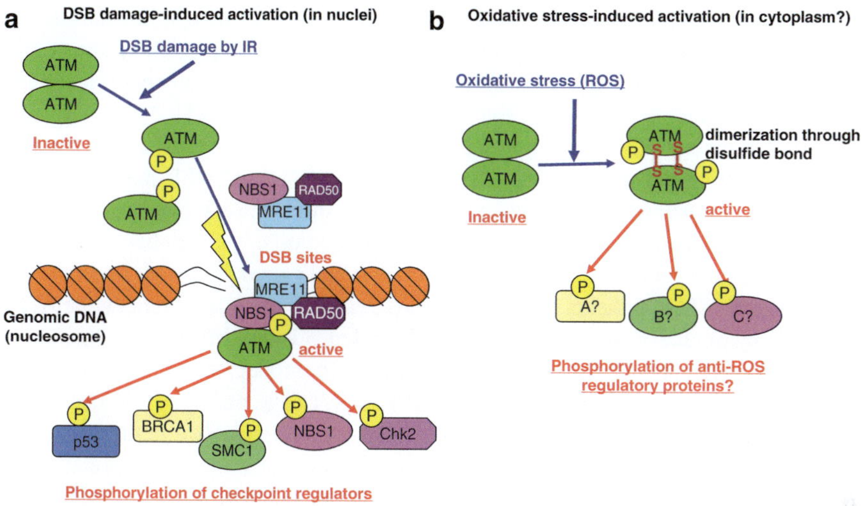

Fig. 13.2 Mechanisms of ATM activation. (**a**) DSB damage induces autophosphorylation of ATM at serine 1981, which allows the interaction of ATM monomer with MRN complex and subsequent recruitment to DSB sites. As a result, ATM could sufficiently phosphorylate checkpoint regulators. (**b**) Oxidative stress induces ATM dimerization through disulfide bonds and subsequent activation. Such activated ATM could phosphorylate anti-ROS regulatory proteins

13.3 Distinct Functions of ATM and NBS1 and Their Relevance to Clinical Phenotypes

Although cellular phenotypes are similar in A-T and NBS, their clinical manifestations, especially neurodegeneration, are different. Thus, A-T patients show progressive cerebellar ataxia, while NBS patients have microcephaly. In A-T patients, progressive cerebellar ataxia mostly leads to gait imbalance at the end of the first decade [36, 37], accompanied with oculomotor abnormalities, swallowing difficulty, and speech defects. Pathology studies indicate that a loss of Purkinje and granular neurons may cause progressive cerebellar cortical degeneration in A-T [38, 39]. Although *Atm*-deficient mice do not display the typical ataxia phenotype [39], neural stem cells derived from induced pluripotent cells of A-T patients had abnormal responses to hypoxia and attenuated neuronal differentiation, especially to GABAergic neurons and oligodendrogenesis [40]. Moreover, ATM-deficient human and mouse cells showed increased generation of endogenous reactive oxygen species (ROS) and accumulation of oxidative damage, suggesting that ATM could have an important role in antioxidative responses [41, 42]. Guo *et al.* reported that ROS directly induced ATM kinase activation independently of MRN complex through stabilization of ATM dimers by the formation of disulfide bonds (Fig. 13.2) [43]. Moreover, several recent studies have described ATM roles in cytoplasmic organelles such as peroxisomes and mitochondria [44–46]. Thus, Walker *et al.* showed that ROS activated peroxisomal ATM, which phosphorylated serine-threonine kinase LKB1 and induced

5′-AMP-activated protein kinase (AMPK) and tuberin (TSC2), dampening mTORC1-mediated signaling. This cascade of events led to a decrease in protein synthesis and enhancement of autophagy (pexophagy) via ATM-dependent phosphorylation of PEX5, which interacts with p62 (an autophagy adaptor) [44, 45]. ATM was also found in mitochondria, where it was suggested to play a role in mitochondrial autophagy (mitophagy) and stabilization of mitochondrial DNA [46–48]; however, ATM substrates in mitochondria and mechanistic aspects of its activity there remain to be investigated. Overall, these data indicate that deregulation of ATM activation in anti-oxidative stress signaling may lead to distinct neuronal phenotypes such as cerebellar ataxia in A-T. ATLD patients show similar progressive cerebellar ataxia, while ATM activation by oxidative stress is suggested to be independent of MRE11; another common role of ATM and MRE11 may be related to the neurodegenerative phenotype.

Immunodeficiency is also a typical feature in A-T. The patients have decreased levels of IgA, IgE, and IgG and are defective in B- and T-cell differentiation [49, 50]. Lymphocytes from A-T patients often contain clonal translocations, mainly in chromosome 7 and 14 loci of the T-cell receptor and Ig heavy chain genes [51, 52]. V(D)J recombination occurs only at the G0 and G1 phases of the cell cycle because of the G1-/G0-specific expression of V(D)J recombination-specific nuclease RAG2; therefore, abnormal regulation of cell cycle checkpoints in A-T may induce incomplete V(D)J recombination and consequently lead to immunodeficiency [53]. Another prominent clinical hallmark of A-T is predisposition to cancer, especially to that of hematopoietic origin, including Hodgkin's and non-Hodgkin's lymphoma, and several forms of leukemia [54, 55], which can be triggered by incomplete differentiation of immune organ. Furthermore, accumulating evidence suggests that ATM-mediated cell cycle checkpoint regulation is important for the prevention of cancer progression [56]; consequently, ATM-inactivating mutations may promote cancer development.

A-T is often accompanied by the deficiency in glucose metabolism and insulin-related responses; the patients often show poor weight gain, progressive decrease in body mass index (BMI), and dystrophy [57]. In fact, ATM was found to participate in several signaling pathways regulated by insulin [58, 59], and ApoE-deficient mice heterozygous for the *Atm* null allele had glucose intolerance and insulin resistance similar to those in type 2 diabetes [60]. Furthermore, $Atm^{-/-}$ mice showed reduction of insulin secretion and impaired differentiation of fibroblasts to adipocytes due to insufficient induction of C/EBPα and PPARγ [61, 62]. These findings suggest an important role of ATM in glucose metabolism mediated via adipocyte differentiation; however, the signals triggering ATM kinase activation in metabolic regulation remain unclear.

13.4 Novel Role of NBS1 and Its Contribution to NBS-Specific Symptoms

As mentioned above, NBS1 forms the MRN complex with MRE11 and RAD50 functioning in HR repair. The C-terminus of NBS1 contains motifs responsible for the binding to MRE11 and ATM, which are conserved among eukaryotes. Another conserved C-terminal sequence was found to be critical for the binding to E3 ubiquitin ligase RNF20 (aa 704–708; Fig. 13.1) [63], which ubiquitinates histone H2B at lysine 119 and is important for transcriptional regulation. RNF20 accumulates at DSB damage sites, and RNF20-dependent ubiquitination of H2B induced by IR is required for the recruitment of BRCA1 and RAD51 to DSB damage sites and HR repair. Furthermore, the interaction between NBS1 and RNF20 is essential for IR-induced focus formation of RPA complex (their accumulation to damaged chromatin) following DSB end resection, suggesting that NBS1 association with RNF20 is involved in the regulation of HR repair through end resection. Thus, while the NBS1 function in HR repair through assembly with MRE11 and RNF20 has been established, the role of its binding partner ATM in HR is unclear. There is evidence to suggest that ATM might be dispensable for HR repair, while some findings indicate the opposite [64–66]; therefore, further investigation is required.

In NBS, the characteristic clinical feature is microcephaly, which is not observed in A-T and ATLD patients, suggesting that physiological functions of NBS1 are distinct from those of ATM and MRE11. Microcephaly is also common for Seckel syndrome patients deficient in ATR, which is a member of the ATM kinase family and regulates cell cycle checkpoint response to replication stress. ATR is recruited to DNA damage sites induced by replication stress through the binding with ATRIP, which interacts with the RPA complex recognizing and coating ssDNA ends generated on stalled forks [67]. In parallel, TopBP1 binds to the RAD9-RAD1-HUS1 (9-1-1) checkpoint clamp, which is loaded onto DNA by the RAD17-replication factor C clamp loader complex [17]. These events result in the catalytic activation of ATR through direct binding of the ATR-ATRIP complex to the activation domain of TopBP1. Shiotani *et al.* reported that NBS1 could activate ATR through binding with RPA32, one of the three RPA subunits (aa 549–561; Fig. 13.1) [68]. NBS1, via its N-terminus, is likely to interact with TopBP1 and facilitate TopBP1 recruitment to DNA damage sites [69]. Overall, these data indicate that NBS1 has a critical role in ATR activation and that both ATR and NBS1 are important for cell cycle checkpoint induction by replication stress. In addition, accumulating evidence suggests that ATR and NBS1 are involved in centrosome maintenance. Centrosome is an organelle with a key role in equal chromosome segregation to daughter cells, which prevents aneuploidy. The *MCPH1* gene responsible for autosomal recessive primary microcephaly disorder has a role in centrosome maintenance, and its defects induce abnormal centrosome amplification [70]. MCPH1 is also involved in ATR-dependent DNA damage responses [71], suggesting a role of ATR in the centrosome cycle. Cumulatively, these data implicate NBS1, ATR, and MCPH1 in normal brain development and prevention of microcephaly [70, 72].

Yanagihara *et al.* reported that another conserved region (aa 639–669; Fig. 13.1) at the NBS1 C-terminus contributes to its binding with RAD18, an E3-ubiqutin ligase for proliferating cell nuclear antigen (PCNA) [73]. PCNA ubiquitination is implicated in DNA polymerase η (Polη)-dependent DNA translesion synthesis (TLS) after UV-induced DNA damage, which is regulated by RAD18. Depletion of NBS1 by siRNA increased UV sensitivity and reduced RAD18-dependent PCNA ubiquitination and subsequent accumulation of Polη at UV-induced DNA damage sites. These data suggest the involvement of NBS1 in RAD18/Polη-dependent TLS, while MRE11 and ATM are known to be dispensable for the TLS pathway. However, cells derived from most NBS patients are not sensitive to UV, because a major NBS1 mutation does not affect its RAD18-binding site and thus, does not prevent the activation of the RAD18/Polη-dependent TLS pathway.

NBS as well as A-T patients have predisposition to cancer, particularly to T-cell and B-cell lymphomas, which is likely due to genomic instability as evidenced by translocation between chromosomes 7 and 14 observed in T lymphocytes [9, 51]. As NBS1 is essential to ATM activation in response to DSB damage, it could also be involved in immune development through the ATM-dependent cell cycle checkpoint. Therefore, NBS1 defects can also lead to incomplete differentiation in immune organs and consequent lymphoid malignancy. Considering NBS1 role in the RAD18/Polη-dependent TLS pathway, deficiency of NBS1 may compromise TLS activation and lead to nucleotide misincorporation during regular DNA synthesis, causing genetic defects. Thus, NBS patients may suffer increased predisposition to malignancy due to deregulation of the RAD18/Polη-dependent TLS pathway.

13.5 Conclusion

ATM, NBS1, and MRE11 exert significant effects on IR-induced DNA damage responses through direct physical associations with and functional activation of each other. Therefore, mutations in the encoding genes underlie radiation hypersensitive genetic disorders such as A-T, ATLD, and NBS, which share common features of chromosomal instability and radioresistant DNA synthesis. However, clinical phenotypes of A-T, ATLD, and NBS are different, especially concerning neurodegeneration. While A-T and ATLD are characterized with progressive cerebellar ataxia, NBS has a typical microcephaly phenotype, suggesting distinct functional roles of the involved genes. Thus, ATM could be activated by oxidative stress as well as DSB damage, while NBS1 and perhaps the MRN complex may be dispensable for oxidative stress responses. At the same time, NBS1 has a role in TLS following UV-induced DNA damage, but ATM and MRE11 do not. Therefore, distinct functional activities of ATM, NBS1, and MRE11 may underlie the differences in clinical phenotypes of A-T, ATLD, and NBS, including neurodegeneration. However, future studies focusing on distinct functions of these genes are required.

References

1. Boder E, Sedgwick RP. Ataxia-telangiectasia; a familial syndrome of progressive cerebellar ataxia, oculocutaneous telangiectasia and frequent pulmonary infection. Pediatrics. 1958;21:526–54.
2. Gotoff SP, Amirmokri E, Liebner EJ. Ataxia telangiectasia. Neoplasia, untoward response to x-irradiation, and tuberous sclerosis. Am J Dis Child. 1967;114:617–25.

3. McKinnon PJ. Ataxia-telangiectasia: an inherited disorder of ionizing-radiation sensitivity in man. Progress in the elucidation of the underlying biochemical defect. Hum Genet. 1987;75:197–208.
4. Shiloh Y. Ataxia-telangiectasia: closer to unraveling the mystery. Eur J Hum Genet. 1995;3:116–38.
5. Shiloh Y. Ataxia-telangiectasia and the Nijmegen breakage syndrome: related disorders but genes apart. Annu Rev Genet. 1997;31:635–62.
6. Hustinx TW, Scheres JM, Weemaes CM, et al. Karyotype instability with multiple 7/14 and 7/7 rearrangements. Hum Genet. 1979;49:199–208.
7. Weemaes CM, Hustinx TW, Scheres JM, et al. A new chromosomal instability disorder: the Nijmegen breakage syndrome. Acta Paediatr Scand. 2000;70:557–64.
8. The International Nijmegen Breakage Syndrome Study Group. Nijmegen breakage syndrome. Arch Dis Child. 2000;82:400–6.
9. Chrzanowska KH, Gregorek H, Dembowska-Bagińska B, et al. Nijmegen breakage syndrome (NBS). Orphanet J Rare Dis. 2012;7:e13.
10. Stewart GS, Maser RS, Stankovic T, et al. The DNA double-strand break repair gene hMRE11 is mutated in individuals with an ataxia-telangiectasia-like disorder. Cell. 1999;99: 577–87.
11. Taylor AM, Groom A, Byrd PJ. Ataxia-telangiectasia-like disorder (ATLD)-its clinical presentation and molecular basis. DNA Repair (Amst). 2004;3:1219–25.
12. Hernandez D, McConville CM, Stacey M, et al. A family showing no evidence of linkage between the ataxia telangiectasia gene and chromosome 1q22-23. J Med Genet. 1993;30:135–40.
13. Savitsky K, Bar-Shira A, Gilad S, et al. A single ataxia telangiectasia gene with a product similar to PI-3 kinase. Science. 1995;268:1749–53.
14. Shiloh Y, Ziv Y. The ATM protein kinase: regulating the cellular response to genotoxic stress, and more. Nat Rev Mol Cell Biol. 2013;14:197–210.
15. Lovejoy CA, Cortez D. Common mechanisms of PIKK regulation. DNA Repair. 2009;8:1004–8.
16. Kong X, Shen Y, Jiang N, et al. Emerging roles of DNA-PK besides DNA repair. Cell Signal. 2011;23:1273–80.
17. Nam EA, Cortez D. ATR signalling: more than meeting at the fork. Biochem J. 2011;436:527–36.
18. Gobbini E, Cesena D, Galbiati A, et al. Interplays between ATM/Tel1 and ATR/Mec1 in sensing and signaling DNA double-strand breaks. DNA Repair (Amst). 2013;12:791–9.
19. Matsuura S, Tauchi H, Nakamura A, et al. Positional cloning of the gene for Nijmegen breakage syndrome. Nat Genet. 1998;19:179–81.
20. Tauchi H, Matsuura S, Kobayashi J, et al. Nijmegen breakage syndrome gene, NBS1, and molecular links to factors for genome stability. Oncogene. 2002;21:8967–80.
21. Chamankhah M, Wei YF, Xiao W. Isolation of hMRE11B: failure to complement yeast mre11 defects due to species-specific protein interactions. Gene. 1998;225:107–16.
22. Tauchi H, Kobayashi J, Morishima K, et al. The forkhead-associated domain of NBS1 is essential for nuclear foci formation after irradiation but not essential for hRAD50 hMRE11 NBS1 complex DNA repair activity. J Biol Chem. 2001;276:12–5.
23. Barbi G, Scheres JMJC, Schindler D, et al. Chromosome instability and X-ray hypersensitivity in a microcephalic and growth-retarded child. Am J Med Genet. 1991;40:44–50.
24. Thompson LH. Recognition, signaling, and repair of DNA double-strand breaks produced by ionizing radiation in mammalian cells: the molecular choreography. Mutat Res. 2012;751:158–246.
25. Fujimori A, Tachiiri S, Sonoda E, et al. Rad52 partially substitutes for the Rad51 paralog XRCC3 in maintaining chromosomal integrity in vertebrate cells. EMBO J. 2001;20:5513–20.
26. Tauchi H, Kobayashi J, Morishima K, et al. Nbs1 is essential for DNA repair by homologous recombination in higher vertebrate cells. Nature. 2002;420:93–8.
27. Khanna KK, Beamish H, Yan J, et al. Nature of G1/S cell cycle checkpoint defect in ataxia-telangiectasia. Oncogene. 1995;11:609–18.

28. Beamish H, Williams R, Chen P, et al. Defect in multiple cell cycle checkpoints in ataxia-telangiectasia postirradiation. J Biol Chem. 1996;271:20486–93.
29. Banin S, Moyal L, Shieh S, et al. Enhanced phosphorylation of p53 by ATM in response to DNA damage. Science. 1998;281:1674–7.
30. Canman CE, Lim DS, Cimprich KA, et al. Activation of the ATM kinase by ionizing radiation and phosphorylation of p53. Science. 1998;281:1677–9.
31. Matsuoka S, Ballif BA, Smogorzewska A, et al. ATM and ATR substrate analysis reveals extensive protein networks responsive to DNA damage. Science. 2007;316:1160–6.
32. Lim DS, Kim ST, Xu B, et al. ATM phosphorylates p95/nbs1 in an S-phase checkpoint pathway. Nature. 2000;404:613–7.
33. Zhao S, Weng YC, Yuan SS, et al. Functional link between ataxia-telangiectasia and Nijmegen breakage syndrome gene products. Nature. 2000;405:473–7.
34. Bakkenist CJ, Kastan MB. DNA damage activates ATM through intermolecular autophosphorylation and dimer dissociation. Nature. 2003;421:499–506.
35. Falck J, Coates J, Jackson SP. Conserved modes of recruitment of ATM, ATR and DNA-PKcs to sites of DNA damage. Nature. 2005;434:605–11.
36. Crawford TO. Ataxia telangiectasia. Semin Pediatr Neurol. 1998;5:287–94.
37. Shiloh Y, Lederman HM. Ataxia-telangiectasia (A-T): an emerging dimension of premature ageing. Ageing Res Rev. 2016;33:76–88. pii: S1568-1637(16)30078-30082.
38. Gatti RA, Vinters HV. Cerebellar pathology in ataxia-telangiectasia: the significance of basket cells. Kroc Found Ser. 1985;19:225–32.
39. Lavin MF. The appropriateness of the mouse model for ataxia-telangiectasia: neurological defects but no neurodegeneration. DNA Repair (Amst). 2013;12:612–9.
40. Carlessi L, Fusar Poli E, De Filippis L, et al. ATM-deficient human neural stem cells as an in vitro model system to study neurodegeneration. DNA Repair (Amst). 2013;12:605–11.
41. Reichenbach J, Schubert R, Schindler D, et al. Elevated oxidative stress in patients with ataxia telangiectasia. Antioxid Redox Signal. 2002;4:465–9.
42. Kamsler A, Daily D, Hochman A, et al. Increased oxidative stress in ataxia telangiectasia evidenced by alterations in redox state of brains from Atm-deficient mice. Cancer Res. 2001;61:1849–54.
43. Guo Z, Kozlov S, Lavin MF, et al. ATM activation by oxidative stress. Science. 2010;330:517–21.
44. Zhang J, Kim J, Alexander A, et al. A tuberous sclerosis complex signalling node at the peroxisome regulates mTORC1 and autophagy in response to ROS. Nat Cell Biol. 2013;15:1186–96.
45. Zhang J, Tripathi DN, Jing J, et al. ATM functions at the peroxisome to induce pexophagy in response to ROS. Nat Cell Biol. 2015;17:1259–69.
46. Valentin-Vega YA, Maclean KH, Tait-Mulder J, et al. Mitochondrial dysfunction in ataxia-telangiectasia. Blood. 2012;119:1490–500.
47. Valentin-Vega YA, Kastan MB. A new role for ATM: regulating mitochondrial function and mitophagy. Autophagy. 2012;8:840–1.
48. Ambrose M, Goldstine JV, Gatti RA. Intrinsic mitochondrial dysfunction in ATM-deficient lymphoblastoid cells. Hum Mol Genet. 2007;16:2154–64.
49. Gatti RA, Bick M, Tam CF, et al. Ataxia-telangiectasia: a multiparameter analysis of eight families. Clin Immunol Immunopathol. 1982;23:501–16.
50. Nowak-Wegrzyn A, Crawford TO, Winkelstein JA, et al. Immunodeficiency and infections in ataxia-telangiectasia. J Pediatr. 2004;144:505–11.
51. Taylor AM, Metcalfe JA, Thick J, et al. Leukemia and lymphoma in ataxia telangiectasia. Blood. 1996;87:423–38.
52. Isoda T, Takagi M, Piao J, et al. Process for immune defect and chromosomal translocation during early thymocyte development lacking ATM. Blood. 2012;120:789–99.
53. Dujka ME, Puebla-Osorio N, Tavana O, et al. ATM and p53 are essential in the cell- cycle containment of DNA breaks during V(D)J recombination in vivo. Oncogene. 2010;29:957–65.
54. Murphy RC, Berdon WE, Ruzal-Shapiro C, et al. Malignancies in pediatric patients with ataxia telangiectasia. Pediatr Radiol. 1999;29:225–30.

55. Olsen JH, Hahnemann JM, Borresen-Dale AL, et al. Cancer in patients with ataxia-telangiectasia and in their relatives in the Nordic countries. J Natl Cancer Inst. 2001;93:121–7.
56. Bartkova J, Rezaei N, Liontos M, et al. Oncogene-induced senescence is part of the tumorigenesis barrier imposed by DNA damage checkpoints. Nature. 2006;444:633–7.
57. Schubert R, Reichenbach J, Zielen S. Growth factor deficiency in patients with ataxia telangiectasia. Clin Exp Immunol. 2005;140:517–9.
58. Yang DQ, Kastan MB. Participation of ATM in insulin signalling through phosphorylation of eIF-4E-binding protein 1. Nat Cell Biol. 2000;2:893–8.
59. Miles PD, Treuner K, Latronica M, et al. Impaired insulin secretion in a mouse model of ataxia telangiectasia. Am J Physiol Endocrinol Metab. 2007;293:E70–4.
60. Wu D, Yang H, Xiang W, et al. Heterozygous mutation of ataxia-telangiectasia mutated gene aggravates hypercholesterolemia in apoE-deficient mice. J Lipid Res. 2005;46:1380–7.
61. Schneider JG, Finck BN, Ren J, et al. ATM-dependent suppression of stress signaling reduces vascular disease in metabolic syndrome. Cell Metab. 2006;4:377–89.
62. Takagi M, Uno H, Nishi R, et al. ATM regulates adipocyte differentiation and contributes to glucose homeostasis. Cell Rep. 2015;10:957–67.
63. Nakamura K, Kato A, Kobayashi J, et al. Regulation of homologous recombination by RNF20-dependent H2B ubiquitination. Mol Cell. 2011;41:515–28.
64. Sakamoto S, Iijima K, Mochizuki D, et al. Homologous recombination repair is regulated by domains at the N- and C-terminus of NBS1 and is dissociated with ATM functions. Oncogene. 2007;26:6002–9.
65. Kass EM, Helgadottir HR, Chen CC, et al. Double-strand break repair by homologous recombination in primary mouse somatic cells requires BRCA1 but not the ATM kinase. Proc Natl Acad Sci U S A. 2013;110:5564–9.
66. Kijas AW, Lim YC, Bolderson E, et al. ATM-dependent phosphorylation of MRE11 controls extent of resection during homology directed repair by signalling through exonuclease 1. Nucleic Acids Res. 2015;43:8352–67.
67. Zou L, Elledge SJ. Sensing DNA damage through ATRIP recognition of RPA-ssDNA complexes. Science. 2003;300:1542–8.
68. Shiotani B, Nguyen HD, Håkansson P, et al. Two distinct modes of ATR activation orchestrated by Rad17 and Nbs1. Cell Rep. 2013;3:1651–162.
69. Morishima K, Sakamoto S, Kobayashi J, et al. TopBP1 associates with NBS1 and is involved in homologous recombination repair. Biochem Biophys Res Commun. 2007;362:872–9.
70. Cox J, Jackson AP, Bond J, et al. What primary microcephaly can tell us about brain growth. Trends Mol Med. 2006;12:358–66.
71. Zhang B, Wang E, Dai H, et al. Phosphorylation of the BRCA1 C terminus (BRCT) repeat inhibitor of hTERT (BRIT1) protein coordinates TopBP1 protein recruitment and amplifies ataxia telangiectasia-mutated and Rad3-related (ATR) signaling. J Biol Chem. 2014;289:34284–95.
72. Shimada M, Sagae R, Kobayashi J, et al. Inactivation of the Nijmegen breakage syndrome gene leads to excess centrosome duplication via the ATR/BRCA1 pathway. Cancer Res. 2009;69:1768–75.
73. Yanagihara H, Kobayashi J, Tateishi S, et al. NBS1 recruits RAD18 via a RAD6-like domain and regulates pol η-dependent translesion DNA synthesis. Mol Cell. 2011;43:788–97.

Chapter 14
Management of Xeroderma Pigmentosum

Deborah Tamura, Ryusuke Ono, John J. DiGiovanna, and Kenneth H. Kraemer

Abstract Xeroderma pigmentosum (XP) is a rare, autosomal recessive inherited disease of DNA repair with a high incidence of sunlight-induced cancer of the skin and eye. Approximately half of the patients have marked burning on minimal sun exposure, often resulting in severe blistering in infancy. The other XP patients do not have this exquisite photosensitivity but develop freckle-like pigmentation of the face and other sun exposed sites, frequently before 2 years of age. Overall, XP patients have a more than 10,000-fold increase in frequency of basal cell or squamous cell carcinoma of the skin with a median age of onset of less than 10 years, nearly 60 years younger than in the US general population. There is a similar high frequency of skin melanoma, but the median age is 22 years, which is 33 years younger than in the US general population. About 25% of the XP patients have progressive neurological degeneration, often first appearing as diminished deep tendon reflexes, microcephaly, and high-frequency sensorineural hearing loss. Early recognition of neurological involvement permits use of hearing aids and other assistive devices. Management is based on early diagnosis, rigorous sun protection, and early detection and treatment of skin cancers. Oral retinoids have been demonstrated to prevent new skin cancers but have numerous side effects. Newer therapies such as use of topical bacterial DNA repair enzymes, correction of the DNA repair defect in cultured cells, and topical agents to increase read-through of premature stop codons are under investigation.

Keywords Xeroderma pigmentosum · DNA repair · Skin cancer · Melanoma Ultraviolet radiation · Sun protection · Patient advocate groups · Rare genetic diseases · Neurologic degeneration · Cancer in children · Sun sensitivity · Eye cancer

D. Tamura · R. Ono · J. J. DiGiovanna · K. H. Kraemer (✉)
Laboratory of Cancer Biology and Genetics, Center for Cancer Research, National Cancer Institute, Bethesda, MD, USA
e-mail: kraemerk@helix.nih.gov

C. Nishigori, K. Sugasawa (eds.), *DNA Repair Disorders*,
https://doi.org/10.1007/978-981-10-6722-8_14

14.1 Introduction

Xeroderma pigmentosum (XP) is a rare inherited disease of DNA repair, with an incidence of one in a million in the United States and Europe [1]. XP is more common in Japan and parts of the Middle East [2–4]. People with XP are unable to repair ultraviolet radiation (UV)-induced DNA damage resulting in an increased risk for skin cancer (10,000-fold) and ocular cancer (1000-fold) in patients under 20 years of age [5, 6]. Symptoms of XP include hyper- or hypopigmented macules (freckling/lentigos) on UV-exposed skin surfaces, severe sun burns (often with blistering) after minimal sun exposure (50% of patients), photophobia, and corneal clouding [7]. These ocular and cutaneous symptoms often occur before 2 years of age. Approximately 25% of XP patients develop a progressive neurodegeneration with cognitive decline, progressive sensorineural hearing loss, loss of deep tendon reflexes, ataxia, dysphagia, and dysarthria. This development and progression of the neurodegeneration is believed to be unrelated to UV exposure [6–8]. Currently there are no methods to reverse the UV-induced DNA damage in the cells once it has occurred. Until a definitive treatment is developed, avoidance of UV exposure and early treatment of cancers is the mainstay for management of XP patients [9]. Information about XP for health professionals can be found online (Table 14.1).

14.2 UV Management for XP Patients

The symptoms of XP are a classic example of gene environment interaction. XP patients have mutations in one of any seven genes (XP-A to XP-G) in the nucleotide excision repair (NER) pathway or in pol eta (XP-V), a bypass DNA polymerase involved in translesion synthesis at the site of UV damage [10, 11]. These mutations lead to decreased ability to repair UV-type DNA damage (cyclobutane dimers or 6-4 photoproducts) resulting in cell death, cell senesce, or the development of cancer. If XP patients are exposed to any amount of UV, they can develop the typical poikilodermatous skin changes (dry, parchment-like skin with lentigos and hypopigmented macules, telangiectasias, and atrophy) and cancer [6, 12]. However, if diagnosed early in life and stringently UV protected, they may develop few if any cutaneous symptoms. Since the ocular and cutaneous symptoms of XP are dependent on the amount of UV exposure, by removing the environmental influence, the consequences of the gene mutations can be modified [13, 14].

14.2.1 UV Management Begins with Identification and Diagnosis of XP Patients

In approximately half of XP patients, the first symptom is a severe often blistering sunburn (Fig. 14.1). Infants have burned through window glass several days after birth; others suffered blistering burns so severe that child abuse was suspected.

Table 14.1 Online information about xeroderma pigmentosum for medical personnel and patients

Educational website	Url	Description
GeneReviews: Xeroderma Pigmentosum	https://www.ncbi.nlm.nih.gov/books/NBK1397/	NCBI site curated by physicians from the National Institutes of Health provides fact-based data and scientific/medical information
US National Library of Medicine Genetics Home Reference: Your Guide to Understanding Genetic Conditions Xeroderma Pigmentosum	https://ghr.nlm.nih.gov/condition/xeroderma-pigmentosum	Curated website with accurate fact-based evidence includes information on disease management and inheritance pattern has PDF conversion for printing of handouts
DermNet New Zealand: Xeroderma pigmentosum	http://www.dermnetnz.org/topics/xeroderma-pigmentosum/	Site founded by New Zealand dermatologist. Has brief fact-based information on disease symptoms and management through UV avoidance
US Department of Health & Human Services NIH/NCATS	https://rarediseases.info.nih.gov/diseases/7910/xeroderma-pigmentosum	Basic information on XP including treatment, research, and support organizations. The chart in the symptoms section is not accurate. There are links for US government sites for social security
NORD: National Organization for Rare Diseases: Xeroderma Pigmentosum	https://rarediseases.org/rare-diseases/xeroderma-pigmentosum/	Comprehensive information including symptoms, genetic information, and treatment. Links for additional information and to support group websites. PDF conversion available for handouts
Gene Tests: Xeroderma Pigmentosum	https://www.genetests.org/disorders/?disid=23895	Lists laboratories that test for XP
Understanding Xeroderma Pigmentosum	https://www.cc.nih.gov/ccc/patient_education/pepubs/xp7_17.pdf	Information about XP for patients and family members written in nontechnical English

These burns tend to develop over the course of several hours and take greater than a week to heal [7]. Despite developing blistering and crusting in the burned areas, there is no scar formation as compared to other photosensitive conditions such as porphyria or hydroa vacciniforme [15, 16]. The other half of XP patients do not burn, they may tan; however they often are photophobic. The eyes of XP patients are very light sensitive, and infants and young children may cry and turn away from sunlight. Erythematous conjunctiva and sclera may be one of the first indications of photosensitivity in XP children. Eventually, after repeated UV exposure, all XP patients will develop lentigos and hypopigmented macules in areas of UV exposure. The face, hands, and forearms are often the first areas of the skin to show changes. Most children with XP will develop the pigmentary changes before age 2 years and often in the first year of life [6, 12, 17]. Once XP is suspected, children should be UV protected while the diagnosis is confirmed. Laboratory testing using DNA analysis for mutations in the XP genes or functional testing (UV sensitivity, unscheduled DNA synthesis, host cell reactivation) on cultured fibroblasts from a skin biopsy is available through laboratories in several countries (Tables 14.1 and 14.2).

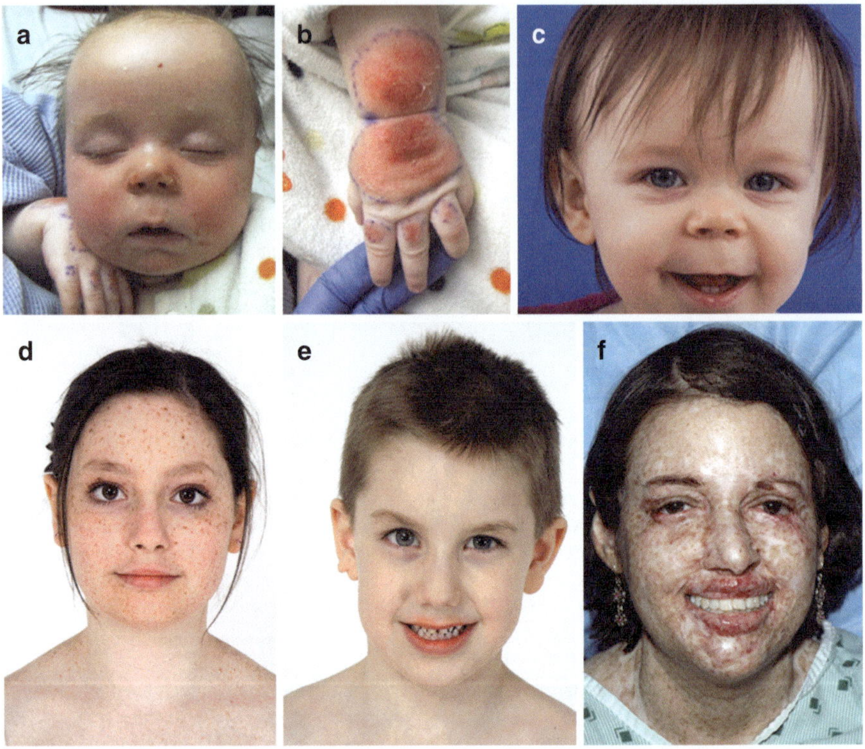

Fig. 14.1 Sun burning in XP-D patient and lack of burning in XP-C patients. (**a–c**) XP-D patient (XP499BE) who sustained severe blistering sunburn following intermittent evening sun exposure. (**a**) Five months of age: day 8 after sun exposure—swelling and burns of the face resolving with delayed erythema on dorsum of the right arm and hand. (**b**) Day 8 after sun exposure—delayed swelling, erythema, blistering, and peeling on dorsum of the left hand and wrist. (**c**) Age 11 months: excellent sun protection and normal skin exam. (**d** and **e**) XP-C siblings with no history of burning on minimal sun exposure. (**d**) Older sister aged 12 (XP198BE)—diagnosed at age 1 secondary to freckling on the face, hands, and arms. She had excellent UV protection since diagnosis and had 1 BCC in her scalp; (**e**) Patient XP338BE, 7-year-old brother of patient XP198BE. He had excellent UV protection since diagnosis at birth and no clinical evidence of XP. (**f**) XP-C patient (XP24BE), aged 32, did not have a history of burning on minimal sun exposure. She had a relatively late age of diagnosis (age 8) and poor UV protection. She has multiple lentigines, chilitis, telangiectasias, and >200 skin cancers. She died at age 35 of a glioblastoma of her brain [76] (Figure from [20])

14.2.2 *Layers of UV Protection Prevent Further UV Damage to Patients*

The first layer of UV protection involves modifying the XP patient's environment. The goal of environmental UV management is to reduce the amount of UV in the XP patient's immediate surroundings. Identification and modification of environmental sources of UV can be difficult. Since the sun is the major source of UV, reduction of sunlight exposure is critical. People with XP should generally avoid

Table 14.2 Clinical evaluation of xeroderma pigmentosum patients

When should I suspect XP?
1. Acute photosensitivity at a young age
2. Early onset of skin freckling <2 years old
3. Severe sun damage including lips and eyes
4. Skin cancer at early age
5. History of large numbers of skin cancers
Initial evaluation
1. Skin exam
(a) Hyper and hypopigmented macules on sun-exposed skin
(b) Cheilitis, freckling, and xerosis of lips
(c) Telangiectasias on sun-exposed skin
(d) Absence of skin lesions on sun-protected skin
(e) Presence of skin cancers in children or young adults
2. General physical exam
(a) Height and weight
(b) Head circumference measurement
(c) Eye exam for evidence of erythema, photophobia, pterygium, or corneal clouding
(d) Assessment of deep tendon reflexes
(e) Evidence of cognitive or developmental delay
(f) Evidence of hearing impairment
3. Laboratory testing
(a) Vitamin D levels
(b) DNA testing for mutations in XP-related genes (genetests.org)
General management guidelines
1. Total body skin exam every 6 months or more frequently if needed for rapidly occurring skin cancers
2. Surgical removal of skin cancers (referral to Mohs surgeon if indicated)
3. Consider field treatment with imiquimod or 5Fluorouracil (5FU) in areas cleared of tumors
4. Referral to ophthalmologist for regular monitoring
5. Referral for audiometry for evidence of high-frequency sensorineural hearing loss
6. Referral to neurologist if indicated

being outside during the day from just after sunup to slightly before sundown (horizon to horizon). XP patients can maintain a safer UV environment by assessing ambient indoor and outdoor UV levels. One important tool many XP patients use is a UV meter. A UV meter measures the level of UV-A (320–400 nm) and UV-B (290–320 nm) in the outdoor or indoor environment permitting the XP patient to identify "UV-safe" areas. It is best that a UV meter be small (handheld), lightweight, sturdy, reasonably priced, and with the UV readings easy to interpret. There are several meters on the commercial market meeting these criteria, and XP families may own more than one meter, so they are readily available for use (Fig. 14.2). There is no "exact" safe UV reading on a meter, and different patients and families will determine what is an acceptable UV level reading on the meter. As an example, in bright sun, a meter may measure UV levels at 2500 mw/cm^2 or greater, and in the

Fig. 14.2 Ultraviolet meters. Small, handheld, portable, battery-operated meters can be used to measure the amount of damaging UV radiation present outdoors or indoors enabling the patient to know when and where to protect themselves

shade on a sunny day, the reading may be 300 mw/cm^2; UV exposure at either of these levels may cause substantial damage to an XP patient. However, indoors, the meter may read 0–3 mw/cm^2, and at this level the potential for damage would be greatly reduced. Many XP patients attempt to keep the UV levels in their homes, cars, schools, and work areas as close to a zero reading on the meter as possible. The meter can be useful to identify areas of high UV levels in public areas such as stores and restaurants. The ability to UV meter their environment permits patients to safely engage in the community and lead active lives [18].

Although window glass blocks about 90% of UV, longer wave UVA can still pass through the glass raising indoor UV levels. Curtains and blinds can be placed over the windows, but they create a darker and potentially isolating environment. The application of clear, UV-blocking window films to windows and glass doors is a better option. The films transmit visible light creating a more atheistically pleasing environment for the family to spend extended amounts of time. UV-blocking film does not need to be darkly tinted to be protective, thus it can be used in cars, in schools, and in workplace areas [19, 20].

UV is also emitted by certain artificial light sources including unshielded florescent bulbs, halogen, and mercury vapor lamps. In contrast, incandescent bulbs, shielded florescent bulbs, and the majority of LED bulbs do not produce appreciable amounts of UV [20]. Since it can be difficult to determine if florescent bulbs are emitting UV, a light meter can be used to determine whether indoor light sources are safe for the XP patient. It may be necessary for schools or work places to replace unsafe bulbs for XP patients. The US ADA (Americans with Disabilities Act—1990) requires that all persons have safe working and learning environments (https://www.ada.gov/cguide.htm).

Other environmental actions should include obtaining handicapped parking permits, so XP patients can park close to the entrances of buildings. XP patients should also carry medical alert IDs (bracelets, necklaces, or wallet identification cards) identifying the person as UV sensitive. So in the event of an accident, first responders will be aware of the condition and can move the XP patient to a UV-safe environment [20].

The second layer of UV management is clothing and other materials that mechanically block UV from reaching the patient's skin. Clothing should generally cover the trunk, arms, legs, and feet. Long-sleeved shirts, gloves, long pants, tights, leggings, long socks, and closed-toed shoes should be the standard wardrobe for XP patients. In areas of high UV exposure (outside or in areas with un-filmed windows), XP patients should have hats, UV-blocking sunglasses, and/or hoods with plastic UV-blocking face shields (Fig. 14.3) and gloves. When XP patients are in known UV-safe areas (e.g., at home), it is not necessary to be as completely covered.

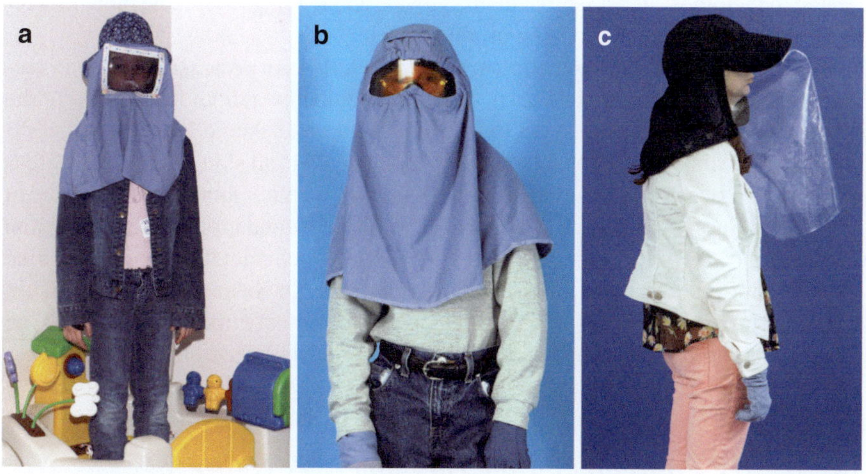

Fig. 14.3 XP patients dressed in UV-protective clothing. Multiple layers of clothing provide good UV protection. The hood is fashioned from a hat with added cloth to shield the neck and shoulders plus clear, UV-absorbing plastic to protect the face. (**a**) XP-D patient (XP84BE) is an 11-year-old girl. She sustained several blistering burns before diagnosis at age 3. She has had excellent UV protection and has never had a skin cancer. (**b**) XP-C patient (XP243BE) is a 13-year-old boy who never sunburned. He was diagnosed with XP at age 3; his first skin cancer a BCC on his nose occurred at 18 months. He began receiving exquisite UV protection at age 5 and had a total of nine skin cancers. He died at age 19 from complications of a bone marrow transplant following the diagnosis of myelodysplastic syndrome. (**c**) XP-D patient (XP116BE) is a 14-year-old girl who was diagnosed with XP at 6 months following two severe blistering burns. She has had excellent UV protection and never had a skin cancer; however, she has developed progressive neurodegeneration

Clothing fabric should be tightly woven and made of a "dense" material; holding clothing up to a light source is a good way to test the weave and density of a material. If visible light is seen coming through the fabric, then UV will also be passing through. Fabrics such as denim, corduroy, canvas, and cotton broad cloth are some examples of tightly woven materials [8, 20, 21]. To enhance the UV-blocking capability of clothing, XP patients can double the layers of fabric or wear several layers of lightweight clothing.

Recently UV-blocking clothing has come onto the commercial market. This specialty clothing is made from very tightly woven fabrics and is treated with a UV-blocking/absorbing chemical to enhance the UPF (ultraviolet protection factor) of the cloth. The types of clothing include shirts, pants, skirts, swimsuits, hats, and gloves. While these clothes come in a variety of styles, colors, and sizes, including children's and infants clothing, they can be expensive.

An alternative to specialty clothing is using one of several UV-blocking laundry additives, rinses, or spray on compounds to add additional UV-blocking capabilities to normal clothing. An XP patient's clothes can be washed or treated with any of these UV-blocking products to add additional protection. The products may last through several washes, but the clothing should be periodically checked to make sure the effectiveness is still at an acceptable level. Not all the clothing needs to be treated, but clothes such as shirts, pants, gloves, and leggings would be more protective if treated with a UV-blocking product.

The third layer of UV protection is sunscreen lotions or creams. There is no recommended "XP" form of sunscreen, but several guidelines should be followed: the SPF (sun protective factor) ought to be 30 or higher and should be UVA and UVB blocking; the first application of sunscreen ideally covers all skin surfaces, and areas of skin not covered by clothing (face, neck, ears, hands) should have sunscreen reapplied at least every 3 h. Since there are multiple formulations of sunscreens on the market with a wide variety of ingredients such as cinnamates, salicylates, zinc and titanium oxides, and the benzones (avobenzone, oxybenzone), finding a sunscreen the patient likes and uses on a regular basis is most important. UV-blocking lip balm should be reapplied frequently, since it is easily licked off [8, 22, 23].

14.2.3 UV Protection for the Face and Eyes Is Critical

The face and eyes of XP patients receive the largest amount of UV exposure; often burning and freckling on the face and photophobia are the first symptoms of the condition and can lead to disabling and disfiguring tumors. In addition, UV damage-related vision loss is a major cause of morbidity and disability in XP patients [24–27]. The lips and tip of the tongue also receive extensive UV exposure, and tumors can occur in these areas [12, 27, 28]. Thus, UV protection of the face and head is extremely important as soon as the diagnosis is made. Patients can protect their faces and eyes by avoiding outdoor activities during the day thus reducing overall UV exposure; however, staying indoors is socially isolating and not practical for children who need to go to school and adults who travel to work. High-SPF sunscreen applied frequently (every 2–3 h) to the face, neck, and ears and wearing large brim hats with neck-shielding drapes made of UV-blocking cloth can significantly reduce UV damage to the facial skin. Despite diligent use of sunscreen, some skin surfaces may not be adequately protected and can sustain more UV damage. To prevent ocular damage, XP patients need to consistently wear UVA- and UVB-blocking "wraparound" sunglasses that protect the globes, supporting tissues, lids, and surrounding ocular skin. A more thorough protective agent for the skin of the head is a UV-blocking hood. The hood is made of lightweight UV-blocking material attached to a UV-blocking hat. The material covers the XP patient's neck and ears and the hat covers the head. A clear plastic UV-blocking face shield is attached to the brim of the hat. The UV-blocking cloth is attached to the sides of the face shield (Fig. 14.3). The hood cannot be considered completely protective, and XP patients still need to avoid being outdoors for extended periods of time. However, stigmatization of XP patients may occur by wearing the hood. Children have been teased and bullied in school settings, and XP adults have been stopped by the police while wearing the hoods and were told they looked "suspicious" [20, 22].

14.3 Management of XP-Related Dermatologic and Ophthalmologic Abnormalities

UV protection is the mainstay for the prevention of dermatologic and ophthalmologic abnormalities in XP patients. There is a 10,000-fold increased risk for developing nonmelanoma skin cancers (NMSC) and a similar elevated risk for developing melanoma skin cancers. Most XP patients will develop skin neoplasms during their life time [7, 11]. When a child (or adult) is first diagnosed, it is important to teach the patient and families to do skin exams at home; pictures can be used to teach patients and families the characteristics between common and concerning skin lesions. Dermatologic care is critical and XP patients need regular full body skin exams by a dermatologist. The interval of time between exams depends on the rate of skin cancer. A child or adult who is developing very few skin cancers may only need to be seen every 4–6 months; however, XP patients who have had poor UV protection or are developing many tumors may need to be seen every 4–6 weeks [29–31].

14.3.1 Field Treatment for XP Patients

Prior to diagnosis most XP patients will have had some amount of UV exposure. Depending on the length of time to diagnosis and the amount of UV exposure, some XP patients will have significant areas of UV damage. There are several treatment options for XP patients with significant photodamage. Field treatment for areas of photodamage and actinic keratoses with imiquimod 5% cream (Aldara),

5 fluorouracil (Effudex) and cryotherapy is effective in preventing the progression of actinic keratoses to SCC (squamous cell carcinoma). In addition, small superficial BCCs (basal cell carcinoma) and areas of severe photodamage that are at risk for tumor development can also be treated [31–38]. The side effects of treatment may be significant (skin redness, swelling, itching, irritation, dryness, burning, pain, tenderness, thickening/hardening of the skin, peeling/flaking/scabbing/crusting, or leaking a clear fluid) and in some cases not well tolerated by the patient. Several reports have noted poor patient compliance, when there are severe side effects or with prolonged use [33, 34]. Side effects may be lessened by using reduced dosing schedules applied over longer time courses.

Photodynamic therapy (PDT) with blue light or lasers and various types of photoactive chemicals (5-aminolevulinic acid, methylaminolevulinate) have been used to treat XP patients [39]; however, a study of mouse fibroblasts found that 8-oxo-7,8-dihydro-2¢-deoxyguanosine (8-oxo-dG) adducts were generated at the site of treatment; these lesions are repaired by the NER pathway. It is possible that PDT therapy may place the XP patient at risk for new lesions at the site of treatment [40, 41].

Chemoprevention and tumor treatment with oral retinoids has been used successfully in XP patients [42]. However, multiple side effects including dry skin with irritation and redness, liver toxicity, abnormalities of serum lipid profiles, and bone abnormalities like hyperostoses and premature epiphyseal closure have occurred. The severity of the side effects is dose-related and is most commonly seen at the higher treatment doses and when taken over a longer time. Retinoids are known teratogens and premenopausal women must use an effective method of birth control while taking these medications. Some retinoid drugs may be slowly eliminated from the body after therapy is discontinued. For example, after discontinuation of treatment with acitretin, effective contraception must be continued for 3 years posttreatment [32, 43, 44].

Other field treatments include dermabrasion and chemical peels. These treatments remove shallow surface lesions and improve cosmetic appearance. However, there is a risk of not treating residual deeper lesions with resulting in the development of deeply invasive lesions [45].

14.3.2 Surgical Management of Skin Tumors

Management and treatment of skin cancers in XP patients can be complicated (Table 14.2). Some tumors can be treated with standard surgical treatments such as excision or electrodessication and curettage. However, cancers on the face, including the lips and tip of the tongue, and head can lead to significant disfigurement, disability, and mortality [28, 46]. The large numbers of skin tumors and the extensive field damage with premalignant changes can make it difficult to obtain a clear margin. Mohs micrographic surgery can offer tissue sparing for XP patients with NMSC on the face, head, or other areas. Mohs surgery assures complete histologic assessment of surgical margins resulting in complete removal of the tumor without removing uninvolved tissue. In non-XP skin, chronic photodamage results in lax, wrinkled skin which is easily moved for surgical repair. In contrast, UV-damaged XP skin becomes severely atrophic and tight, augmenting the need for tissue sparing surgery. Large or deeply invasive tumors may require a team approach with several surgical specialties involved in the management of patients [47, 48]. Rarely, in patients with severe involvement of facial skin, total resurfacing with skin grafts from sun-protected skin has been performed [49, 50].

14.3.3 Ocular Management

The eyes of XP patients often have sustained substantial UV damage. Photophobia (crying, turning away from sunlight, rubbing the eyes) may be the first suggestion that a child has a photosensitive condition [6]. Due to the erythema and irritation caused by UV, some young XP children are misdiagnosed as allergic conjunctivitis. XP ocular damage includes dry eye, visual impairment, keratitis, band-like keratopathy, pterygium, corneal scarring and ulceration, neovascularization, and neoplasms of the ocular surface. Brooks et al. reported on 87 XP patients finding that 91% of the patients had at least one ocular abnormality [26]. Other ophthalmologic studies of XP patients also found high rates of abnormalities including cancer in the periocular skin and ocular surfaces of patients [25, 51].

Initial ocular management begins with strict UV protection. The UV-blocking hood and face shield protect the eyes and face from UV (Fig. 14.3); however some adult XP patients and children find the hood too cumbersome or stigmatizing to wear on a regular basis. UVA- and UVB-blocking sunglasses that wrap around the eyes and protect not just the globes but the lids and sides of the palpebral areas and sides of the face near the eyes can provide good protection. The sunglasses should be worn on both sunny and cloudy days and until the sun is below the horizon. Some XP patients will obtain clear UV-blocking glasses (nonprescription for some patients) to wear indoors as an additional level of protection [52].

Benign, ocular neoplasms associated with UV exposure include pterygium, pinguecula, pannus, and abnormal blood vessel development with corneal clouding. These lesions can progress over time, leading to reduced acuity or vision loss. Surgical removal is necessary in some cases [26]. Corneal transplant has been used to treat vision loss in some XP patients, but in many cases, the graft is rejected due to the presence of abnormal cornea vessels [53]. Malignancies including, squamous cell, basal cell carcinomas carcinoma and melanoma occur on the lids and any UV-exposed areas of the eye and can lead to significant morbidity and mortality. Surgical procedures on the lids can lead to lagophthalmos exacerbating dry eye and leading to corneal scarring [24, 26, 27].

Regular ophthalmologic exams, at least annually, are an important part of care for XP patients. Since dry eye and ocular surface abnormalities can be seen even in young XP patients, tear film and ocular lubrication should be regularly monitored

[26]. Patients should be reminded about the continual need for UV protection and the use of ocular lubricants to treat dry eye.

14.4 Management of the XP Patient at Risk for XP Neurologic Disease

Approximately 25% of XP patients in the US and Europe develop a progressive neurodegeneration. Most of the patients will have mutations in the *XP-A*, *XP B (ERCC3)*, *XP-D (ERCC2)*, *XP-F (ERCC 4)*, or *XP-G (ERCC5)* genes; XP patients who experience severe burning on minimal sun exposure are most at risk for developing neurologic degeneration. Neurodegeneration is thought to be a result of accumulation of unrepaired oxidative damage in the central nervous system and is unrelated to external UV exposure [54, 55]. The symptoms of neurodegeneration can occur at any age beginning in early childhood through adulthood [6, 10, 56, 57].

Understandably, parents of XP children are very concerned about the risk for their children developing neurologic disease; many parents inquire about ways to determine if or when it will develop [58]. Currently it is usually not possible to predict with certainty which XP patients will develop neurologic disease; however there are several methods to monitor for early symptoms of neurodegeneration (Table 14.2). Prompt identification of neurodegeneration is important so appropriate supportive interventions can be started as soon as symptoms appear. High-frequency neurosensory hearing loss is one of the earliest symptoms of neurodegeneration, and monitoring for hearing loss is simple and noninvasive. Routine (every 1–2 years) audiology exams can help detect early neurosensory hearing abnormalities; hearing aids are an effective means of treatment, when hearing loss becomes clinically significant [6, 10, 59]. Loss of deep tendon reflexes is another early symptom of neurodegeneration. Reflex assessment can be instituted as regular part of an XP patient's skin exam. If sensorineural hearing loss or areflexia are noted, referral to a neurologist for a complete neurologic exam and additional testing is indicated. Most XP patients will have nerve conduction testing and brain imaging (CT or MRI) as part of the neurologic evaluation. Peripheral neuropathy with neuronal loss and brain atrophy are commonly found in XP patients with neurologic disease [7, 57, 60].

As a consequence of the neurodegeneration, XP patients may lose cognitive abilities; monitoring for cognitive decline can be done through neuropsychological testing. Tests including the Wechsler Adult Intelligence Scales, Wechsler Intelligence Scale for Children, Wechsler Memory Scales, and WAIS-III are commonly used to determine intelligence, memory, language, and executive function. If school-age children develop XP with neurologic disease, they will require special educational support. An Individualized Education Plan (IEP) to include requirements for UV protection, hearing support, and special educational services should be developed as soon as neurodegeneration is suspected. Adult patients with XP neurodegeneration will develop a progressive loss of abilities; eventually they may require assistance in self-care activities including bathing, toileting, eating, and ambulation. As the neurodegeneration progresses, XP patients can also develop dysarthria and dysphagia and may require feeding tubes to prevent aspiration. Ambulation becomes increasingly difficult and a broad-based gait is commonly seen in XP patients with neurologic disease; walkers are commonly used to help maintain balance and ambulatory independence. As ataxia and balance abnormalities progress, wheelchairs may be needed for safe transport [6, 7, 11]. Bradford et al. reported that XP patients with neurologic disease had a significantly shortened life span with a median age at death of 29 years [7].

14.5 Research Studies for Treatment of XP

Patients with XP have defective nucleotide excision repair. Bacteria also have an enzyme system that removes DNA damage. Unrepaired DNA damage may lead to formation of skin cancers. A study of 30 XP patients treated with a topical formulation of the bacterial DNA repair enzyme T4 endonuclease in a liposome formulation reported reduction in the frequency of precancerous actinic keratoses and of basal cell carcinomas [61]. No side effects were reported. However, this treatment has not been approved by the US Food and Drug Administration (FDA).

Gene transfer therapy by use of retrovirus-based transduction of the normal XPC gene was reported to increase DNA repair in cultured skin keratinocytes from three XPC patients [62]. Another approach was designed to correct the defect in XPC cells with a common TG dinucleotide deletion found in North African XPC patients. They used meganuclease and TALEN (transcription activator-like effector nuclease) to form a double-strand break in the DNA near the mutation and promote correction by homologous recombination [63].

About 10–20% of human genetic disorders involve inherited mutations that lead to premature termination codons (PTC) during the process of translating the mRNA to protein. This results in low levels of mRNA due to nonsense-mediated message decay and to low levels of protein. XP patients with premature termination codon could be targeted for read-through therapy by use of aminoglycosides. Aminoglycosides bind to 30S ribosomal subunit and modify fidelity between mRNA and tRNA. At premature termination codon site, another amino acid could be inserted in the presence of aminoglycosides. This would result in increased mRNA and protein. Long-term use of systemic aminoglycosides may induce kidney and nerve toxicity. In contrast, topical use of aminoglycosides would be expected to have lower toxicity than systemic use. Kuschal et al. [64] reported that the aminoglycosides geneticin and gentamicin successfully induced full-length XP-C protein in six of eight skin fibroblast cell lines from XP-C patients with premature termination codons. This treatment resulted in the reduction of UV-induced DNA damage in the XP-C fibroblasts. Similar results were found by use of other aminoglycosides (paromomycin, neomycin, and kanamycin) in cultured XP-C cells [65]. The effect depended on the 5′ and 3′ context of PTC site and the type of aminoglycoside. A drug that inhibits nonsense-mediated message decay was reported to increase the level of XPC mRNA in cultured XP-C cells alone or in combination with aminoglycosides [66]. Therapy of PTC bypass for XP patients needs careful in vitro and in vivo assessment.

14.6 XP Family Social Support

The diagnosis of a rare and potentially fatal disease, particularly in a young child, is devastating [18, 22, 58]. One of most important aspects of XP patient management is education and social support. When the diagnosis is confirmed, it is important to have a sensitive discussion with the patient and family allowing ample time for questions from the family. Although young children may not comprehend detailed explanations, they can understand simple concepts such as *the sun is not your friend.* The family should be provided factual, clearly written, educational material on XP at the time of diagnosis. This helps the families explain the disorder to the extended family and provides concrete information for them to reference when they have questions [67, 68] (Tables 14.1 and 14.3).

There are many pictures on the Internet of poorly protected children and adults with XP, often with devastating skin cancers, neurologic disease, and early death, and this can be very upsetting to families [58, 69]. Thus, during the early discussions of the disease, it is important to stress the relationship between consistent UV protection and favorable patient outcomes [18, 22]. Families also may express guilt over UV exposure the child has already had, since the pregnancy and early infancy were normal with few symptoms to alert the family to the presence of a disease [9]. Many families were unaware that the freckling was abnormal, until the diagnosis was suggested. Remarkably, when a child sustains severe sunburns after minimal

Table 14.3 Xeroderma pigmentosum support groups worldwide

Country/ location	XP patient support group	URL
United States	XP Family Support Group	www.xpfamilysupport.org
United States	XP Society	www.xps.org
North America	XP Grupo Luz De Esperanza (in Spanish)	www.xpgrupoluzdeesperanza.org
Great Britain	XP Support Group	xpsupportgroup.org.uk
France	Les Enfantes De La Lune (in French)	www.enfantsdelalune.org
Germany	Xeroderma Pigmentosum (in German)	www.xerodermapigmentosum.de/
Tunisia	Association d'aide aux Enfants Atteints de Xeroderma Pigmentosum	www.xp-tunisie.org.tn
Japan	Japanese National Network of Xeroderma Pigmentosum (in Japanese)	www.xp-japan.net
South Africa	Xeroderma Society South Africa	www.xpsociety.co.za

sun exposure, families may have had long standing concerns about their child's health and finally having a diagnosis can be a relief. Whether the parents suspected health issues or not, the guilt over previous UV exposure may intensify when the child is diagnosed with skin cancers [68, 70].

Adjustment to strict UV protection may be easier for younger children. Younger children will not have come to expect extensive outdoor day activity. For an older child, UV protection is more complicated; the child may be used to extensive outdoor time and may not want to conform to UV-safe activities and restrictions [18, 22, 71]. They may also have worries over peer group acceptance. Referral to an XP patient support group at the time of diagnosis can be very beneficial. Members of the support group can help newly diagnosed patients and their families adjust to "UV-safe" living. The experiences of support group members living the "UV-safe lifestyle" can help newly diagnosed patients and families, by providing helpful suggestions regarding day to day adjustments needed for UV avoidance. As one experienced XP mother said "In a year UV protection will be just second nature." Many of the support groups have educational meetings and UV-safe social events, where the patients and families can share experiences and companionship. These interactions help families normalize the changes required for UV-safe living [18, 70, 72, 73].

Families need to work with schools, workplaces, and other extracurricular sites (churches and recreational organizations) to provide UV-safe environments. XP is a recognized disabling condition by the US Social Security Administration, and people with XP are covered by The Americans with Disabilities Act (ADA). The ADA prohibits discrimination against people with disabilities and requires schools and employers to provide reasonable accommodations for people with disabilities. Schools should provide safe accommodations with as little restriction to normal activity for XP students' education as possible. For example, the windows in classrooms and other areas of the school (where the child will spend extensive time) can be covered in UV-protective films and light bulbs checked for UV emission to provide a safe environment. The school should provide substitute activities for outdoor physical education classes, recess, and disaster drills for XP children so as not to exclude them from normal activities. XP adults need to work with their employers to develop UV-safe work environments [8, 18, 21, 22].

Families who have unaffected children can face additional stress. Balancing the needs of the non-XP child to be outside and participate in organized sports activities while simultaneously accepting the restrictions of UV safety for the XP child can be a daunting task. Although studies have found that siblings of children with disabilities adjust to the situation, parents can still find it difficult to manage the requirements for all children equally [74, 75]. These adjustments are easier in families with good social support systems such as positive family and community assistance [18, 68, 70]. However, not all families have good social support systems, or they may have significant stressors including poverty, addiction, mental illness, and domestic violence that make adapting to UV-safe living more difficult. In cases where children are not well UV protected or their health seems compromised, social service referrals should be made.

In summary, XP is a severe disorder caused by mutations in genes responsible for nucleotide excision repair of UV-induced DNA damage. XP patients may develop hundreds of cancers in UV-exposed areas of the skin and eyes. However, with early diagnosis, stringent UV protection, and prompt management of cancers, XP patients can live longer active lives, attend schools, work at a wide variety of jobs, and raise families. As research and new treatments evolve, the outlook for XP patients' health and longevity will continue to improve.

Acknowledgments This research was supported by the Intramural Research Program of the NIH, National Cancer Institute, Center for Cancer Research. We thank the XP patients and their families for participation in our research studies.

References

1. Kleijer WJ, Laugel V, Berneburg M, Nardo T, Fawcett H, Gratchev A, et al. Incidence of DNA repair deficiency disorders in western Europe: xeroderma pigmentosum, Cockayne syndrome and trichothiodystrophy. DNA Repair (Amst). 2008;7(5):744–50.
2. Cartault F, Nava C, Malbrunot AC, Munier P, Hebert JC, N'Guyen P, et al. A new XPC gene splicing mutation has lead to the highest worldwide prevalence of xeroderma pigmentosum in black Mahori patients. DNA Repair (Amst). 2011;10(6):577–85.
3. Nakano E, Masaki T, Kanda F, Ono R, Takeuchi S, Moriwaki S, et al. The present status of xeroderma pigmentosum in Japan and a tentative severity classification scale. Exp Dermatol. 2016;25(Suppl 3):28–33.
4. Doubaj Y, Laarabi FZ, Elalaoui SC, Barkat A, Sefiani A. Carrier frequency of the recurrent mutation c.1643_1644delTG in the XPC gene and birth prevalence of the xeroderma pigmentosum in Morocco. J Dermatol. 2012;39(4):382–4.
5. Kraemer KH, Lee MM, Andrews AD, Lambert WC. The role of sunlight and DNA repair in melanoma and nonmelanoma skin cancer. The xeroderma pigmentosum paradigm. Arch Dermatol. 1994;130(8):1018–21.
6. DiGiovanna JJ, Kraemer KH. Shining a light on xeroderma pigmentosum. J Invest Dermatol. 2012;132(3 Pt 2):785–96.
7. Bradford PT, Goldstein AM, Tamura D, Khan SG, Ueda T, Boyle J, et al. Cancer and neurologic degeneration in xeroderma pigmentosum: long term follow-up characterises the role of DNA repair. J Med Genet. 2011;48(3):168–76.
8. Emmert S, Ueda T, Zumsteg U, Weber P, Khan SG, Oh KS, et al. Strict sun protection results in minimal skin changes in a patient with xeroderma pigmentosum and a novel c.2009delG mutation in XPD (ERCC2). Exp Dermatol. 2009;18(1):64–8.
9. Tamura D, Khan SG, Merideth M, DiGiovanna JJ, Tucker MA, Goldstein AM, et al. Effect of mutations in XPD(ERCC2) on pregnancy and prenatal development in mothers of patients with trichothiodystrophy or xeroderma pigmentosum. Eur J Hum Genet. 2012;20(12):1308–10.
10. Fassihi H. Importance of genotype-phenotype correlation in xeroderma pigmentosum. Br J Dermatol. 2015;172(4):859–60.
11. Fassihi H, Sethi M, Fawcett H, Wing J, Chandler N, Mohammed S, et al. Deep phenotyping of 89 xeroderma pigmentosum patients reveals unexpected heterogeneity dependent on the precise molecular defect. Proc Natl Acad Sci U S A. 2016;113(9):E1236–45.
12. Kraemer KH, Lee MM, Scotto J. Xeroderma pigmentosum. Cutaneous, ocular, and neurologic abnormalities in 830 published cases. Arch Dermatol. 1987;123(2):241–50.
13. Giarelli E, Jacobs LA. Modifying cancer risk factors: the gene-environment interaction. Semin Oncol Nurs. 2005;21(4):271–7.

14. Kraemer KH. Xeroderma pigmentosum. A prototype disease of environmental-genetic interaction. Arch Dermatol. 1980;116(5):541–2.
15. Almeida HL, Kopp J, Jorge VM, Sartori DS, Velloso CD. Extensive hydroa vacciniforme. An Bras Dermatol. 2013;88(4):620–2.
16. Andersen J, Gjengedal E, Sandberg S, Raheim M. A skin disease, a blood disease or something in between? An exploratory focus group study of patients' experiences with porphyria cutanea tarda. Br J Dermatol. 2015;172(1):223–9.
17. Fassihi H. Spotlight on xeroderma pigmentosum. Photochem Photobiol Sci. 2013;12(1):78–84.
18. Milota M, Jones DL, Cleaver J, Jamall IS. Xeroderma pigmentosum family support group: helping families and promoting clinical initiatives. DNA Repair (Amst). 2011;10(7):792–7.
19. Duarte I, Rotter A, Malvestiti A, Silva M. The role of glass as a barrier against the transmission of ultraviolet radiation: an experimental study. Photodermatol Photoimmunol Photomed. 2009;25(4):181–4.
20. Tamura D, DiGiovanna JJ, Khan SG, Kraemer KH. Living with xeroderma pigmentosum: comprehensive photoprotection for highly photosensitive patients. Photodermatol Photoimmunol Photomed. 2014;30(2–3):146–52.
21. Davis BE, Koh HK, Rohrer TE, Gonzalez E, Cleaver JE. Sunlight avoidance and cancer prevention in xeroderma pigmentosum. Arch Dermatol. 1994;130(6):806–8.
22. Webb S. Xeroderma pigmentosum. BMJ. 2008;336(7641):444–6.
23. Lim HW, Arellano-Mendoza MI, Stengel F. Current challenges in photoprotection. J Am Acad Dermatol. 2017;76(3S1):S91–S9.
24. Ramkumar HL, Brooks BP, Cao X, Tamura D, Digiovanna JJ, Kraemer KH, et al. Ophthalmic manifestations and histopathology of xeroderma pigmentosum: two clinicopathological cases and a review of the literature. Surv Ophthalmol. 2011;56(4):348–61.
25. Dollfus H, Porto F, Caussade P, Speeg-Schatz C, Sahel J, Grosshans E, et al. Ocular manifestations in the inherited DNA repair disorders. Surv Ophthalmol. 2003;48(1):107–22.
26. Brooks BP, Thompson AH, Bishop RJ, Clayton JA, Chan CC, Tsilou ET, et al. Ocular manifestations of xeroderma pigmentosum: long-term follow-up highlights the role of DNA repair in protection from sun damage. Ophthalmology. 2013;120(7):1324–36.
27. Mahindra P, DiGiovanna JJ, Tamura D, Brahim JS, Hornyak TJ, Stern JB, et al. Skin cancers, blindness, and anterior tongue mass in African brothers. J Am Acad Dermatol. 2008;59(5):881–6.
28. Olson MT, Puttgen KB, Westra WH. Angiosarcoma arising from the tongue of an 11-year-old girl with xeroderma pigmentosum. Head Neck Pathol. 2012;6(2):255–7.
29. Behan JW, Sutton A, Wysong A. Management of skin cancer in the high-risk patient. Curr Treat Options Oncol. 2016;17(12):60.
30. Cerio R. The importance of patient-centred care to overcome barriers in the management of actinic keratosis. J Eur Acad Dermatol Venereol. 2017;31(Suppl 2):17–20.
31. Lambert WC, Lambert MW. Development of effective skin cancer treatment and prevention in xeroderma pigmentosum. Photochem Photobiol. 2015;91(2):475–83.
32. Giannotti B, Vanzi L, Difonzo EM, Pimpinelli N. The treatment of basal cell carcinomas in a patient with xeroderma pigmentosum with a combination of imiquimod 5% cream and oral acitretin. Clin Exp Dermatol. 2003;28(Suppl 1):33–5.
33. Goldenberg G. Treatment considerations in actinic keratosis. J Eur Acad Dermatol Venereol. 2017;31(Suppl 2):12–6.
34. Gupta AK, Paquet M, Villanueva E, Brintnell W. Interventions for actinic keratoses. Cochrane Database Syst Rev. 2012;12:CD004415.
35. Roseeuw D. The treatment of basal skin carcinomas in two sisters with xeroderma pigmentosum. Clin Exp Dermatol. 2003;28(Suppl 1):30–2.
36. Stockfleth E. The importance of treating the field in actinic keratosis. J Eur Acad Dermatol Venereol. 2017;31(Suppl 2):8–11.
37. Walker JL, Siegel JA, Sachar M, Pomerantz H, Chen SC, Swetter SM, et al. 5-fluorouracil for actinic keratosis treatment and chemoprevention: a randomized controlled trial. J Invest Dermatol. 2017;137(6):1367–70.

38. Yang JQ, Chen XY, Engle MY, Wang JY. Multiple facial basal cell carcinomas in xeroderma pigmentosum treated with topical imiquimod 5% cream. Dermatol Ther. 2015;28(4):243–7.
39. Larson DM, Cunningham BB. Photodynamic therapy in a teenage girl with xeroderma pigmentosum type C. Pediatr Dermatol. 2012;29(3):373–4.
40. Brooks PJ, Wise DS, Berry DA, Kosmoski JV, Smerdon MJ, Somers RL, et al. The oxidative DNA lesion 8,5′-(S)-cyclo-2′-deoxyadenosine is repaired by the nucleotide excision repair pathway and blocks gene expression in mammalian cells. J Biol Chem. 2000;275(29):22355–62.
41. Besaratinia A, Bates SE, Synold TW, Pfeifer GP. Similar mutagenicity of photoactivated porphyrins and ultraviolet A radiation in mouse embryonic fibroblasts: involvement of oxidative DNA lesions in mutagenesis. Biochemistry. 2004;43(49):15557–66.
42. Kraemer KH, DiGiovanna JJ, Moshell AN, Tarone RE, Peck GL. Prevention of skin cancer in xeroderma pigmentosum with the use of oral isotretinoin. N Engl J Med. 1988;318(25):1633–7.
43. Campbell RM, DiGiovanna JJ. Skin cancer chemoprevention with systemic retinoids: an adjunct in the management of selected high-risk patients. Dermatol Ther. 2006;19(5):306–14.
44. DiGiovanna JJ. Retinoid chemoprevention in the high-risk patient. J Am Acad Dermatol. 1998;39(2 Pt 3):S82–5.
45. Nelson BR, Fader DJ, Gillard M, Baker SR, Johnson TM. The role of dermabrasion and chemical peels in the treatment of patients with xeroderma pigmentosum. J Am Acad Dermatol. 1995;32(4):623–6.
46. Sibar S, Findikcioglu K, Erdal AI, Barut I, Ozmen S. Technical aspects and difficulties in the management of head and neck cutaneous malignancies in xeroderma pigmentosum. Arch Plast Surg. 2016;43(4):344–51.
47. Hoorens I, Batteauw A, Van Maele G, Lapiere K, Boone B, Ongenae K. Mohs micrographic surgery for basal cell carcinoma: evaluation of the indication criteria and predictive factors for extensive subclinical spread. Br J Dermatol. 2016;174(4):847–52.
48. van Leeuwen AC, The A, Moolenburgh SE, de Haas ER, Mureau MA. A retrospective review of reconstructive options and outcomes of 202 cases large facial Mohs micrographic surgical defects, based on the aesthetic unit involved. J Cutan Med Surg. 2015;19(6):580–7.
49. Tayeb T, Laure B, Sury F, Lorette G, Goga D. Facial resurfacing with split-thickness skin grafts in xeroderma pigmentosum variant. J Craniomaxillofac Surg. 2011;39(7):496–8.
50. Cox SE, Roberts LJ, Bergstresser PR. Prevention of skin cancer in xeroderma pigmentosum: the physician as advocate. J Am Acad Dermatol. 1993;29(6):1045–6.
51. Alfawaz AM, Al-Hussain HM. Ocular manifestations of xeroderma pigmentosum at a tertiary eye care center in Saudi Arabia. Ophthal Plast Reconstr Surg. 2011;27(6):401–4.
52. Lim R, Fedele F, Patel P, Morley AM. Ocular solar protection in xeroderma pigmentosum: the role of untinted lenses in blocking ultraviolet radiation. Br J Dermatol. 2016;175(3):625–7.
53. Jalali S, Boghani S, Vemuganti GK, Ratnakar KS, Rao GN. Penetrating keratoplasty in xeroderma pigmentosum. Case reports and review of the literature. Cornea. 1994;13(6):527–33.
54. Brooks PJ. DNA repair in neural cells: basic science and clinical implications. Mutat Res. 2002;509(1–2):93–108.
55. Brooks PJ. The cyclopurine deoxynucleosides: DNA repair, biological effects, mechanistic insights, and unanswered questions. Free Radic Biol Med. 2017;107:90–100.
56. Robbins JH. A childhood neurodegeneration due to defective DNA repair: a novel concept of disease based on studies xeroderma pigmentosum. J Child Neurol. 1989;4(2):143–6.
57. Robbins JH, Brumback RA, Mendiones M, Barrett SF, Carl JR, Cho S, et al. Neurological disease in xeroderma pigmentosum. Documentation of a late onset type of the juvenile onset form. Brain. 1991;114(Pt 3):1335–61.
58. Davies H. Living with dying: families coping with a child who has a neurodegenerative genetic disorder. Axone. 1996;18(2):38–44.
59. Totonchy MB, Tamura D, Pantell MS, Zalewski C, Bradford PT, Merchant SN, et al. Auditory analysis of xeroderma pigmentosum 1971-2012: hearing function, sun sensitivity and DNA repair predict neurological degeneration. Brain. 2013;136(Pt 1):194–208.
60. Kraemer KH, DiGiovanna JJ. Forty years of research on xeroderma pigmentosum at the US National Institutes of Health. Photochem Photobiol. 2015;91(2):452–9.

61. Yarosh D, Klein J, O'Connor A, Hawk J, Rafal E, Wolf P. Effect of topically applied T4 endonuclease V in liposomes on skin cancer in xeroderma pigmentosum: a randomised study. Xeroderma Pigmentosum Study Group. Lancet. 2001;357(9260):926–9.
62. Warrick E, Garcia M, Chagnoleau C, Chevallier O, Bergoglio V, Sartori D, et al. Preclinical corrective gene transfer in xeroderma pigmentosum human skin stem cells. Mol Ther. 2012;20(4):798–807.
63. Dupuy A, Valton J, Leduc S, Armier J, Galetto R, Gouble A, et al. Targeted gene therapy of xeroderma pigmentosum cells using meganuclease and TALEN. PLoS One. 2013;8(11):e78678.
64. Kuschal C, DiGiovanna JJ, Khan SG, Gatti RA, Kraemer KH. Repair of UV photolesions in xeroderma pigmentosum group C cells induced by translational read-through of premature termination codons. Proc Natl Acad Sci U S A. 2013;110(48):19483–8.
65. Kuschal C, Khan SG, Enk B, DiGiovanna JJ, Kraemer KH. Readthrough of stop codons by use of aminoglycosides in cells from xeroderma pigmentosum group C patients. Exp Dermatol. 2015;24(4):296–7.
66. Ono R, Khan S, Kuschal C, Tamura D, Chen J, Luo X, DiGiovanna JJ, Kraemer KH. Induced DNA repair in xeroderma pigmentosum group C cells by read-through of premature termination codons. J Investig Dermatol. 2017;137(5S):S117.
67. Levenson D. Communication with parents counts. Am J Med Genet A. 2010;152A(5):fmvii–fmix.
68. Riley C, Rubarth LB. Supporting families of children with disabilities. J Obstet Gynecol Neonatal Nurs. 2015;44(4):536–42.
69. Rupp K, Davies PS, Newcomb C, Iams H, Becker C, Mulpuru S, et al. A profile of children with disabilities receiving SSI: highlights from the National Survey of SSI Children and Families. Soc Secur Bull. 2005;66(2):21–48.
70. Joachim G, Acorn S. Living with chronic illness: the interface of stigma and normalization. Can J Nurs Res. 2000;32(3):37–48.
71. Austin A, Herrick H, Proescholdbell S, Simmons J. Disability and exposure to high levels of adverse childhood experiences: effect on health and risk behavior. N C Med J. 2016;77(1):30–6.
72. Deatrick JA, Knafl KA, Murphy-Moore C. Clarifying the concept of normalization. Image J Nurs Schloarsh. 1999;31(3):209–14.
73. Silverman AM, Molton IR, Smith AE, Jensen MP, Cohen GL. Solace in solidarity: disability friendship networks buffer well-being. Rehabil Psychol. 2017;62(4):525–33.
74. Janson K, Law M. Siblings of children with special needs. Phys Occup Ther Pediatr. 2002;22(1):73–8.
75. Pit-Ten Cate IM, Loots GM. Experiences of siblings of children with physical disabilities: an empirical investigation. Disabil Rehabil. 2000;22(9):399–408.
76. Lai JP, Liu YC, Alimchandani M, Liu Q, Aung PP, Matsuda K, et al. The influence of DNA repair on neurological degeneration, cachexia, skin cancer and internal neoplasms: autopsy report of four xeroderma pigmentosum patients (XP-A, XP-C and XP-D). Acta Neuropathol Commun. 2013;1:4.

MIX
Papier aus verantwortungsvollen Quellen
Paper from responsible sources
FSC® C105338

FSC
www.fsc.org

If you have any concerns about our products,
you can contact us on
ProductSafety@springernature.com

In case Publisher is established outside the EU,
the EU authorized representative is:
Springer Nature Customer Service Center GmbH
Europaplatz 3, 69115 Heidelberg, Germany

Printed by Libri Plureos GmbH
in Hamburg, Germany